BIOCHEMISTRY RESEARCH TRENDS

OLIGOSACCHARIDES

FOOD SOURCES, BIOLOGICAL ROLES AND HEALTH IMPLICATIONS

BIOCHEMISTRY RESEARCH TRENDS

Additional books in this series can be found on Nova's website
under the Series tab.

Additional e-books in this series can be found on Nova's website
under the e-book tab.

BIOCHEMISTRY RESEARCH TRENDS

OLIGOSACCHARIDES

FOOD SOURCES, BIOLOGICAL ROLES AND HEALTH IMPLICATIONS

LORI S. SCHWEIZER
AND
STANLEY J. KREBS
EDITORS

New York

For permission to use material from this book please contact us:
Telephone 631-231-7269; Fax 631-231-8175
Web Site: http://www.novapublishers.com

Library of Congress Cataloging-in-Publication Data

ISBN: 978-1-62948-328-3

Library of Congress Control Number: 2013950463

Published by Nova Science Publishers, Inc. † New York

Contents

Preface **vii**

Chapter I Oligosaccharides: A Key for Gut Health **1**
Haydé A. Vergara-Castañeda, B. Dave Oomah, Guadalupe Loarca-Piña, Hector Pool and Rocio Campos-Vega

Chapter II Prebiotic Oligosaccharides, Resistant Starch, and Sugar Alcohols in Lentils: Implications for Obesity **49**
Casey Johnson, Dil Thavarajah, Pushparajah Thavarajah and Shiv Kumar

Chapter III Behaviour of Dietary Oligosaccharides in the Human Gastrointestinal Tract **85**
Fraser L. Courts, Liz K. Wells, Stephanie K. Allen and Huw S. Jones

Chapter IV The Role of Human Milk Oligosaccharides in Preventing Respiratory Infections in Infants **115**
Mariarosaria Marotta and Rita M. Hickey

Chapter V Alterations in the Glycosylation Pattern of Alpha-1-Glycoprotein May be Diagnostic for the Detection of Breast Cancer and/or Reducing the Effect of Chemotherapy *In Vivo* **143**
Kate Doak, Jodi Flaws, Yvonne Cruickshank and Kevin D. Smith

Chapter VI Production of Oligosaccharides with Low Molecular Weights, Secoiridoids and Phenolic Glycosides from Thermally-Treated Olive By-Products **173**
Juan Fernández-Bolaños, Fátima Rubio-Senent, Antonio Lama-Muñoz, Aránzazu García and Guillermo Rodríguez-Gutiérrez

Chapter VII Supramolecular Analytical Applications of Cyclodextrins and their Derivatives Using Fluorescence Spectroscopy **209**
Abdalla A. Elbashir

Chapter VIII Utilization of Soy Oligosaccharides during Fermentation by Lactobacilli **233**
Subrota Hati, Shilpa Vij, Surajit Mandal, Kaushik Khamrui and J. B. Prajapati

Chapter IX Impact of Low Oligosaccharide Soybean Meal on Nutrient Digestibility, Growth Performance, and Carcass Characteristics of Broiler Chickens **251**
W. A. Dozier, III and K. R. Perryman

Chapter X Lipo-Oligosaccharides in Miller Fisher Syndrome **267**
Kyung Yeon Lee

Chapter XI Oligosaccharides from Ancient Beverages Fermented by Symbiotic Cultures **287**
Jose Mauricio Schneedorf

Index **343**

Preface

In this book, the authors present current research in the study of the food sources, biological roles and health implications of oligosaccharides. Topics discussed in this compilation include functional oligosacchardes and their positive health benefits on the digestive system; prebiotic oligosaccharides; behaviour of dietary oligosaccharides in the human gastrointestinal tract; the role of human milk oligosaccharides in preventing respiratory infections in infants; alterations in the glycosylation patterns of alpha-1 glycoprotein in the diagnostic detection of breast cancer; production of oligosaccharides with low molecular weights, secoiridoids and phenolic glycosides from thermally treated olive by-products; supramolecular analytical applications of cyclodextrins and their derivatives using fluorescence spectroscopy; utiization of soy oligosaccharides during fermentation by lactobacilli; impact of low oligosaccharide soybean meal on nutrient digestibility, growth performance, and carcass characteristics of broiler chickens; lipo-oligosaccharides in Miller Fisher Syndrome; and oligosaccharides from ancient beverages fermented by symbiotic cultures.

Chapter I – The recent fermentable fiber revolution has emerged on the prospect of providing digestive health benefits such as enhanced immune system, mineral aborption, and general colonic health. These benefits accrue from dietary component, primarily 'non-digestible' (ND) dietary carbohydrates that reach the colon intact or partially digested by the stomach or small intestine. The fructo-, galacto-, and xylo-oligosaccharides, as well as fermentable carbohydrates such as pectin, arabinoxylans, and resistant starch form part of this diverse group of compounds also commonly referred to as 'prebiotic'. These prebiotic modulate the composition and metabolism of the gut microbiota thereby improving long-term human health status. Numerous studies have shown that gut microbiota is primarily influenced by the type and quantity of nondigestible carbohydrate in the diet, hence the increased awareness of food industries and consumers in incorporating a wide range of pre- and pro-biotics in the daily diet. Oligosaccharides present important physicochemical, functional and beneficial physiological properties for consumers, including anti-carcinogenic effect, low caloric value and the ability to stimulate the growth of beneficial bacteria in the colon. Thus, oligosaccharides from various sources, such as dairy products, bacteria, algae, fungi and higher plants, have been widely used to formulate food ingredients and pharmacological supplements. In the food industries, the non-digestible oligosaccharides have been implicated as dietary fiber, sweetener, and weight control agent, as a humectant in confectioneries, bakeries and breweries. Keeping in view the importance of the functional oligosaccharides, we present an overview of their positive health benefits, focused on the digestive system and their potential applications.

Chapter II – Imbalanced energy intake, coupled with low concentrations of bioactive compounds in foods, has fueled the rising epidemic of obesity and related non-communicable diseases. Currently, over 35% of American adults are obese, including nearly 17% of children and adolescents. In relation, chronic non-communicable diseases result in an estimated 36 million deaths globally each year. To combat these diet-related disorders, support health, and contribute to overall sustainability of health care, it is important to develop foods that supply proper amounts of energy and nutrients. Lentil (*Lens culinaris* L.), a cool season food legume, is a good source of protein (20-30%), essential fatty acids, micronutrients, and prebiotic carbohydrates and elicits a low glycemic index. Carbohydrate profiles of lentil that contribute to healthful effects include prebiotics: raffinose-family oligosaccharides, fructooligo-saccharides, sugar alcohols, and resistant starch. The book chapter will provide insights on prebiotic carbohydrates in lentil, including genetic diversity and growing environmental and processing effects. Additionally, the chapter will include a synopsis of health consequences associated with consumption of these important prebiotic carbohydrates, and, in closure, will discuss avenues of further research.

Chapter III – Dietary oligosaccharides are classically assumed to remain inert during passage through the human gastrointestinal (GI) tract, being resistant to both hydrolysis and uptake into systemic circulation. Advances in the fields of paediatric nutrition, gut microbiology and animal nutrition have more recently provided intriguing evidence for diverse physiological and physicochemical behaviours of oligosaccharides in the mammalian GI tract however, including mucosal adhesive properties, prebiotic activity, and modulation of intestinal nutrient uptake. The GI tract is also now known to process some dietary oligosaccharides via acidic and enzymatic hydrolysis of specific glycosidic linkages, forming shorter and less-branched saccharides, the latter of which may be actively transported across the intestinal mucosa to participate in glycolytic metabolism. This evidence, fuelled by the advent of advanced analytical methodologies for carbohydrate analysis and the global rise in demand for functional food product innovation has resulted in interest surrounding the functional behaviour of dietary oligosaccharides through the human GI tract, and downstream effects on health.

In this chapter, current and future experimental strategies for modelling oligosaccharide behaviour in the mammalian GI tract are summarised, focusing on the biological models and methods of oligosaccharide analysis used in the literature. The application of these techniques towards elucidating dietary oligosaccharide pharmacokinetics is described, revealing often stark differences between oligosaccharide families in susceptibility to hydrolysis and transmucosal uptake into systemic circulation. Evidence for the ability of surviving oligosaccharide structures in the lumen to prevent pathogen adhesion to the intestinal mucosa, act as excipients for nutrient uptake and as prebiotic substrates for the colonic microflora, selectively encouraging the growth of particular microbial genera or species is also detailed, with implications for intestinal bacterial invasion and gut health discussed. Overall, we present evidence for the importance of oligosaccharides in maintaining a healthy GI tract and the modulatory role of the GI tract in determining the potential protective effects of oligosaccharides against disease and dysfunction.

Chapter IV – Breast milk plays a major role in protecting infants from respiratory and gastrointestinal infections. In addition to IgA, molecules such as free oligosaccharides present in human milk have been implicated in this protective role. When considering respiratory tract infection, pathogen adhesion and colonization is often a prelude to infection, and intervention at this early stage can help prevent disease. Bacteria have evolved a multitude of colonisation mechanisms commonly targeting surface carbohydrate structures of the host. Many human milk oligosaccharides are considered to be soluble receptor analogs of epithelial cell surface carbohydrates, displaying structural homology to host cell receptors. Thus, they function as receptor decoys, which pathogens can bind to instead of the host. Here, the authors review the ability of human milk oligosaccharides and related structures to prevent colonisation of pathogens to host respiratory cells. In particular, they examine the use of *in vitro* assays to study milk oligosaccharides anti-infective effect. The authors also access the limited data, which is available from infant studies involving the use of human milk oligosaccharides. Finally, intervention studies assessing the effects of galactooligosaccharides and fructooligosaccharides when incorporated into infant formula on respiratory infections in developed countries are also reviewed.

Chapter V– Alpha-1-acid glycoprotein AGP is an important binder of drugs in the plasma. Although the binding site is protein in nature, the glycosylation pattern present could still affect the overall conformation of the glycoprotein and could also have implications on access to the binding site. Drugs bound by plasma proteins have reduced efficacy as pharmacological effects are exerted by free unbound drug.

The present study aimed to determine if AGP could bind to tamoxifen, the gold standard of treatment for breast cancer, and if, through this binding, AGP could represent a mechanism of resistance to tamoxifen.

The glycosylation of AGP was also examined to determine if any disease-specific changes were present in breast cancer and if these disease-specific glycoforms displayed altered binding properties for tamoxifen.

The study used different cohorts of patients - healthy individuals, those with untreated breast cancer, those at risk of breast cancer and those with breast cancer being treated with tamoxifen therapy. AGP was isolated by a technique specifically chosen because it did not degrade the glycoprotein structure, thus reflecting the situation *in vivo*. Drug binding was examined by the use of an intrinsic fluorescence assay. Patients with untreated and tamoxifen treated breast cancer had significantly increased AGP levels compared to healthy individuals. The monosaccharide fucose, which is normally not present on AGP glycosylation in healthy plasma, was found in untreated breast cancer patients. Fucose was also found in some tamoxifen treated samples, which correlated with tamoxifen treatment, with those patients at the end of treatment displaying no fucose indicating no inflammation or disease, thus backing up the standard 5 year treatment period with tamoxifen. Patients with untreated breast cancer displayed a disease specific shift to bisiaylated glycan chains compared with healthy AGP. Tamoxifen treated breast cancer patients showed changes in the oligosaccharide structures over the time course of tamoxifen therapy, reverting to normal at a later time point – also indicated by the monosaccharide analysis. Untreated breast cancer patients were able to bind tamoxifen to a similar extent as healthy AGP. In contrast those at risk and those with tamoxifen treated breast cancer had significantly reduced binding. Although binding was minimal at therapeutic concentrations, the decrease in binding by breast cancer patients could lead to issues regarding toxicity as there would be more free drug in the plasma and therefore this concentration could exceed the therapeutic level.

Chapter VI – Alperujo was hydrothermally pre-treated to solubilize the oligosaccharides from the cell wall material of olive pulp. A protocol for the isolation of oligosaccharides of low molecular weight has been developed, which involves thermal treatment, mild acid hydrolysis, separation by ultra-filtration, adsorption chromatography and size exclusion chromatography using Superdex Peptide HR. Three groups of fractions with common characteristics were obtained and studied in two samples of alperujo from different seasons, origin and variety of olives. In the first group, neutral and acidic oligosaccharides with a low DP (4-10) bound or not to phenol-containing compounds and with antioxidant activity were identified. In the second group, several secoiridoid glycosides esterified to phenolic compounds and their aglycones were also detected. In the third group, a significant fraction of di- and trisaccharides of glucose or glucose-xylose bound to phenol residues (hydroxytyrosol and tyrosol) that could be considered antioxidant-soluble fiber with bioactive properties was also isolated. All of them were identified by HPLC, CG, and ESI-MS and in certain cases by ^{1}H and ^{13}C MNR. The similar behavior obtained for both samples confirms the protocol followed as a useful tool for bioactive compound isolation.

Chapter VII – Cyclodextrins (CDs) are a family of structurally related natural products formed during bacterial digestion of cellulose. Theses cyclic oligosaccharides composed of α-(1, 4) linked glucopyranose subunits. Due to the chair conformation of the glucopyranose units, the cyclodextrins are shaped like a truncated cone rather than perfect cylinders. The most important feature of cyclodextrins is their ability to form inclusion complexes (host–guest complexes) with a very wide range of solid, liquid and gaseous compounds by a molecular complexation, that lead to enhancement of fluorescence intensity. During the last decade a considerable number of research works have been focused on the use of CDs to enhance fluorescence intensity of different analytes and to develop CD-induced spectrofluorimetric method. This chapter provides an overview of spectrofluorimetric methods based on host-inclusion complex for determination of various compounds.

Chapter VIII – Soymilk is the water extract of soybean. The major carbohydrates present in soy milk are sucrose, raffinose and stachyose and is free of lactose and cholesterol. The main problem of limiting the widespread consumption of soy products is their objectionable beany flavour and high levels of oligosaccharides, raffinose and stachyose, which can lead to flatulence when consumed in large amounts. An alternative approach would be the utilization of microorganisms possessing α-galactosidase activity which can hydrolyze the oligosaccharides during fermentation. This would provide simple sugars as substrate for the growth of lactic acid bacteria as well as reduce the level of oligosaccharides present in soy milk. Growth of lactic cultures in soymilk has been reported as a good growth medium. They produce small amounts of lactic acid, acetic acid etc during fermentation and also lower down the pH, making the environment acidic against pathogens. Soy products are considered as a good substrate for functional foods since fermentation by lactic acid bacteria has the potential to reduce the levels of some carbohydrates which can be responsible for gas production in the intestinal system, increase free isoflavone levels and flavor desirable changes in bacterial populations in the gastrointestinal tract, antioxidant activity, anticarcinogenic activity, antimicrobial activity and phytoestrogen activity in post menopausal women. Some researchers have reported the utilization of oligosaccharides during fermentation of soymilk using either pure or mixed culture of the following probiotic lactobacilli for the development of functional fermented soy foods: *Lactobacillus cellobiosis, L. plantarum, L. fermentum, L. fermenti, L. pentosaceus, L casei, L. rhamnosus.*

Chapter IX – Previous research has reported that removal of raffinose and stachyose from SBM can increase the energy utilization of SBM by broilers. Several strategies have been implemented to ameliorate the adverse effects of raffinose and stachyose on nutrient utilization, which include ethanol extraction of SBM, galactosidase supplementation, and developing new varieties of low galactooligosaccharide SBM. Soybean meals stemming from novel genetic varieties of soybeans have been developed to contain either a 75% (low oligosaccharide SBM [LOSBM]) or 90% (ultra-low oligosaccharide SBM [ULSBM]) reduction in GAL concentrations. Broilers fed diets containing LOSBM had higher AME_n compared with birds fed CSBM and ULSBM. Both ULSBM and LOSBM had higher digestible AA content compared with CSBM, with ULSBM having the highest AA digestibility. Broilers fed diets formulated with ULSBM had superior feed conversion from 1 to 28 and 1 to 42 days of age compared with birds consuming LOSBM and CSBM based diets. Diets formulated with LOSBM and ULSBM had supplemental fat concentrations reduced by up to 70% compared with CSBM based diets. Due to the better nutrient profile of LOSBM and ULSBM, less supplemental fat can be utilized in diet formulation without any adverse effects on growth performance or carcass characteristics.

Chapter X – Miller Fisher syndrome is a clinical variant of Guillain-Barré syndrome, characterized by the acute-onset of ophthalmoplegia, ataxia, and areflexia, with a relative sparing of strength in the extremities and trunk. In 1992, the discovery of immunoglobulin G antibodies against GQ1b ganglioside in patients with Miller Fisher syndrome inaugurated a new era in research on the pathogenesis of Miller Fisher syndrome. Subsequent studies revealed that the oculomotor, trochlear, and abducens nerves have significantly higher percentages of the GQ1b than the other cranial nerves. Because of the molecular similarity between lipo-oligosaccharides on GQ1b ganglioside of the human neuronal cell surfaces and lipo-oligosaccharides on the outer membranes of certain bacteria, e.g., *Campylobacter jejuni*, anti-GQ1b antibody generated in the bacterially infected patients cross-reacts with these lipo-oligosaccharides, which can cause Miller Fisher syndrome. This antibody is now widely used as a diagnostic marker for Miller Fisher syndrome. However, a small number of anti-GQ1b-negative Miller Fisher syndrome cases have also been documented, although they presented with clinical features nearly identical to those of patients who were anti-GQ1b-positive. Despite the enormous advancement of knowledge on the pathogenesis of Miller Fisher syndrome thanks to the progress of comprehension about the structure of lipo-oligosaccharide in human gangliosides and bacterial cell membranes and the role of anti-GQ1b antibody, we still have unresolved questions about Miller Fisher syndrome.

Chapter XI – Symbiotic colonies of bacteria and yeasts (Scoby) are known to ferment dairy or non-dairy foods (sugar, fruits, cereals, vegetables, tubers, roots), leading to traditional or indigenous drinks distributed worldwide. These Scoby drinks remain today as a house art, as no special condition is required for their formulation. The fermented beverages have different microbial strains (sometimes probiotics) together with bioactive compounds (sometimes prebiotics), including oligosaccharides related to alleged health properties of the drinks. These oligosaccharides are known to play a role in gastrointestinal equilibrium, pathogen inhibition and immunity enhancement, among others. This chapter exploits the origin, structure, and health properties of oligosaccharides of Scoby beverages, either inherent to the carbon source used for souring or released from microbial enzyme activities (e.g., glucosyltransferase, dextransucrase, fructofuranosidase, fructosyltransferase, inulinase). The oligosaccharides found in ancient Scoby drinks, e.g., kefir, kvass, pulque, kombucha, boza, koumiss, and borde, among others, as well as non-Scoby indigenous drinks (as miki and koji) are discussed. These carbohydrates, e.g., isomaltotriose, rhamnoarabinan, fructans, and inulin, have typically less than 20 degree of polymerization. The main strains involved in their production belongs to *Lactobacillus, Leuconostoc, Lactococcus, Saccharomyces, Kluyveromyces*, among others. Complementarily, some activities of an oligosaccharide isolated from Kefir d'aqua and named AK (aqueous kefiran) on biomimetic systems and living cells are presented by means of a combined electrochemical impedance spectroscopy and cyclic voltammetry data.

In: Oligosaccharides
Editors: L. S. Schweizer, S. J. Krebs
ISBN: 978-1-62948-328-3

Chapter I

Oligosaccharides: A Key for Gut Health

Haydé A. Vergara-Castañeda[1], B. Dave Oomah[2], Guadalupe Loarca-Piña[3], Hector Pool[1] and Rocio Campos-Vega[3*]

[1]Nucitec, S. A. de C. V., Departamento de Investigación, Querétaro, México

[2]National Bioproducts and Bioprocesses Program, Pacific Agri-Food Research Centre, Agriculture and Agri-Food Canada, Summerland, British Columbia, Canada

[3]Programa de Posgrado en Alimentos del Centro de la República (PROPAC), Research and Graduate Studies in Food Science, School of Chemistry, Universidad Autónoma de Querétaro, Querétaro, México

Abstract

The recent fermentable fiber revolution has emerged on the prospect of providing digestive health benefits such as enhanced immune system, mineral aborption, and general colonic health. These benefits accrue from

[*] Corresponding author: Telephone +52 (442) 1921200 Ext 5568; Email: chio_cve@yahoo.com.mx.

dietary component, primarily 'non-digestible' (ND) dietary carbohydrates that reach the colon intact or partially digested by the stomach or small intestine. The fructo-, galacto-, and xylo-oligosaccharides, as well as fermentable carbohydrates such as pectin, arabinoxylans, and resistant starch form part of this diverse group of compounds also commonly referred to as 'prebiotic'. These prebiotic modulate the composition and metabolism of the gut microbiota thereby improving long-term human health status. Numerous studies have shown that gut microbiota is primarily influenced by the type and quantity of nondigestible carbohydrate in the diet, hence the increased awareness of food industries and consumers in incorporating a wide range of pre- and pro-biotics in the daily diet. Oligosaccharides present important physicochemical, functional and beneficial physiological properties for consumers, including anti-carcinogenic effect, low caloric value and the ability to stimulate the growth of beneficial bacteria in the colon. Thus, oligosaccharides from various sources, such as dairy products, bacteria, algae, fungi and higher plants, have been widely used to formulate food ingredients and pharmacological supplements. In the food industries, the non-digestible oligosaccharides have been implicated as dietary fiber, sweetener, and weight control agent, as a humectant in confectioneries, bakeries and breweries. Keeping in view the importance of the functional oligosaccharides, we present an overview of their positive health benefits, focused on the digestive system and their potential applications.

Introduction

Oligosaccharides are low molecular weight carbohydrates, containing sugar moieties with 3 to 10 degrees of polymerization, intermediate in nature between simple sugars and polysaccharides. In the non-digestible oligosaccharides, the anomeric C atoms (C1 or C2) of the monosaccharide units have configuration that makes their osidic bonds non-susceptible to the hydrolytic activity of human digestive enzymes. The indigestible oligosaccharides possess important physicochemical and physiological properties and serve as dietary fibers and prebiotics (Patel and Goyal, 2011).

Prebiotics are non-digestible but fermentable oligosaccharides that are specifically designed to change the composition and activity of the intestinal microbiota aimed at promoting the health of the host. Dietary fiber and non-digestible oligosaccharides are the main growth substrates of gut microorganisms. Their fermentation results in the acidification of the colonic contents and the formation of short chain fatty acids, which serve as fuels in different tissues and may play a role in the regulation of cellular processes (Blaut, 2002).

The intestinal microbiota is of critical importance to our health and wellbeing. Its modulation by exogenous (oligosaccharides) and endogenous substrates can be expected to improve various physiological functions of our body, not just by those in the intestine. Recently, new targets such as immune function and areas outside the colon have been considered to be influenced by the intestinal microbiota. Microbiota is involved in the production and absorption of essential vitamins and micronutrients. It has been established that the levels of essential vitamins, such as folate and biotin, may be higher in individuals than their diet would suggest/support. This can be attributed to microbial production of folate and biotin, as well as a number of vitamins, including other water-soluble B vitamins and vitamin K. It was assumed that, as these vitamins were produced in the colon, they would not be absorbed, but excreted. This concept was challenged with the discovery of colonic transporters for biotin, thiamine, riboflavin, pyridoxine and folate (Jeferry and O'Toole, 2013). Additionally, data suggest that oligosaccharides can stimulate the cecal bolus passage through the colon, shortening its transit time and reducing the time available for water reabsorption. All these factors increase the weight of the faeces and impart a softer composition (Sabater-Molina et al., 2009).

Interest in functional oligosaccharides is motivated by their diverse industrial applications in nutraceutical, pharmaceutical, prebiotics, cosmetics, animal feed and agriculture sector. The health benefits of functional oligosaccharides are well-known and far reaching. Furthermore, oligosaccharides have been the subject of intensive and cutting-edge research during the last three decades in response to their ever-increasing demand (Patel and Goyal, 2011).

Beneficial Properties of Oligosaccharides in the Small and Large Intestine

Recent rapid changes in diet and lifestyle due to industrialization, urbanization and market globalization have significantly impacted the health and nutritional status of population in the entire world (Evans et al., 2001; Hu, 2011). The lack of physical activity and over consumption are main factors that increase the risk of developing nutrition related diseases, such as obesity, hypertension, cardiovascular diseases, osteoporosis, type II diabetes, and several cancers (Amine et al., 2002; James, 1988; Walker, 1995). Therefore, it is not surprising that, over the last decade, drastic changes have taken place in a rapid expansion of numerous relevant scientific fields; particularly, population-based epidemiological evidence clearly demonstrating the role of diet in preventing and controlling morbidity and premature mortality resulting from diet related diseases (Korhonen, 2002; Malinow et al., 1999). Thus, food is no longer judged only in terms of taste or nutritional needs, but also in terms of their ability to improve the health and well-being of consumers (Schneeman, 2003). Diet is becoming the major modifiable determinant of nutrition related diseases, with scientific evidence that supports the positive effects of diet and its components on health throughout life (Brouns, 2002; Estruch et al., 2006; Mustalahti et al., 2002). Moreover, functional foods and functional food ingredients exert beneficial effect on host health and/or reduce the risk of chronic disease beyond their nutritive value (Charalampopoulos et al., 2002; Puupponen-Pimiä et al., 2002). Recently, the concept of functional foods has changed in terms of supplementation, the first generation was based on the implementation of food components, such as minerals (mainly calcium) and vitamins (Ziemer and Gibson, 1998); but in the last years food additives, as pro- and pre-biotics have gained great interest, due to their positive effect on the gut microbiota (Puupponen-Pimiä et al., 2002; Roberfroid, 2002). Since some probiotics are susceptible to oxygen and heat, their applications in food has been limited; therefore, the prebiotics market is expanding rapidly in order to formulate novel food products and improve the intestinal health of consumers (Roberfroid et al., 2010; Tuohy et al., 2005).

Definitions and Effect on Intestinal Microbiota (*in vitro* Studies) of Non-Digestible Oligosaccharides

Prebiotics are non-digestible food ingredients that target certain components within the microbiota of the human large intestine. Efficient prebiotics require specific fermentation and thereby have the ability to alter the faecal microflora composition towards a more 'beneficial' community structure (Kolida et al., 2002; Roberfroid et al., 2010; Tuohy et al., 2005). This implies that they resist hydrolysis by digestive enzymes and/or are not absorbed in the upper part of the gastrointestinal tract (Gibson et al., 2004b;

Roberfroid, 2002). Indeed, prebiotics must pass into the large bowel where they modify the colonic microbiota in such a way that potential health-promoting bacteria (especially *lactobacilli* and *bifidobacteria*) become numerically and/or metabolically predominant (Roberfroid, 1998). While some peptides, proteins, and certain lipids are potential prebiotics, non-digestible carbohydrates, particularly non-digestible oligosaccharides (NDOs), have received the most attention, mainly due to their beneficial effects lead by their low absorption in the gastrointestinal tract (GI) (Gibson et al., 2004a; Kaur et al., 2002). Studies have shown that the indigestibility of NDOs results from the configuration of the glycosidic bond between monomeric sugar units, (Kaur and Gupta, 2002; Sako et al., 1999). Most NDOs have a β-configuration and cannot be degraded by human gastrointestinal digestive enzymes which are specific for α-glycosidic bonds (Swennen et al., 2006). However, there are NDOs with *α*-configuration that in principle can be degraded by human *α*-glycolytic enzymes. Besides, the *β*-galactosidase-enzyme, localized at the brush border membrane of the small intestine, has the potential to degrade *β*-galactooligosaccharides. However, degradation of NDOs with *α*-glycosidic bonds or *β*-galactooligosacharides by the above– mentioned gastrointestinal digestive enzymes is low because of weak activities and/or substrate selectivity (Sako et al., 1999). NDOs are water soluble and highly likely to be fermented in the colon; their utilization is mediated by hydrolytic enzymes of colon bacteria (Cummings and Macfarlane, 2002; Sako et al., 1999). The resistance of oligosaccharides to digestive enzyme hydrolysis and digestive conditions were studied recently by Wichienchot et al., (2010). Extracts of oligosaccharides obtained from pitaya (dragon fruit) were evaluated for their resistance to the hydrolysis by artificial human gastric juice and human α-amylase. The degree of hydrolysis in artificial human gastric at pH 1, 2, 3, 4 and 5 increased with pH reduction (4.07%, 2.43%, 1.66%, 0.85% and 0.02%, respectively). Food is usually retained in the human stomach where gastric juice (pH 2-4) is released within 2h; thus, when pitaya oligosaccharides showed a high resistance to the gastric hydrolysis and consumed, 96% are estimated to reach the intestine. Human salivary α-amylase hydrolysis of pitaya oligosaccharides at pH 4-8 showed that higher pH gave a significantly higher degree of hydrolysis ($P<0.05$), with high enzymatic resistance compared to control (inulin). These results suggested that about 50% of the pitaya oligosaccharides consumed would reach the colon since some of them are hydrolyzed by α-amylase, by acid stomach and by brush-border enzymes (i.e. isomaltase, glucoamylase, maltase, sucrose and lactase) in the small intestine. The partial or total-resistance of oligosaccharides, and their end-products, may help the growth of the intestinal microbiota demonstrated by this study. Pitaya oligosaccharides stimulated the growth of *Lactobacilli* and *Bifidobacterium* for 48h. These results have spurred the interest of several researchers to investigate the prebiotic effect of oligosaccharides from different natural sources on the growth of intestinal bacteria species, which can lead to beneficial properties and enhance intestinal health.

NDOs are metabolized by bacterial species (**Figure 1**) producing glycolytic enzymes which hydrolyze them into mono- or disaccharides that are transported into the cell where they can be further metabolized to short chain fatty acids (SCFAs), gases and water (Gibson et al., 2004a; Schneeman, 2002). These SCFAs, particularly acetate, propionate, and butyrate, are the major end-products of bacterial fermentation reactions, acidifying the colon environment (Miller and Wolin, 1996). This is beneficial for the development of bacteria such as *bifidobacteria* and *lactobacilli*, and detrimental to the growth of potential pathogenic species (Blaut, 2002; Sako et al., 1999; Tungland, 2003). All the SCFAs are rapidly absorbed from the hindgut and then metabolized by various tissues: butyrate by the colonic epithelium, propionate and acetate (partly) by the liver, and acetate (partly) by muscle and other peripheral tissues (Miller et al., 1996; Priebe et al., 2002). Some authors claim that the absorbed SCFAs impact the host metabolism (Gibson et al., 2004a; Topping, 1996). Acetate and propionate influence cholesterol production (Kontula et al., 1998), propionate reduces hepatic gluconeogenesis and inhibits ureagenesis in the liver (Roberfroid and Delzenne, 1998) and butyrate is a growth and differentiation regulator for colonic epithelial cells besides being a major energy source (Tungland and Meyer, 2002; Velázquez et al., 2000). Butyrate as a major SCFA is associated with many others beneficial biological functions in the colon. One of the important effects of butyrate on the DNA methylation is probably ascribed to modified gene expression, and the yet unclear action mechanism, especially in the context of colon cancer (Pool-Zobel et al., 1996; Wollowski et al., 2001). However, butyrate may enhance cell proliferation of normal cells. Furthermore, apoptosis in the presence of butyrate may be enhanced in transformed cells but inhibited in normal cells (Hass et al., 1997; Hsu et al., 2004). Furthermore, The ultimate aim of supplementation of the human diet with prebiotics is the beneficial management of the gut microbiota (Blaut et al., 2002).

mixtures of fecal slurries carefully controlling variables such as temperature, pH and nutrient composition. This technique enables the study of NDOs influence on the bacterial composition of mixed bacterial faecal slurries as well as the fermentation of NDOs by pure bacterial cultures. The different compartments of the colon can be simulated with continuous culture experiments with (sequences of) chemostats (Macfarlane et al., 1998). One limitation of this type of experiment is that metabolites are not withdrawn from the environment as they are in the colon. The accumulation of these products affects the equilibrium of many ongoing biochemical processes (inhibition). It is not possible to quantify the influence of the artificial environment of a bioreactor. Nevertheless, the relative comparisons that can be studied are considered important tools in prebiotic research, as they allow the demonstration (qualitatively) of certain mechanisms, leading to the formulation of hypotheses that can be tested by well-designed human feeding studies (Macfarlane et al., 1998; Van Loo et al., 1999). In this way, different studies have been carried out by *in vivo* or *in vitro* experiments.

Prebiotic Effects of NODs on Intestinal Microbiota (*in vivo Studies*)

Researchers evaluated the increase in bifidobacteria in the gut of individuals fed a diet containing 15 g of oligofructose or inulin during 15 days. Results showed an increase of 0.7 and 0.9 log colony forming units (CFU) per gram stool for oligofructose and inulin, respectively (Gibson et al., 1995). Bouhnik et al., (1999) tested the effect of different doses (0, 2.5, 5, 10 and 20 g per day) of oligofructose on bifidobacteria in a 7-day study with 40 volunteers. The lowest dose (2.5 g per day) had no effect; a dose of 5 g per day showed an increase of 0.9 log CFU per g, whereas the two largest doses led to a maximum increase between 1.3 and 1.5 log CFU per gram.

In another study mice were given gavage of selected fructo-oligosaccharides (FOS), galactosaccharides (GOS), mannanoligosaccharides (MOS), and chitooligosaccharides (COS) to elucidate the effects of oligosaccharides on SCFAs concentrations (Pan et al., 2009). Cecal contents were collected, pH-measured and processed for SCFAs analysis (by gas chromatography). Results showed that oligosaccharides increased the cecal concentrations of total SCFAs compared with the control group. In addition, the FOS and GOS groups had higher ($P<0.05$) SCFAs concentrations than the

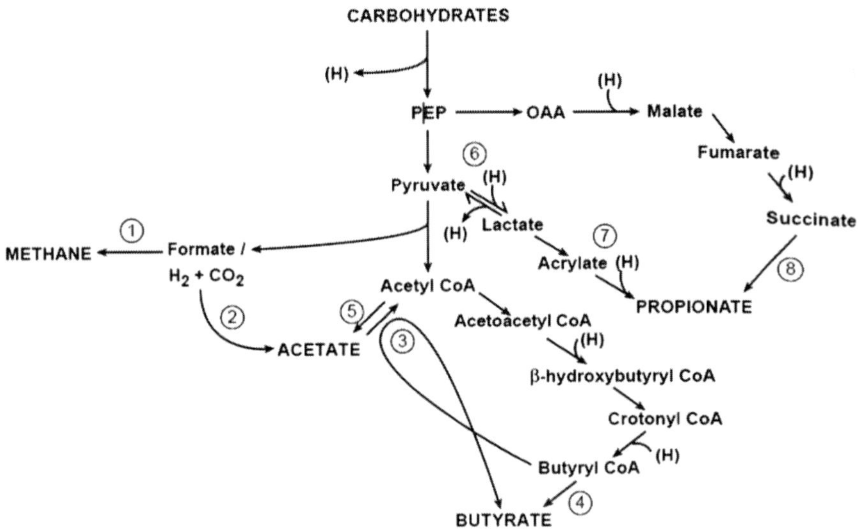

Figure 1. Carbohydrate fermentation pathways in the large intestine. 1=Methanogenesis, 2= reductive acetogenesis, 3= butyryl CoA:acetate CoA transferase, 4 =phosphotransbutyrylase/butyrate kinase, 5= phosphotransacetylase/ acetate kinase, 6 =lactate dehydrogenase, 7 =acrylate pathway, 8= succinate decarboxylation. Adapted from Pryde et al., (2002).

Many NDOs beneficially affect the host by selectively stimulating the growth and/or activity of one or a limited number of beneficial bacteria in the colon, thereby improving host health (Gibson et al., 1995). These prebiotics effects can be evaluated on the basis of growth of bacteria, such as *lactobacilli* and *bifidobacteria*, the decrease of intestinal pathogens and the increase or decrease in production of health related bacterial metabolites (Grootaert et al., 2007).

Changes in the microflora of GI using NODs increase the protection of the human intestine reducing the antagonist effects of pathogenic bacteria, decrease the intestinal infections, diarrhoea disease, stimulate the immune system and antigenotoxicity activities (Kirjavainen et al., 2002; Pool-Zobel et al., 1996; Saavedra et al., 1994).

Because of their positive attributes bifidobacteria and lactobacilli are the most frequent target organisms (Kolida et al., 2002). In this way, several studies have focused their efforts in investigating the effect of some NDOs in the growth of intestinal microbiota. Most *in vitro* experimental studies use mainly small bioreactors inoculated with pure bacterial strains, reconstituted

MOS and the COS groups. Acetate, propionate, and the butyrate rate in the cecum increased with administration of oligosaccharides, while the concentrations of valerate were unaffected. Similarly, all oligosaccharides significantly increased ($P<0.05$) the cecal concentration of lactate. Bacterial concentrations (*bifidobacteria* and *lactobacilli*) also increased in mice cecum after 14 days of gavage. In addition, the concentrations of *Enterococcus* and *Enterobacteriaceae* were significantly reduced ($P<0.05$). These findings demonstrated that the use of selected oligosaccharides increase beneficial bacterial concentrations and decrease potential pathogens, which play an important role in the eco-physiology of the intestinal microbiota, such as resistance to infections and diarrhea (Saavedra et al., 1994). This study demonstrate that the increase of SCFAs concentrations is associated with beneficial changes in enteric functions, lowering the colon pH, which has been related to promoting resistance/protection of colorectal cancer (Hsu et al., 2004). NDOs promote the growth of the bacterial (bifidobacteria and lactobacilli) population in GI, activating signal pathways associated with the infection response (Winkler et al., 2007). Bifidobacteria are responsible for synthesizing antimicrobial substances that inhibit the growth and adherence of pathogen microorganisms, simultaneously modulating the production of nitric oxide and pro-inflammatory cytokines, such as interleukins Il-1β, Il-6, and Il-12 and the tumor necrosis factor (Tnf-α) in murine macrophages (Trebichavsky et al., 2009; Vulevic et al., 2008).

Various strategies have been developed to extract oligosaccharides from different natural sources, and investigate the prebiotics effects of these compounds, mainly in the intestinal microbiota. Prebiotic effects on the *Bifidobacterium bifidum* and *Lactobacillus sp.* concentrations using COS (prepared with three degree of polymerization) were investigate by Lee et al., (2002). In this study, FOS only stimulated the growth of three strains: *Bifidobacterium bifidum, B. infantis* and *Lactobacillus casei*, whereas COS stimulated the growth of all *Lactobacillus sp.* and *B. bifidum* KCTC 3440. Furthermore, the growth rate of *B. bifidum* increased with increasing COS concentrations. The authors indicated that bifidobacteria are unable to use FOS as nutrient due to a relatively high amount of β-fructosidase (De Vries et al., 1967). It is known that the amine moiety and the molecular weight are important for the biological activity of chitosan, especially, in terms of antimicrobial activity (Choi et al., 2001; Tsai et al., 2004). However, only the chitosan polymer but not the COS exerted an inhibitory effect on bifidobacteria and lactic acid bacteria. These findings suggested that COS lack antimicrobial effect, although they were fully deacetylated.

Recent efforts focused on improving infection resistance in infants, using infant's formulas supplemented with mixtures of NDOs. Human milk contains approximately 7g/100 ml lactose, 1 g/100 ml neutral oligosaccharides and 0.1 g/100 ml acidic oligosaccharides. More than 100 different oligosaccharide structures have been identified, and most of them with a low molecular weight (Thurl et al., 1997; Thurl et al., 1996). Also, low amounts of high molecular weight oligosaccharides have been detected in human milk, further confirming the extreme complexity of human milk oligosaccharides (HMO) (Stahl et al., 1994). HMO are important components of the defense system of human milk, having both the prebiotic potential and the direct interaction with the immune cells (Eiwegger et al., 2004; Schumacher et al., 2006). Thus, identical structures are not available for use in infant formulas. Searching for alternatives to mimic the prebiotic effect of human milk, standard formulas supplemented with mixtures of NDOs have been formulated in the last years, in order to exert favorable effects in the intestinal health and immune system of infants. It is well known that the intestinal microbiota composition is regarded as an important factor for infant health and well-being (Ouwehand et al., 2002). In this context, a specific prebiotic mixture of GOS and FOS has been described for infant formulas, which can stimulate the growth of bifidobacteria and lactobacilli similar to HMO (Weaver, 2003). Several reports showed that the supplementation of infant formulas with this specific mixture of GOS and FOS increases the numbers of *Bifidobacterium* (Weaver, 2003) and the total numbers of *Lactobacillus* (Moro et al., 2002), reduces the numbers of pathogens, and induces a short-chain fatty acid profile similar to that found in breastfed infants (Moro et al., 2002; Weaver, 2003). Addition of the specific prebiotic mixture of GOS and FOS also results in a distribution of the different *Bifidobacterium* species similar to that found in breast-fed infants (Haarman and Knol, 2005).

Several studies have demonstrated that using mixtures of oligosaccharides increases the number of bacteria species. However, bacteria species are stimulated only when they obtain energy from a specific carbon source. Many different types of carbohydrate are present in the human colon at any given time, particularly in the proximal bowel. Consequently, nutritionally versatile bacteria that can grow on a variety of carbon sources and adapt rapidly to changing substrate availability have a competitive advantage in the ecosystem. Carbohydrate utilization in colonic micro-organisms is to a large extent controlled by catabolite regulatory mechanisms, which determine the synthesis of hydrolytic enzymes and substrate transport systems, as well as their physiological activities. Catabolite regulatory mechanisms enable colonic micro-organisms to selectively use certain substrates to the exclusion of others, enabling some species to compete more efficiently for a restricted range of preferred substrates in relatively specialized metabolic niches. Because different bacteria form distinct patterns of metabolic end products, catabolite regulatory mechanisms ultimately affect the types and amounts of these substances that can be produced from individual substrates. Consequently, these control processes are ecologically important, and are of considerable physiological significance to the host (Macfarlane and Macfarlane, 2003). Hopkins et al., (2001) compared the growth of several bifidobacterial strains on a range of oligosaccharide substrates. Results showed that as with monosaccharide utilization, considerable inter-species differences existed in their abilities to utilize these substrates. GOS and oligofructose supported the best growth of the test bacteria, while pyrodextrins and xylo-oligosaccharides were poor substrates for the test strains. Growth was often better on oligosaccharides compared to their component monosaccharides, supporting previous suggestions that some bifidobacterial species may be particularly adapted to growth on oligosaccharides in the large gut (Macfarlane et al., 2003). Ohtsuka et al., (1989) found that several pure cultures of bifidobacteria (*B. bifidum, B. adolescentis, B. longum, B. infantis, B. breve*) grew well on GOS, but only one of four lactobacillus species tested (*Lactobacillus acidophilus*) utilized the substrate. No growth was seen with *L. gasseri, L. salivarius or L. casei*. Substrate preferences occur in the utilization of GOS by intestinal bacteria. This was shown in a study where *B. adolescentis* was able to degrade oligosaccharides with a d.p. > 3, while *B. Infantis* and *L. acidophilus* could only use those with a d.p. < 3 (Van Den Broek et al., 2003). In another study, substrate preferences were seen with respect to GOS utilization, where *Bif. lactis*, unlike *L. rhamnosus* was able to metabolize tri and tetrasaccharide fractions (Gopal et al., 2001). As mentioned earlier, organisms other than bifidobacteria and lactobacilli are able to grow on GOS, including species of clostridia and bacteroides, which can utilize the trisaccharides 4´ galactosyllacose and 6´ galactosyllactose (Gopal et al., 2001; Ohtsuka et al., 1989), and it seems that the pattern of glycosidic linkages in different GOS products influences the abilities of individual bacteria to grow on them.

Currently, in order to enumerate some bacteria species, such as lactobacilli, several methods have been developed and used to accomplish this task. However, some disadvantages of many techniques increase the need for the use and the development of novel methods that assure the real growth of bacteria species. Traditional plating methods have insufficient selectivity and the presence of non culturable bacteria in fecal samples; the fluorescent in situ hybridization (FISH) has a high detection limit, which disables the quantification of very low bacterial numbers present in fecal samples; the conventional PCR assay, is only used for semiquantitative assessment, due to endpoint analysis limitations such as the plateu phase (O'Sullivan, 2000; Walter et al., 2001).

A new method to demonstrated that supplemented infant formulas with a mixture of GOS-FOS can stimulate the growth of species of Lactobacilli was developed recently and compared with the efficiency of the detection of conventional PCR assays (Haarman and Knol, 2006). Duplex 5´nucelase assay was more sensitive than the conventional PCR assays (1,000- to 10,000-fold) and a detection limit ranging from 0.75 to 1.25 CFU/ml compared to the conventional plating techniques. These results showed that Duplex 5´ nuclease assays were designed, optimized, validated, and used to study the distribution of *Lactobacillus* species in fecal samples of infants obtained from a nutritional intervention study. With these accurate assays, it was demonstrated that after an intervention with a mixture of galacto- and fructo-oligosaccharides the *Lactobacillus* species distribution in the feces of formula-fed infants closely resembles the distribution in breast-fed infants. A major finding of this study is that GOS-FOS supplemented in a standard formula results in a *Lactobacillus* distribution with relatively high levels of *L. acidophilus*, *L. casei*, and *L. paracasei*, which is rather similar to that of breastfed infants.

Several studies and efforts have shown that the use of oligosaccharides stimulating the bifidobacteria and lactobacilli is useful in promoting GI health. Thus, providing different types of oligosaccharides and their mixtures as ingredients to enrich formulas or to develop novel functional foods will increase the consumer's health and wellness.

General Gastro-Intestinal Health

Naturally, non-digestible oligosaccharides are found in breast milk and in fermented milk products. Supplementation with prebiotic non-digestible oligosaccharides has been referred to as an important step in improving infant milk formula. In preterm infants formula feeding is commonly associated with hard stools, delayed gastrointestinal transport, and constipation. As these problems considerably hamper tolerance to enteral feeding, it would be desirable to attain reduction of stool viscosity and acceleration of gastrointestinal transport. Boehm et al., (2002) demonstrated that preterm infants fed with mother's milk had lower stool consistency and higher stool frequency than infants fed a preterm bovine milk formula. Supplementation of the same formula with a mixture of short chain GOS and long chain FOS resulted in a reduction in stool consistency and an increase in stool frequency. While, Mihatsch et al., (2006) demonstrated a clinically relevant reduction in the gastrointestinal transit time in preterm infants fed a formula supplemented with these prebiotics. Later, Indrio et al., (2009) showed that prebiotic oligosaccharides can modulate the electrical activity and the gastric emptying may improve the intestinal tolerance of enteral feeding in preterm infants. Although breast milk is the ideal source of nutrition for infants, advancements in infant formula research are necessary to provide the best possible alternative for infants who cannot receive human milk. In a recent study (Ashley et al., 2012) routine infant formula with adjustments in fat, carbohydrate, and calcium composition and supplemented with 4 g/L of either the prebiotic blend of polydextrose and GOS or GOS alone was well-tolerated and supported normal growth. Compared to infants who received the unsupplemented control formula, infants receiving prebiotic supplementation experienced a softer stooling pattern similar to that reported in breastfed infants. In this regard, Mihatsch et al. (2007) investigating a mixture of prebiotic non-digestible oligosaccharides (GosFos; referring to galacto- and fructo-oligosaccharides) showed that GosFos significantly reduces both stool viscosity and gastrointestinal transit time without adverse effects in preterm infants. One benefit of breast milk over infant formulas is greater gastrointestinal comfort. The effect of a whey-predominant starter formula containing LCPUFAs and oligosaccharides (FOS/GOS) on gastrointestinal comfort in infants was evaluated by Vivatvakin et al., (2010). The data suggested that in infants the prebiotic-containing whey-based formula provides superior gastrointestinal comfort than a control formula. Neutral and acidic oligosaccharides (small-chain galacto-oligosaccharides (scGOS)/long-chain fructo-oligosaccharides (lcFOS)) decreases stool viscosity and stool pH with a trend towards increased stool frequency in preterm infants; however, the inclusion of pectin-derived acidic oligosaccharides (pAOS) in a formula containing a mixture of scGOS/lcFOS does not add specific advantages to the formula in terms of stool viscosity, frequency, pH as well as feeding tolerance (Westerbeek et al., 2011).

Any carbohydrate fermented or not that reaches the large bowel should have a laxative effect. Macfarlane et al., (2006) summarized the results of seven published investigations in which mean daily faecal weight was measured, and the response to a prebiotic determined. A significant increase in stool output was observed in only two of the seven studies with 1.3 g of stool/g of prebiotic for oligofructose (134–154 g of stool/day) in the study of Gibson et al., (1995) and 2.4 g/g for inulin (129–204 g/day) in the study of Castiglia-Delavaud et al., (1998). Four studies recorded virtually no change in bowel habit.

The positive effects of a daily intake of FOS can contribute to an increase in the microbial mass and gas production, increasing the cecal content of the colon. SCFAs, the end products of FOS fermentation by the bacteria of the colon, are absorbed and used by the cells of the human colonic epithelium, stimulating their growth as well as the absorption of salts and water increasing the humidity of the cecal bolus through osmotic pressure, resulting in increased intestinal motility. This increase in the cecal bolus stimulates its passage through the colon, shortening its transit time and reducing the time available for the reabsorption of water. These factors increase the weight of the faeces and impart a softer composition (Sabater-Molina et al., 2009). FOS are an important source of intestinal gas and, for some people, rapid gas generation may produce abdominal pain, eructation, flatulence, stomach swelling, and intestinal cramps. Furthermore, the high intake of FOS can cause diarrheic processes due to the osmotic retention of liquids in the small and large intestine. However, the appearance and intensity of these effects depends on the dose, daily consumption of the product, and can vary significantly from one individual to another (Sabater-Molina et al., 2009). Hond et al., (2000) observed that the time of intestinal transit is not affected by a daily dose of 4-15 g/day, although such a dose would increase deposition frequency and fecal volume in healthy individuals. In older persons suffering constipation the consumption of inulin at 20-40 g/day increases deposition frequency although with great variability between subjects and accompanied in some cases by flatulence (Kleessen et al., 1997).

An unexpected finding from a traveller's diarrhoea study was the significantly greater proportion of subjects on FOS group (12.9% vs. 4.7%, $P < 0.04$) responding affirmatively to the post study questionnaire, by ticking the box affirming a general improvement in well-being. Well-being is a state of body and mind that is very difficult to define and measure (Cummings et al., 2001). It is, however, a core principle of the functional food concept that wellness is improved rather than treating disease or symptoms. Paienau et al., (2008) reported that regular consumption of sc-FOS improve digestive comfort in a working population not undergoing medical treatment. Using the functional digestive disorders quality of life questionnaire, the discomfort item scores increased in the sc-FOS group ($P = 0{\cdot}03$). However, daily activities expressed as change in quality of life (improvement, decline or unchanged), improved significantly in the sc-FOS group (P = 0.02). scFOS ingestion in elderly healthy humans produced no observable effect on stool composition and oro-fecal transit time (Bouhnik et al., 2007). However, in a placebo-controlled, double blind study, yacon, a root crop containing high FOS, was administered to 16 healthy individuals (8 males, 8 females) in a dose of 20 g daily (equal to 6.4 g FOS) in a 2-week crossover design (Geyer et al., 2008). Yacon markedly accelerates colonic transit in healthy individuals. On the other hand, inulin, a non-digestible oligosaccharide classified as a prebiotic, is a substrate that promotes the growth of certain beneficial gut microorganisms. Slavin and Feirtag (2011) examined the effect of a 20 g day^{-1} supplement of chicory inulin on stool weight, intestinal transit time, stool frequency and consistency, among other parameters, in twelve healthy male volunteers; inulin was well tolerated, but had minimal effects on measures of laxation in healthy human subjects. The consumption of breads with in situ-produced arabinoxylan oligosaccharides (AXOS) may favorably modulate intestinal fermentation and overall gastrointestinal properties, according to a healthy human cross-over study (Damen et al., 2012).

Effects of Oligosaccharides on Mineral Absorption

Absorption of an adequate amount of calcium and other minerals is particularly important during early adolescence to help achieve peak bone mass. The current recommended daily intakes of calcium are largely based on dose-effect relations to maximize its net retention, which is primarily determined by intestinal calcium absorption. This is especially important, given the large disparity between recommended and typical intakes of calcium (Yates et al., 1998). Experimental evidence indicates that NDOs have a role to play in increasing bioavailability of the absorption of both calcium and magnesium in growing animals and in adolescents. Numerous animal studies have shown that NDOs significantly increase calcium absorption and bone mineralization (Coudray et al., 2003; Roberfroid et al., 2002). Inulin and oligofructose improve mineral absorption, which is associated with the production of SCFAs and a lower pH in the intestinal lumen. Compared to inulin, oligofructose stimulated calcium absorption slightly more effectively, while the effect on magnesium was similar. The production of total SCFAs was not different but lactate was significantly higher following xylo-oligosaccharides and butyrate and was highest in oligofructose (derived both from inulin and sucrose) fed rats (Scholz-Ahrens et al., 2001). Moreover, it has been claimed the dose of NDOs and the background dietary calcium used in different studies have their own impact on mineral absorption (Scholz-Ahrens and Schrezenmeir, 2002).

In young growing rats NDOs exerted their most prominent effects on calcium and magnesium availability. In aged ovariectomized rats 5% oligofructose added to a diet high in calcium (1 %) persistently reduced faecal calcium loss after 4 and 8 weeks. This effect became significant after 16 weeks. Calcium retention was slightly but persistently higher after 4, 8, and 16 weeks. This effect failed significance because increased absorption was associated with increased urinary calcium. Higher urinary calcium however, reflected metabolic reaction on higher calcium influx from the gut and not calcium mobilization from the skeleton because bone mineral content was higher on oligofructose (Scholz-Ahrens et al., 2002).

The effect of NDOs (inulin and FOS) on calcium metabolism was realized using an ovariectomized rat model to mimic postmenopausal women (Zafar et al., 2004). The absorption, excretion, bone formation, and bone resorption were measured by combining a metabolic balance study. Results showed that control (no NDOs diet) and NDOs groups had similar weight, food intake and total calcium intake (Table 1). Fecal calcium excretion was lower in the NDOs group, and they absorbed and retained more calcium than the control group. Diet supplemented with NDOs not only increased the calcium content, but bone mineral density (BMD) in the distal part of the femurs and the suppression of bone resorption. However, BMD in proximal and midshaft did not differ, perhaps due to the relatively short feeding period. This study demonstrated that the protective effect of NDOs on bone in a rat model to mimic menopausal women was established through the following: 1) increased calcium absorption, 2) calcium balance, 3) bone mineralization, and 4) decreased bone turnover rate.

Most studies evaluating the beneficial effect of NDOs on mineral absorption are carried out in rat models. However, during the last years, some human studies have been carried (adolescents and adults) to investigate the potential beneficial effect of FOS on mineral intestinal absorption, particularly that of Ca (Coudray et al., 1997; Van Den Heuvel et al., 1999). One mineral that plays an important role in human health is magnesium (Mg). Mg has an essential role in a wide range of fundamental cellular reactions and it is not surprising that an increasing number of clinical disorders such as diabetes, cardiovascular disease, and osteoporosis have been associated with Mg deficiency (Durlach et al., 1999).

The effect of a 5-week lasting intake of a moderate daily dose (10 g) of scFOS on intestinal Mg absorption was investigated in postmenopausal women (Tahiri et al., 2001). The results showed that the addition of 10 g sc-FOS to the diet increased Mg absorption by 12.3%, (from 30.2 ± 5.0% placebo treatment to 33.9 ± 7.2% sc-FOS treatment; $p < 0.02$). This increase in intestinal Mg absorption was accompanied by an increase in plasma Mg level resulting in higher urinary Mg excretion. Since many animal studies clearly report a beneficial effect of fermentable carbohydrates including FOS on intestinal Mg absorption in different animal models, this was one of the first studies in humans demonstrating a significant effect of a nutritional dose of sc-FOS on intestinal Mg absorption.

Table 1. Effect of feeding NDOs on body weight, calcium balance, and femur measurements in ovariectomized rats. Adapted from Zafar et al., (2004)

Variable	Control	NDOs
N	13	13
Final body weight, g	344 ± 24	337 ± 23
Food intake, g/d	12 ± 2	11 ± 2
Ca intake, mg/d	62 ± 9	60 ± 11
Urinary Ca, mg/d	0.9 ± 0.4	1.2 ± 0.4
Fecal Ca, mg/d	57 ± 9	41 ± 10***
Ca balance[1], mg/d	4 ± 13	19 ± 9**
Apparent Ca absorption, %	8 ± 2	32 ± 4**
Femur wet weight, g	1.04 ± 0.07	1.07 ± 0.6
Femur length, cm	37.2 ± 0.6	37.4 ± 0.8
Femoral total Ca, mg	145 ± 10	155 ± 10*
BMD of distal femur, g/cm^2	0.203 ± 0.02	0.221 ± 0.02*
Peak force of femur breaking, kg	12 ± 3	4 ± 1

Values are means ± SD. Asteriks indicate difference from control: * P< 0.05; ** P < 0.005; *** P< 0.001.

[1] Calcium balance data: Ca intake – fecal Ca- urinary Ca were collected over 4 d.

The mechanism for increased calcium retention is not known, but has been related to enhanced absorption efficiency. It is hypothesized that NDOs is fermented by the intestinal microflora in the colon and converted into SCFAs that lower the luminal pH. The insoluble, unabsorbed calcium coming from the upper part of the intestine is converted to the ionic form in the lower pH medium. Both low pH and SCFAs result in the hypertrophy of the mucosal cells, leading to an enlargement of the surface area of the intestine and enhanced calcium absorption (Remesy et al., 1993; Younes et al., 1996). Others have proposed increased paracellular calcium diffusion in the ileum due to an enhanced permeability of the tight junctions in the presence of NDOs (Kishi et al., 1996). An increase in the ceco-colorectal calbindin-D9K protein expression in the presence of inulin has also been reported (Takasaki et al., 2000). In addition, butyrate, a substrate for cell growth and proliferation (Lupton and Kurtz, 1993) has given prominence to be a potent candidate for enhancing mineral absorption. By increasing the floral butyrate production, FOS might indirectly induce cell growth and growth of the gut's absorptive area, another way to contribute to the enhanced mineral absorption. Moreover, FOS may stimulate active calcium transport via butyrate production. Sodium butyrate increased the concentration of calbindin and of 1, 25[OH]2D3 receptor activity in chick primary culture cells (Maiyar and Norman, 1992). Results reported by Le Blay et al., (1999) support the hypothesis that butyrate may have a marked impact on the effect of oligofructose. They observed that oligofructose reduced the caecal lumen pH after 2 weeks and this was mainly due to stimulated synthesis of lactate (63·5mmol/g wet contents at oligofructose and 3·6 mmol/g at basal) > acetate > butyrate. However, after 8 weeks, lactate production declined to concentration of basal diet, while acetate and butyrate were persistently stimulated over 27 weeks. These results indicated a relationship between high luminal butyrate concentrations and increased mineral uptake across the apical membrane.

Different studies have demonstrated that the use of NDOs and their mixtures exert several beneficial properties in human health, such as the prevention of some chronic diseases, increase in resistance to some intestinal pathogens, and mineral absorption, which can be used as future therapies to prevent bone related diseases, i.e. osteoporosis. Overall, the excellent properties of NDOs, make them suitable for the enrichment of different food products or to formulate novel functional foods, with the principal aim to enhance the human health and wellness.

Oligosaccharides as Immune Modulators

Several dietary compounds are known to modulate immune functions in human by multiple ways; many studies demonstrate the potential role of nutrients within the immune system. However, there are also many non-essential food components whose functions may be associated with immunomodulation, but have not yet been extensively studied. These constituents include phytochemicals, microbial products of food fermentation (probiotics and their metabolites) and prebiotics. Recent results indicate that prebiotics may modulate the systemic immune system as well as the local immune system in the gut (Watzl et al., 2005).

Mechanisms contributing to altered *in vivo* immune function induced by functional foods may include microflora modulation, improved barrier function and direct effects of bacteria on different epithelial and immune cell types (monocytes/macrophages, B cells, T cells and NK cells) (Romeo et al., 2010).

The supplementation with prebiotic oligosaccharides has been extensively studied in infants and children. The beneficial effect on the immune system of preterm infants due to the specific conditions in the luminal part of the developing gut wall after administration of the combination of neutral oligosaccharides with acidic oligosaccharides (maximal dose of 1.5 g/kg/day added to breast milk or preterm formula), have been suggested by Westerbeek et al., (2008). Arslanoglu et al., (2007) investigated the prevention of infections during the first 6 mo of life by oligosaccharides supplementation (8 g/L scGOS/lcFOS) in term infants. After 6 mo of supplementation, infants in the GOS/FOS group had fewer episodes of all types of infections combined, such as overall, upper respiratory tract infections (URTI), otitis media, gastrointestinal infections, and urinary tract infections compared to control group. Timing of this immune modulation coincides with the critical period of gut colonization.

Bakker-Zierikzee et al., (2006) studied the effects of prebiotic and probiotic infant formulas, compared with a standard diet, on the faecal secretory immunoglobulin A (SIgA) levels, which plays an important role in the gastrointestinal tract defense. Prebiotic diet was formed by a mixture of 0.6 g GOS/FOS/100 ml formula and probiotic diet by $6x10^9$ cfu *Bifidobacterium animalis* (Bb-12)/100 ml formula. Although many factors may influence SIgA survival in the large intestine, measuring SIgA levels in faeces gives a good representation of the amount available in the colon (Prentice et al., 1987). The results of the study indicated that adding non-digestible oligosaccharides to infant formula results in higher levels of faecal SIgA compared with unsupplemented infant formula, while probiotic strain Bb-12 was found to be less effective. This suggests the possibility to stimulate the development of the mucosal immune response, decreasing the episodes of gastrointestinal and respiratory tract infections by non-digestible oligosaccharides even without probiotics synergy (Bakker-Zierikzee et al., 2006).

Moreover, Bruzzese et al., (2009) added galacto and fructooligosaccharides to infant formula and tested the hypothesis that prebiotics reduce the incidence of the intestinal and respiratory infections in healthy infants during 12 months. The results confirmed that GOS/FOS formula significantly decreased diarrheal episode/child compared to control group, and also URTI episodes resulted in a number lower in the GOS/FOS group. URTI was defined as more than 3 episodes of URTI in 12 months. As a result, antibiotic prescription was lower in GOS/FOS group. The results together, provide a new option for disease prevention strategy in the first year of life. Several preventive strategies as vaccination against frequent agents have been used for a long time, and other specific therapies are lacking. This accounts for the importance of further options to support the immune system of our children, in a natural and less invasive way. The European Society for Pediatric Gastroenterology, Hepatology and Nutrition, has made a call for clinical trials on prebiotics in order to produce recommendations on their inclusion in infant formula (Agostoni et al., 2005).

The mechanisms by which oligosaccharides exert their immunoprotective action have not been fully elucidated; however, there have been numerous efforts to establish the basis of its function, through the study of oligosaccharides from human milk and other foods, in both children and adults. The postnatal period is a crucial time for the maturation of the immune system. At birth, T-lymphocytes exhibit a Th2- profile, characterized by a limited ability to produce cytokines. Throughout the first months after birth these Th2-skewed responses are modified into low-level immunity, predominantly expressing Th1-cytokines and IgG-antibodies, particularly of the IgG1 subclass (Holt and Jones, 2000). Human milk contain large amounts of free oligosaccharides, that affect the production of murine IL-10 (Velupillai and Harn, 1994), and is involved in the generation of anti-inflammatory mediators that suppress Th1-type and inflammatory responses in mice (Terrazas et al., 2001). Eiwegger et al., (2004) examined the capability of two different fractions of HMOs and low-molecular-weight fucoidan (LMWF), as an example of plant-derived oligosaccharides, on function and maturation of cord blood mononuclear cells (CBMC) *in vitro*. The presence of the acidic

HMOs and the LMWF severely affected T-cell cytokine production, acidic HMOs also significantly elevated IFN- γ-producing $CD4^+$ and $CD8^+$ T-cells and IL-13 production in $CD8^+$ T-cells. In contrast, LMWF did not affect IFN-γ production in $CD4^+$ T-cells. Moreover, LMWF increased the amount of IL-4–producing $CD3^+CD4^+$ T-cells and the percentage of IL-13–expressing $CD3^+CD8^+$ T-cells significantly. In addition, acidic HMOs and LMWF significantly increased CD25 expression.

In order to elucidate the biochemical mechanism responsible for the immunomodulatory protective effect of the oligosaccharides, Hosono et al., (2003) orally administrated FOS to BALB/c mice for 6 weeks and measured the intestinal mucosal immune responses. FOS given in two doses: 0-7.5% and 2.5% up-regulated IgA secretion in the intestinal mucosa and CD4+T cells derived from Peyer's patch (PP, site of induction of mucosal immune responses) induced Th1 and Th2 cytokine responses when stimulated by the mitogen BP, derived from Gram-positive bacteria. The ability of PP cells to produce IgA was dose-dependent in response to FOS. Moreover, IL-10 and IFN-γ secretions increased by FOS administration, in a dose-dependent manner. Secretions of IL-5 and IL-6 were maintained at high secretion in response to FOS-feeding and systemic immune responses (IgG1 and IgG2a production) were remarkable at 6 weeks by FOS administration. The authors concluded that this is enough information to suggest that FOS acts as a preventive agent with regard to food allergy.

Nakamura et al., (2004) examined whether dietary FOS in infancy can have a beneficial effect on the mucosal immune system. The study conducted with newborn BALB/c mice administered 5% (w/w) of FOS through the diet. FOS supplementation increased IgA levels in intestinal tissue extracts (jejunum, ileum and colon). Moreover, ileal and colonic, but not jejuna, polymeric immunoglobulin receptor (pIgR) expression, which plays a critical role in transepithelial transport of intestinal IgA onto mucosal surface, were significantly higher in FOS group compared to control diet. Thus, dietary FOS affects the mucosal immune system in the small intestine of infant mice as well as in the colon.

Several types of oligosaccharides also have beneficial effects on asthma episodes (Vos et al., 2007). scGOS mixed with lcFOS with or without pAOS were used as supplement in Male BALB/cByJIco SPF mice. Significant effects were observed on airway hyperactivity and the number of inflammatory cells in the BAL fluid, contributing to decreased allergic asthma. Conversely, pAOS exerted more pronounced effects than the scGOS/lcFOS mixture without pAOS, but a direct comparison between the mixtures was not made. The results are very important to design products with oligosaccharide or its combination thereof that can help improve the health of infant or other age groups.

Feruloylated oligosaccharides (FOs), the ferulic acid ester of oligosaccharides, display good antioxidant capacity making it a potential candidate for health food, pharmaceutical and cosmetic industries. Fang et al., (2012) studying the effects of FOs on the anti-inflammatory cytokines and the inflammatory mediators release in RAW264.7 stimulated with LPS, found that FOs stimulated RAW264.7 cells to produce TNF-α, IL-β, IL-6, NO and PGE2, and thereby exhibited immunomodulatory activities in innate immunity. In fact, the authors suggest that the consumption of FOs may enhance innate immunity, and protect against cardiovascular and chronic inflammatory diseases.

Recently, a mixture of R-galacto-oligosaccharides was isolated and characterized from a new source, the roots of *Lycopus lucidus* Turcz. (RL), used for centuries as an oriental traditional medicine and dietary herb for its anti-inflammatory, cardiac, sedative, wound-healing, pain relief, and tonic effects (Slusarczyk et al., 2009; Lee et al., 2008). The plant contains sucrose, raffinose, stachyose, and verbascose. The research opens the possibility of exploiting RL-GOS as a new source of bioactive oligosaccharides with immunostimulatory activity, because RL-GOS intervention significantly enhanced cellular and humoral immune responses (Yang et al., 2010).

Other mechanism of host defense of the oligosaccharides that result in the improvement of immunity response is the mucosal barrier. It has been shown that supplementing diets of chicken, pig, and rats with oligofructose and other oligosaccharides reduces fecal densities of *Salmonella* (Letellier et al. 2000), *Candida* in small intestine (Buddington et al. 2002a), and increases colonization resistance to *Clostridium difficile* after antibiotic treatment (Gaskins et al., 1996). The interactions between the bacteria and host defenses can be direct and involve attachment of gastrointestinal tract bacteria to receptors of host cells, thereby permitting 'cross-talk'. Adherence of *Lactobacilli* to receptors of epithelial cells can induce expression of genes coding for variants of intestinal mucins that inhibit the adherence of enteropathogenic *E. coli* to the epithelium (Buddington et al., 2002b).

Oligosaccharides act as intestinal receptor analogs because enteropathogenic bacteria and their toxins use oligosaccharide part of glycoproteins as a target for their adhesion (Kunz et al., 2000). Mammalian cell membranes contain glycolipids and glycoproteins, and its membrane structure is such that the oligosaccharide chains of the cell membrane glycoconjugates protrude into the extracellular region. Complex carbohydrate chains are involved in cell-cell interactions and may act as receptors for hormones as well as later humoral mediators. They also act as receptor antibodies, toxins and pathogens. There are common sites where viruses and bacteria adhere to target host cells to inhibit the pathogenic process. Oligosaccharides, probably similar in structure to those found on surfaces, can act as soluble homolog to the cell surface receptors, targets of specific pathogens, inhibiting its binding to host cells and therefore the colonization process (Newburg, 2000; Dai et al., 2000; Zhang and Normark, 1996).

Finally, there is another hypothesis claiming that SCFAs, produced by fermentation of oligosaccharides, influence the immune system by their immunomodulatory and anti-inflammatory properties (Kelly-Quagliana and Buddington, 2003). The increased concentrations of SCFAs resulting from bacterial fermentation of non-digestible oligosaccharides, and particularly butyrate enhance NK cytotoxic activity and increase the densities of splenic T-helper cells (Campbell and Fahey, 1997). Therefore, the SCFAs may represent an important set of signals for immunomodulation and triggering a TH1 shift. Moreover, butyrate inhibited *Staphylococcus aureus*-induced TH1 cytokines IL-2 and IFN-g and upregulated TH2 cytokine IL-4 and anti-inflammatory IL-10 production, and prevented NF-κB translocation to the nucleus thereby preventing transcription of the proinflammatory genes TNF-α, IL-1β, and IL-6 (Meijer et al., 2010).

In conclusion, non-digestible oligosaccharides are considered non-caloric agents because of their resistantance to the hydrolytic action of digestive enzymes. Once they reach the colon, oligosaccharides produce short chain fatty acids due to degradation by intestinal bacteria. Oligosaccharides are excellent prebiotics, because they act like growth factor to particular commensal bacteria, which inhibit the adherence and invasion of pathogens in the colonic epithelia by competing for the same glycoconjugates present on the surface of epithelial cells, altering the colonic pH, favoring the barrier function, improving the mucus production, producing short-chain fatty acids and inducing cytokine production, promoting multiple benefits and better health (Mussatto and Mancilha, 2007; Choque-Delgado et al., 2011).

Preventing Colon Cancer by Non-Digestible Oligosaccharides

Many non-digestible carbohydrates are not digested by humans due to lack of enzymes required to hydrolyze the β-links formed among the units of some monosaccharides. Once non-digestible carbohydrates reach the caeco-colon, most are hydrolyzed to small oligomers and monomers by one, a few, or most of the anaerobic bacteria. The hydrolysis occur through a metabolic process, known as fermentation, providing energy for proliferation, producing gases (H_2, CO_2, CH_4), metabolically useless to the host, and SCFAs (Hirayama et al., 2002; Scheppach et al., 2004).

NDOs fermentation in the caeco-colon by the bacteria modulates several physiological functions, primarily reduction of cancer risk of the gut. The anticarcinogenic effect is related to an increase in cellular immunity, the components of the cell wall and the extra-cellular components of bifidobacteria. SCFAs production is also relevant to avoid the development of neoplastic colonic disease (Scheppach et al., 2004; Mussatto and Mancilha 2007). Short chain fatty acids provide protection against cancer development, which is carried out by several mechanisms that include tumor suppression, apoptosis, cell proliferation, and affect proteolytic and antiproteolytic enzymes (Emenaker et al., 2001). Especially butyric acid, have been suggested to play a role in reducing the risk of cancer and acting as a cancer suppressing agent (Kautenburger et al., 2005). Butyric acid at physiological concentrations of 1-10 mmol/ml stops the growth of neoplastic enterocytes the G1 phase of the cell cycle in multiple culture cell lines and *in vivo* models, where experience is more limited (Benzineb-Rouet et al., 2004; Kim et al., 2004; Li and Elsasser, 2005). Butyric acid is the main energy substrate for colonocytes and is involved in cell differentiation, adherence, proliferation, motility and apoptosis, as well as gene regulator of cell cycle (Emenaker et al., 2001; Fung et al., 2012).

The anticarcinogenic potential of butyrate is due to induction of RNAm expression and glutation S-Transferase (GST) activity in cell of human colonic tumors (Knoll et al., 2005). The apoptosis induced by butyrate is caspase-dependent and caspase-independent, and it also increase the expression of pro-apoptotic proteins (p21, p53) (Hague et al., 1993; Chen et al., 2006). Moreover, butyrate induces RNAm and protein levels of p21WAF1/CIP1 for cell cycle block in G1 phase, resulting in inhibition of cell proliferation (Scheppach and Weiler, 2004). Butyrate also dismisses Bcl-2 expression in cell cultures and increases Bax expression, which induce apoptosis (Avivi-Green et al., 2001).

Several authors highlight the potential for nutritional intervention with specific non-digestible carbohydrates, especially oligosaccharides, to selectively increase butyrate production, which is important for benefits to colon health mentioned above. Morrison et al., (2006) showed that addition of oligofructose to faecal batch cultures significantly increased butyrate production and that selectively stimulates the bacterial conversion of acetate and lactate to butyrate. The authors concluded that selectively increasing butyrate production will depend on the regional differences throughout the colon of bacteria expressing butyryl CoA:acetyl CoA transferase and the regional differences in lactate-utilizing, butyrate-producing bacteria. Manderson et al., (2005) studied the prebiotic properties of pectin-oligosaccharides (POS) extracted from the peel by-product of orange juice manufacturing, resulting in increased butyrate production by POS after 10 h of fermentation (10.69 mM of butyrate), peaking at 24 h (18 mM of butyrate). The increase SCFAs concentration, especially butyrate, in mice large bowel was significantly higher for FOS and GOS groups than MOS and COS (chitooligosaccharides) groups (Pan et al., 2009). This information is important because companies producing dietary supplements, particularly infant formulas, could use these types of oligosaccharides as ingredients with potential benefits of the gastrointestinal tract.

The effects of xylo-oligosaccharides and FOS on the inhibition of precancerous colon lesions, called aberrant crypt foci (ACF) in male Sprague-Dawley rats was studied by Hsu et al., (2004). The rats were fed FOS or GOS for 35 days, and a third group without oligosaccharides, and colon cancer was induced with 1,2-dimethylhydrazine (DMH). Data suggested that dietary supplementation with 60 g/kg xilo-oligosaccharides and FOS inhibited the development of ACF precancerous lesions, promoted the growth of bifidobacteria, and lowered the cecal pH in rats. Pierre et al., (1997) observed that scFOS reduced colon tumors associated with the distal colon. In contrast, there were no significant difference between diets in the number of tumors found in the small intestine suggesting strongly that events specific to the colon were involved. scFOS counteract advanced stages of colon carcinogenesis, possibly via stimulation of antitumoral immunity by modulation of the colonic ecosystem.

Recently, Pietro-Femia et al., (2010) probed arabinoxylan (AX)-oligosaccharides (AXOS) to test its effect in a rat colon carcinogenesis model by determining the preneoplastic lesions: ACF and mucin depleted foci

(MDF). The AXOS group had the same number, but different distribution of ACF than control. Furthermore, the AXOS group had significantly less ACF in distal colon than the control group, indicating an interesting protection, because in this zone ACF are more differentiated and most of them develop into adenocarcinoma (Reddy et al., 1976). On the other hand, the number and multiplicity of MDF in the entire colon was significantly lower in the animals fed AXOS than in the control group, resulting in reduced MDF in distal zone in AXOS group, similar to the ACF. These data together, showed a protective effect of AXOS on two types of preneoplastic lesions in a short-term rat colon carcinogenesis model.

Li et al., (2013) performed a deeper analysis of oligosaccharide chempoprotection against colon cancer. The authors prepared oligosaccharide from apple polysaccharide and investigated its antitumor activity and mechanism on HT-29 human colon cancer cells. The apple oligossacharide displayed cytotoxicity on HT29 cells as a result of apoptosis determined by Anenexin V-FITC/PI staining and flow cytometry. Apple oligosaccharide induced more apoptosis than 5 FU and apple polysaccharide group consistent with the cytotoxicity. Furthermore, the percentage of cells in the G1 phase was reduced by incubating with apple oligosaccharide or 5-FU, accompanied by an increase in cells in S phase. Bax expression was enhanced with apple oligosaccharide or 5-FU, while the expression of antiapoptotic proteins Bcl-2 and Bcl-xl was suppressed, and cyclin B1 and Cdk2 were down-regulated. The expression of cyclin A1 was up-regulated with at high oligosaccharide concentration (10μmol/l). Also mRNA expression after apple oligosaccharide treatment was investigated, showing that mRNA levels of *bcl-2, bcl-xl, cdk2* and *cyclin B1* decreased. This study demonstrates that oligosaccharides induce apoptosis and cell cycle arrest by protein and gene modulation, and approaches to elucidate the possible mechanism of action.

Other mechanisms such as the prevention of adenomas, measured by parameters of colorectal carcinogenesis have been suggested for the physiological actions for the protective effect of oligosaccharides. Generally, decrease of faecal enzyme activities (including β-glucuronidase, β-glucosidase, nitroreductase and urease) and a decrease of the concentration of secondary bile acids (lithocholic acid and deoxycholic acid) could reduce the risk for developing colon cancer (Goldin and Gorbach, 1984; Reddy et al., 1987; Pool-Zobel et al., 2002). Faecal enzymes enable the release of carcinogens in the intestinal lumen from phase II enzymes, which is the biologically active form of xenobiotics and other compounds (Rafter, 2002). Fatty acids and secondary bile acids act as carcinogens capable of generating hyper-proliferation in epithelial cells of the crypts, causing DNA damage and selecting for apoptosis-resistant cells (both leading to increased mutation) (Bernstein et al., 2005).

The supplementation period (15 days) of FOS significantly decreased the concentration of deoxycholic acid in dry faeces in a group of healthy male subjects (van Dokkum et al., 1999). Buddington et al., (1996) administered 4 g of FOS (neosugar)/d for 25 days to 20 healthy adults, and observed reduced β-glucuronidase and glycocholic acid hydroxylase activities in 12 subjects without affecting nitroreductase activity. β-glucosidase and β-glucuronidase activities were also reduced by the addition of low (0.1%, w/v) soy oligosaccharides concentration to a two-stage continuous fecal bacterial culture resulting in changes in the microflora and a significant decrease in azoreductase activity (Saito et al., 1992). It should be noted that the main oligosaccharides contained in soybeans are raffinose and stachyose.

Boutron-Ruault et al., (2005) analyzed further biological markers between adenoma and adenoma-free subjects, before and after 3 months of daily intake of 10 g sc-FOS in a multicenter study. 5 g of sc-FOS administered twice daily for 3 months to subjects with small (<10 mm diameter), large, or no adenomas significantly reduced lithocholic acid and increased cholic ($P = 0.02$), chenodeoxycholic, total primary bile and ursodeoxycholic acids. However, sc-FOS consumption had no effect on colonic cell proliferation. In patients with adenomas (especially small adenomas), the consumption of sc-FOS increased butyrate and SCFAs excretion to values similar to those in adenoma-free controls and interestingly, these patients had similar fecal levels but higher serum concentrations of this bile acid than adenoma-free subjects. Other studies, such as those of Bouhnik et al., (1996) demonstrating no effect on nitroreductase, azoreductase, or β-glucuronidase activities in feces report the opposite effect of oligosaccharides. Moreover, Juskiewicz et al., (2006) reported that FOS (0.5, 1 and 2%) administered in a diet for 8 wk to commercial male turkeys produced a slight, but statistically non significant beneficial decrease in β-glucosidase and β-glucuronidase activities. Finally, the adhesion mechanism studied in detail specifically for milk oligosaccharides in colon cancer prevention proposes that the adhesion capacity of oligosaccharides to epithelial cell receptor prevents the adhesion of pathogens by modulating signaling cascades that lead to cell cycle arrest and apoptosis.

As mentioned above, the oligosaccharides may induce or suppress the gene and protein expression involved in cellular processes that prevent carcinogenic progress. Kuntz et al., (2009) observed the influence on cell cycle events via flow cytometry and expression levels using quantitative real-time

RT–PCR of human milk oligosaccharides. The total oligosaccharide fraction was further separated into acidic (acidic HMO) and neutral fractions (neutral HMO). After treatment with neutral or acidic HMO, cells accumulated in the G2/M phase of the cell cycle (44, 30 and 31 %; and 37, 44 and 33% of HT-29, HIEC and Caco-2 cells, respectively, after exposure to neutral and acidic HMO oligosaccharides, respectively). This indicates that the anti-proliferative effect of acidic and neutral oligosaccharides occurs in G2/M arrest in intestinal epithelial cells. On the other hand, the authors investigated the effect of oligosaccharides on mRNA expression of key protein regulating cell cycle. The treatment with oligosaccharides did not alter cyclin D and E mRNA levels, suggesting that cells enter the cell cycle and proceed to G2. However, the treatment increased cyclin B1 expression, which prevents the cells from entering the G2/M stage, reducing cell proliferation with the application of oligosaccharides. In HT-29 and HIEC cells an increase of mRNA expression level of cyclin A was observed after incubation with acidic oligosaccharides. Oligosaccharides from human milk also up-regulated $p21^{cip1}$ mRNA expression, indicating the participation of oligosaccharides on G2/M arrest. Finally, oligosaccharides were able to induce EGFR phosphorylation and the authors confirmed that the GFR/Ras/Raf/ERK pathway is involved.

Oligossacharides Applications

Functional and/or designer foods reach beyond the traditional role of food as nutrition; they provide specific function and/or achieve specific health benefit(s) upon sustained consumption. Functional foods in relation to oligosaccharides and other prebiotics target gut health. GOS and other prebiotics have been suggested for various human food products such as baked goods, sweeteners, yogurts, nutrition bars and meal replacement shakes, as well as pet foods (Lamsal, 2012). The major prebiotics used in human nutrition are inulin, galacto-oligosaccharides, fructo-oligosaccharides, lactulose, malto-oligosaccharides and resistant starch (Singh et al., 2012). These prebiotic ingredients can be utilized for either their nutritional advantages or functional properties, but they are often applied to offer a double benefit: an improved organoleptic quality and a better-balanced nutritional approach. GOS and other prebiotic ingredients are proposed for various food applications; major physicochemical properties that make them useful in various food systems were discussed above. The use of inulin and non-digestible oligosaccharides as fiber ingredients leads to improved taste and texture in food, along with health benefits. In yogurt and desert applications, prebiotic oligosaccharides are used for sugar replacement, texture and mouth feel, and fiber delivery, whereas in beverage applications foam stabilization also becomes important. In bakery and cake applications, moisture retention is the important property imparted by prebiotics, along with sugar replacement and fiber delivery. Other applications of prebiotic oligosaccharides for mouthfeel, fiber and sugar replacement include baby food, fillings, confectionery and sauces (Lamsal, 2012). Soluble inulin fibers influence dough and bread quality (Hager et al., 2011).

Rodrigues et al., (2012) investigated the influence of FOS and inulin (50:50) on the free fatty acid profile of cheese, with special emphasis on the conjugated linoleic acid. Increase of the conjugated linoleic acid content during the ripening stage suggests the addition of prebiotics in probiotic cheese manufacture, for better quality product with lower atherogenicity index. Fermentable dietary fibers such as short-chain FOS can be supplemented in foods to induce satiety and thus prevent obesity (Hess et al., 2011). Wong et al., (2010) studied the effect of soy food diet enriched with prebiotic in hyperlipidemic adults. Results showed that intake of soy-fortified prebiotic resulted in greater reductions in low-density lipoprotein cholesterol and increase in high-density lipoprotein. Co-ingestion of a prebiotic may potentiate the effectiveness of soy foods in improving serum cholesterol profile.

Presad and Sherkat, (2013) examined the influence of GOS and modified waxy maize starch (MWMS) addition on the growth of starter cultures, and syneresis and firmness of low-fat yogurt during storage, finding a significant ($P< 0.05$) increase in proteolysis and lactic and acetic acid concentrations. Prebiotic soluble fibre (fructooligosaccharides)-incorporated whey protein crisps have been produced by low-shear supercritical fluid extrusion (SCFX), which utilizes supercritical CO_2 as an expansion agent instead of steam (Paraman et al., 2013). The process produces an expanded protein matrix with unique internal microstructure of uniformly distributed closed cells. Also, the abundance of arabinoxylan (AX) in bran offered excellent possibilities for manufacturing AXOS, which meet the criteria for prebiotics (Delcour et al., 2012). Other study has opened a new avenue for preparation of commonly consumed FOS incorporated in Indian food products (Sheth, 2013). Wheat flour in chapati and thepla, and bengal gram in dhokla and patra substituted with FOS significantly reduced all organoleptic attributes of chapati and thepla, while for chapatis the most affected attribute were chewability, breakability and overall acceptability.

Additionally, a juice containing prebiotic oligosaccharide is a potential new functional beverage. Da Silva et al., (2012) produced prebiotic oligosaccharides (104.73 g/L) by enzymatic process with addition of the dextransucrase enzyme in a clarified cashew apple juice.

Ice creams supplemented with *L. casei* and 2.5% inulin showed good nutritional and sensory properties (Criscio et al., 2010). Rößle et al., (2010) developed potentially synbiotic fresh- cut apple wedges by applying probiotic bacteria (*L. rhamnosus GG*) and prebiotics oligofructose and inulin. Fructan analysis showed that all prebiotics remained stable over the 14-day storage period and an intake of 100 g of apple supplied 2–3 g of prebiotics. Different dehydrated prebiotic fibers viz. oat bran, β-glucan and green banana flour provided substratum for adherence and trehalose acted as a cell protectant prolonging the viability of *L. casei*. In the sensory evaluation, the prebiotic oat bran added to a dairy fruit beverage has been well accepted by consumers (Guergoletto et al., 2010). Apple purees enriched with two commercial FOS prebiotics (Beneo GR®(inulin) and HSI®) showed stability for 30 days' storage at 4 °C (Keenan et al., 2011). However, ample quantities of prebiotics were required to deliver prebiotic effect and high hydrostatic pressure posed the risk of certain prebiotics hydrolysis. Lyophilized inulin syrup made from Jerusalem artichoke tuber was used as a prebiotic ingredient in the small-scale manufacture of wafer crackers. Sensory analysis revealed a significant influence of product formulation on appearance, flavor and texture of the crackers (Hempel et al., 2007). Also, when baking fat is replaced (50%, 75% or 100%) by inulin in a muffin, the increasing amounts of added inulin, increase product moisture and crumb density significantly, whereas muffin volume decreased (Zahn et al., 2010).

On the other hand, the utilization of oligosaccharides by oral microorganisms and intestinal enzymes are important factors in determining their effectiveness as alternative sweeteners. Hodoniczky et al., (2012) analyzed oligosaccharide properties in structural classes; oligosaccharides containing a (1 → 6)-β-Glc group, including gentiobiose and gentiobiitol, together with melezitose, a sucrose-derived oligosaccharide containing a α-Glc-(1 → 3)-Fru moiety were resistant to digestion by both *S. mutans* and mammalian intestinal enzymes, highlighting their potential as dietary sugar substitutes. Protonotariou et al., (2013) substituted (10–50%) sucrose with FOS in cream caramel dessert. Panellists did not find significant difference in terms of sweetness when sucrose was substituted up to 30%.

Conclusion

According to a global consensus non-digestible oligosaccharides play an important role in nutrition and health of individuals. Non-digestible oligosaccharides are obtained from the diet by extracts of natural sources such as inulin, followed by an enzymatic hydrolysis, oligosaccharides in food or synthesized by subjecting disaccharides such as lactose or sucrose to the transferases action. The quality of the oligosaccharides escaping the enzymatic action in the gastrointestinal tract, position them as powerful prebiotics, because they reach the colon and act as excellent substrates for beneficial bacteria in such organ. Moreover, its ability to enhance mineral absorption, immune function, and protect against cancer development, makes them attractive to the pharmaceutical and nutrition industries, reason for their frequent use and various applications, but mainly as functional ingredients in food. This has stimulated the research regarding the discovery of new oligosaccharides. This chapter shows that the oligosaccharides in low doses, exhibit great benefits and contribute to the health of individual, and will be exploited by many industries for a common purpose, the general health of the population and disease prevention of global prevalence.

References

Agostoni, C., Axelsson, I., Goulet, O., Koletzko, B., Michaelsen, K. F., Puntis, J. W., ... & Turck, D. (2004). Prebiotic oligosaccharides in dietetic products for infants: a commentary by the ESPGHAN Committee on Nutrition. *Journal of pediatric gastroenterology and nutrition*, *39*(5), 465-473.

Amine, E., Baba, N., Belhadj, M., Deurenbery-Yap, M., Djazayery, A., Forrester, T., ... & Yoshiike, N. (2002). Diet, nutrition and the prevention of chronic diseases: report of a Joint WHO/FAO Expert Consultation.

Arslanoglu, S., Moro, G. E. & Boehm, G. (2007). Early supplementation of prebiotic oligosaccharides protects formula-fed infants against infections during the first 6 months of life. *The Journal of nutrition*, *137*(11), 2420-2424.

Ashley, C., Johnston, W. H., Harris, C. L., Stolz, S. I., Wampler, J. L. & Berseth, C. L. (2012). Growth and tolerance of infants fed formula supplemented with polydextrose (PDX) and/or galactooligosaccharides

(GOS): double-blind, randomized, controlled trial. *Nutrition Journal, 11*(1), 38.

Avivi-Green, C., Polak-Charcon, S., Madar, Z. & Schwartz, B. (2002). Different molecular events account for butyrate-induced apoptosis in two human colon cancer cell lines. *The Journal of nutrition, 132*(7), 1812-1818.

Bakker-Zierikzee, A. M., Tol, E. V., Kroes, H., Alles, M. S., Kok, F. J. & Bindels, J. G. (2006). Faecal SIgA secretion in infants fed on pre-or probiotic infant formula. *Pediatric allergy and immunology, 17*(2), 134-140.

Bernstein, H., Bernstein, C., Payne, C. M., Dvorakova, K. & Garewal, H. (2005). Bile acids as carcinogens in human gastrointestinal cancers. *Mutation Research/Reviews in Mutation Research, 589*(1), 47-65.

Blaut, M. (2002). Relationship of prebiotics and food to intestinal microflora.*European Journal of Nutrition, 41*(1), i11-i16.

Blaut, M., Collins, M. D., Welling, G. W., Dore, J., Van Loo, J. & De Vos, W. (2002). Molecular biological methods for studying the gut microbiota: the EU human gut flora project. *British journal of nutrition, 87*(2), 203-212.

Boehm, G., Lidestri, M., Casetta, P., Jelinek, J., Negretti, F., Stahl, B. & Marini, A. (2002). Supplementation of a bovine milk formula with an oligosaccharide mixture increases counts of faecal bifidobacteria in preterm infants. *Archives of Disease in Childhood-Fetal and Neonatal Edition, 86*(3), F178-F181.

Bouhnik, Y., Achour, L., Paineau, D., Riottot, M., Attar, A. & Bornet, F. (2007). Four-week short chain fructo-oligosaccharides ingestion leads to increasing fecal bifidobacteria and cholesterol excretion in healthy elderly volunteers. *Nutr J, 6*, 42.

Bouhnik, Y., Flourié, B., Riottot, M., Bisetti, N., Gailing, M. F., Guibert, A., ... & Rambaud, J. C. (1996). Effects of fructo-oligosaccharides ingestion on fecal bifidobacteria and selected metabolic indexes of colon carcinogenesis in healthy humans.

Bouhnik, Y., Vahedi, K., Achour, L., Attar, A., Salfati, J., Pochart, P., ... & Rambaud, J. C. (1999). Short-chain fructo-oligosaccharide administration dose-dependently increases fecal bifidobacteria in healthy humans. *The Journal of nutrition, 129*(1), 113-116.

Boutron-Ruault, M. C., Marteau, P., Lavergne-Slove, A., Myara, A., Gerhardt, M. F., Franchisseur, C., ... & The Eripolyp Study Group. (2005). Effects of a 3-mo consumption of short-chain fructo-oligosaccharides on parameters of colorectal carcinogenesis in patients with or without small or large colorectal adenomas. *Nutrition and cancer*, *53*(2), 160-168.

Brouns, F. (2002). Soya isoflavones: a new and promising ingredient for the health foods sector. *Food Research International*, *35*(2), 187-193.

Bruzzese, E., Volpicelli, M., Squeglia, V., Bruzzese, D., Salvini, F., Bisceglia, M., ... & Guarino, A. (2009). A formula containing galacto-and fructo-oligosaccharides prevents intestinal and extra-intestinal infections: an observational study. *Clinical Nutrition*, *28*(2), 156-161.

Buddington, K. K., Donahoo, J. B. & Buddington, R. K. (2002). Dietary oligofructose and inulin protect mice from enteric and systemic pathogens and tumor inducers. *The Journal of nutrition*, *132*(3), 472-477.

Buddington, R. K., Kelly-Quagliana, K., Buddington, K. K. & Kimura, Y. (2002). Non-digestible oligosaccharides and defense functions: lessons learned from animal models. *British Journal of Nutrition*, *87*(2), 231-240.

Buddington, R. K., Williams, C. H., Chen, S. C. & Witherly, S. A. (1996). Dietary supplement of neosugar alters the fecal flora and decreases activities of some reductive enzymes in human subjects. *The American journal of clinical nutrition*, *63*(5), 709-716.

Campbell, J. M., Fahey, G. C. & Wolf, B. W. (1997). Selected indigestible oligosaccharides affect large bowel mass, cecal and fecal short-chain fatty acids, pH and microflora in rats. *The Journal of nutrition*, *127*(1), 130-136.

Castiglia-Delavaud, C., Verdier, E., Besle, J. M., Vernet, J., Boirie, Y., Beaufrere, B., ... & Vermorel, M. (1998).Net energy value of non-starch polysaccharide isolates (sugarbeet fibre and commercial inulin) and their impact on nutrient digestive utilization in healthy human subjects. *British journal of nutrition*, *80*(04), 343-352.

Charalampopoulos, D., Wang, R., Pandiella, S. S. & Webb, C. (2002). Application of cereals and cereal components in functional foods: a review.*International Journal of Food Microbiology*, *79*(1), 131-141.

Chen, J., Ghazawi, F. M., Bakkar, W. & Li, Q. (2006). Valproic acid and butyrate induce apoptosis in human cancer cells through inhibition of gene expression of Akt/protein kinase B. *Molecular cancer*, *5*(1), 71.

Choi, B. K., Kim, K. Y., Yoo, Y. J., Oh, S. J., Choi, J. H. & Kim, C. Y. (2001). In vitro antimicrobial activity of a chitooligosaccharide mixture against< i> Actinobacillus</i>< i> actinomycetemcomitans</i> and< i> Streptococcus mutans</i>. *International journal of antimicrobial agents*, *18*(6), 553-557.

Choque Delgado, G. T., Tamashiro, W. M. D. S. C., Junior, M. R. M., Moreno, Y. M. F. & Pastore, G. M. (2011). The putative effects of prebiotics as immunomodulatory agents. *Food Research International*, *44*(10), 3167-3173.

Coudray, C., Bellanger, J., Castiglia-Delavaud, C., Remesy, C., Vermorel, M. & Rayssignuier, Y. (1997). Effect of soluble or partly soluble dietary fibres supplementation on absorption and balance of calcium, magnesium, iron and zinc in healthy young men. *European Journal of Clinical Nutrition*, *51*(6), 375-380.

Coudray, C., Tressol, J. C., Gueux, E. & Rayssiguier, Y. (2003). Effects of inulin-type fructans of different chain length and type of branching on intestinal absorption and balance of calcium and magnesium in rats. *European journal of nutrition*, *42*(2), 91-98.

Cummings, J. H. & Macfarlane, G. T. (2002). Gastrointestinal effects of prebiotics. *British Journal of Nutrition*, *87*(S2), S145-S151.

Cummings, J. H., Christie, S. & Cole, T. J. (2001).A study of fructo oligosaccharides in the prevention of travellers' diarrhoea. *Alimentary pharmacology & therapeutics*, *15*(8), 1139-1145.

da Silva, I. M., Rabelo, M. C. & Rodrigues, S. (2012). Cashew juice containing prebiotic oligosaccharides.*Journal of Food Science and Technology*, 1-7.

Dai, D., Nanthkumar, N. N., Newburg, D. S. & Walker, W. A. (2000). Role of oligosaccharides and glycoconjugates in intestinal host defense. *Journal of pediatric gastroenterology and nutrition*, *30*, S23-S33.

Damen, B., Cloetens, L., Broekaert, W. F., François, I., Lescroart, O., Trogh, I., ... & Courtin, C. M. (2012).Consumption of Breads Containing In Situ–Produced Arabinoxylan Oligosaccharides Alters Gastrointestinal Effects in Healthy Volunteers. *The Journal of nutrition*, *142*(3), 470-477.

De Vries, W., Gerbrandy, S. J. & Stouthamer, A. H. (1967). Carbohydrate metabolism in< i> Bifidobacterium bifidum</i>. *Biochimica et Biophysica Acta (BBA)-General Subjects*, *136*(3), 415-425.

Delcour, J. A., Rouau, X., Courtin, C. M., Poutanen, K. & Ranieri, R. (2012). Technologies for enhanced exploitation of the health-promoting potential of cereals. *Trends in Food Science & Technology*, *25*(2), 78-86.

Den Hond, E., Geypens, B. & Ghoos, Y. (2000). Effect of high performance chicory inulin on constipation. *Nutrition Research*, *20*(5), 731-736.

Di Criscio, T., Fratianni, A., Mignogna, R., Cinquanta, L., Coppola, R., Sorrentino, E. & Panfili, G. (2010). Production of functional probiotic, prebiotic, and synbiotic ice creams. *Journal of dairy science*, *93*(10), 4555-4564.

Durlach, J., Bac, P., Bara, M. & Guiet-Bara, A. (1999). Cardiovasoprotective foods and nutrients: possible importance of magnesium intake. *Magnesium research: official organ of the International Society for the Development of Research on Magnesium*, *12*(1), 57.

Eiwegger, T., Stahl, B., Schmitt, J., Boehm, G., Gerstmayr, M., Pichler, J., ... & Szépfalusi, Z. (2004). Human milk–derived oligosaccharides and plant-derived oligosaccharides stimulate cytokine production of cord blood T-cells in vitro. *Pediatric research*, *56*(4), 536-540.

Emenaker, N. J., Calaf, G. M., Cox, D., Basson, M. D. & Qureshi, N. (2001). Short-chain fatty acids inhibit invasive human colon cancer by modulating uPA, TIMP-1, TIMP-2, mutant p53, Bcl-2, Bax, p21 and PCNA protein expression in an in vitro cell culture model. *The Journal of nutrition*, *131*(11), 3041S-3046S.

Estruch, R., Martínez-González, M. A., Corella, D., Salas-Salvadó, J., Ruiz-Gutiérrez, V., Covas, M. I., ... & Ros, E. (2006). Effects of a Mediterranean-Style Diet on Cardiovascular Risk Factors A Randomized Trial. *Annals of internal medicine*, *145*(1), 1-11.

Evans, M., Sinclair, R. C., Fusimalohi, C. & Liava'a, V. (2001). Globalization, diet, and health: an example from Tonga. *Bulletin of the World Health Organization*, *79*(9), 856-862.

Fang, H. Y., Chen, Y. K., Chen, H. H., Lin, S. Y. & Fang, Y. T. (2012). Immunomodulatory effects of feruloylated oligosaccharides from rice bran. *Food chemistry*, *134*(2), 836-840.

Femia, A. P., Salvadori, M., Broekaert, W. F., François, I. E., Delcour, J. A., Courtin, C. M. & Caderni, G. (2010). Arabinoxylan-oligosaccharides (AXOS) reduce preneoplastic lesions in the colon of rats treated with 1, 2-dimethylhydrazine (DMH). *European journal of nutrition*, *49*(2), 127-132.

Fung, K. Y., Cosgrove, L., Lockett, T., Head, R. & Topping, D. L. (2012). A review of the potential mechanisms for the lowering of colorectal oncogenesis by butyrate. *British Journal of Nutrition*, *108*(5), 820.

Gaskins, H.R., Mackie, R.I., May, T., Garleb, K.A. (1996) Dietary fructo-oligosaccharides modulate large intestinal inflammatoryresponses to Clostridium difficile in antibiotic-compromised mice. *Microbial ecology in health and disease, 9*, 157–166.

Geyer, M., Manrique, I., Degen, L. & Beglinger, C. (2008). Effect of yacon (Smallanthus sonchifolius) on colonic transit time in healthy volunteers. *Digestion*, *78*(1), 30-33.

Gibson, G. R., Beatty, E. R., Wang, X. & Cummings, J. H. (1995).Selective stimulation of bifidobacteria in the human colon by oligofructose and inulin. *Gastroenterology*, *108*(4), 975-982.

Gibson, G. R., Probert, H. M., Van Loo, J., Rastall, R. A. & Roberfroid, M. B. (2004). Dietary modulation of the human colonic microbiota: updating the concept of prebiotics. *Nutr Res Rev*, *17*(2), 259-275.

Goldin, B. R. & Gorbach, S. L. (1984). The effect of milk and lactobacillus feeding on human intestinal bacterial enzyme activity. *The American journal of clinical nutrition*, *39*(5), 756-761.

Gopal, P. K., Sullivan, P. A. & Smart, J. B. (2001). Utilisation of galacto-oligosaccharides as selective substrates for growth by lactic acid bacteria including< i> Bifidobacterium lactis</i> DR10 and< i> Lactobacillus rhamnosus</i> DR20. *International Dairy Journal*, *11*(1), 19-25.

Grootaert, C., Delcour, J. A., Courtin, C. M., Broekaert, W. F., Verstraete, W. & Van de Wiele, T. (2007). Microbial metabolism and prebiotic potency of arabinoxylan oligosaccharides in the human intestine. *Trends in Food Science & Technology*, *18*(2), 64-71.

Guergoletto, K. B., Magnani, M., Martin, J. S., Andrade, C. G. T. D. J. & Garcia, S. (2010). Survival of< i> Lactobacillus casei</i>(LC-1) adhered to prebiotic vegetal fibers. *Innovative Food Science & Emerging Technologies*, *11*(2), 415-421.

Haarman, M. & Knol, J. (2005). Quantitative real-time PCR assays to identify and quantify fecal Bifidobacterium species in infants receiving a prebiotic infant formula. *Applied and environmental microbiology*, *71*(5), 2318-2324.

Haarman, M. & Knol, J. (2006). Quantitative real-time PCR analysis of fecal Lactobacillus species in infants receiving a prebiotic infant formula. *Applied and environmental microbiology*, *72*(4), 2359-2365.

Hager, A. S., Ryan, L. A., Schwab, C., Gänzle, M. G., O'Doherty, J. V. & Arendt, E. K. (2011). Influence of the soluble fibres inulin and oat β-glucan on quality of dough and bread. *European Food Research and Technology*, *232*(3), 405-413.

Hague, A., Manning, A. M., Hanlon, K. A., Hart, D., Paraskeva, C. & Huschtscha, L. I. (1993). Sodium butyrate induces apoptosis in human colonic tumour cell lines in a p53-independent pathway: implications for the possible role of dietary fibre in the prevention of large-bowel cancer. *International Journal of Cancer*, *55*(3), 498-505.

Hass, R. A. L. F., Busche, R. O. G. E. R., Luciano, L. I. L. I. A. N. A., Reale, E. N. R. I. C. O. & Engelhardt, W. V. (1997). Lack of butyrate is associated with induction of Bax and subsequent apoptosis in the proximal colon of guinea pig. *Gastroenterology*, *112*(3), 875-881.

Hempel, S., Jacob, A. & Rohm, H. (2007). Influence of inulin modification and flour type on the sensory quality of prebiotic wafer crackers. *European Food Research and Technology*, *224*(3), 335-341.

Hess, J. R., Birkett, A. M., Thomas, W. & Slavin, J. L. (2011).Effects of short-chain fructooligosaccharides on satiety responses in healthy men and women. *Appetite*, *56*(1), 128-134.

Hirayama, M. (2002). Novel physiological functions of oligosaccharides. *Pure and applied chemistry*, *74*(7), 1271-1279.

Hodoniczky, J., Morris, C. A. & Rae, A. L. (2012). Oral and intestinal digestion of oligosaccharides as potential sweeteners: A systematic evaluation. *Food Chemistry*, *132*(4), 1951-1958.

Holt, P. G. & Jones, C. A. (2000). The development of the immune system during pregnancy and early life. *Allergy*, *55*(8), 688-697.

Hopkins, M. J., Sharp, R. & Macfarlane, G. T. (2001). Age and disease related changes in intestinal bacterial populations assessed by cell culture, 16S rRNA abundance, and community cellular fatty acid profiles. *Gut*, *48*(2), 198-205.

Hosono, A., Ozawa, A., Kato, R., OHNISHI, Y., Nakanishi, Y., Kimura, T. & Nakamura, R. (2003). Dietary fructooligosaccharides induce immunoregulation of intestinal IgA secretion by murine Peyer's patch cells. *Bioscience, biotechnology, and biochemistry*, *67*(4), 758-764.

Hsu, C. K., Liao, J. W., Chung, Y. C., Hsieh, C. P. & Chan, Y. C. (2004). Xylooligosaccharides and fructooligosaccharides affect the intestinal microbiota and precancerous colonic lesion development in rats. *The Journal of nutrition*, *134*(6), 1523-1528.

Hu, F. B. (2011). Globalization of Diabetes The role of diet, lifestyle, and genes. *Diabetes Care*, *34*(6), 1249-1257.

Indrio, F., Riezzo, G., Raimondi, F., Francavilla, R., Montagna, O., Valenzano, M. L., ... & Boehm, G. (2009). Prebiotics improve gastric motility and gastric electrical activity in preterm newborns. *Journal of pediatric gastroenterology and nutrition*, *49*(2), 258-261.

James, W. P. T. (1988). *Healthy nutrition. Preventing nutrition-related diseases in Europe*. WHO Regional Office for Europe.

Jeffery, I. B. & O'Toole, P. W. (2013). Diet-Microbiota Interactions and Their Implications for Healthy Living. *Nutrients*, *5*(1), 234-252.

Juskiewicz, J., Jankowski, J., Zdunczyk, Z. & Mikulski, D. (2006). Performance and gastrointestinal tract metabolism of turkeys fed diets with different contents of fructooligosaccharides. *Poultry science*, *85*(5), 886-891.

Kaur, I. P., Chopra, K. & Saini, A. (2002). Probiotics: potential pharmaceutical applications. *European Journal of Pharmaceutical Sciences*, *15*(1), 1-9.

Kaur, N. & Gupta, A. K. (2002). Applications of inulin and oligofructose in health and nutrition. *Journal of Biosciences*, *27*(7), 703-714.

Kautenburger, T., Beyer-Sehlmeyer, G., Festag, G., Haag, N., Kühler, S., Küchler, A., ... & Pool-Zobel, B. L. (2005). The gut fermentation product butyrate, a chemopreventive agent, suppresses glutathione S-transferase theta (hGSTT1) and cell growth more in human colon adenoma (LT97) than tumor (HT29) cells. *Journal of cancer research and clinical oncology*, *131*(10), 692-700.

Keenan, D. F., Brunton, N., Butler, F., Wouters, R. & Gormley, R. (2011). Evaluation of thermal and high hydrostatic pressure processed apple purees enriched with prebiotic inclusions. *Innovative Food Science & Emerging Technologies*, *12*(3), 261-268.

Kelly-Quagliana, K. A., Nelson, P. D. & Buddington, R. K. (2003). Dietary oligofructose and inulin modulate immune functions in mice. *Nutrition Research*, *23*(2), 257-267.

Kim, Y. H., Park, J. W., Lee, J. Y. & Kwon, T. K. (2004). Sodium butyrate sensitizes TRAIL-mediated apoptosis by induction of transcription from the DR5 gene promoter through Sp1 sites in colon cancer cells. *Carcinogenesis*, *25*(10), 1813-1820.

Kirjavainen, P. V., Arvola, T., Salminen, S. J. & Isolauri, E. (2002). Aberrant composition of gut microbiota of allergic infants: a target of bifidobacterial therapy at weaning?. *Gut*, *51*(1), 51-55.

Kishi, K., Goda, T. & Takase, S. (1996). Maltitol increases transepithelial diffusional transfer of calcium in rat ileum. *Life sciences*, *59*(14), 1133-1140.

Kleessen, B., Sykura, B., Zunft, H. J. & Blaut, M. (1997). Effects of inulin and lactose on fecal microflora, microbial activity, and bowel habit in elderly constipated persons. *The American journal of clinical nutrition*, *65*(5), 1397-1402.

Knoll, N., Ruhe, C., Veeriah, S., Sauer, J., Glei, M., Gallagher, E. P. & Pool-Zobel, B. L. (2005). Genotoxicity of 4-hydroxy-2-nonenal in human colon tumor cells is associated with cellular levels of glutathione and the modulation of glutathione S-transferase A4 expression by butyrate. *Toxicological Sciences*, *86*(1), 27-35.

Kolida, S., Tuohy, K. & Gibson, G. R. (2002). Prebiotic effects of inulin and oligofructose. *British Journal of Nutrition*, *87*(S2), S193-S197.

Kontula, P., von Wright, A. & Mattila-Sandholm, T. (1998). Oat bran β-gluco- and xylo-oligosaccharides as fermentative substrates for lactic acid bacteria. *International journal of food microbiology*, *45*(2), 163-169.

Korhonen, H. (2002). Technology options for new nutritional concepts. *International Journal of Dairy Technology*, *55*(2), 79-88.

Kuntz, S., Kunz, C. & Rudloff, S. (2009). Oligosaccharides from human milk induce growth arrest via G2/M by influencing growth-related cell cycle genes in intestinal epithelial cells. *British Journal of Nutrition*, *101*(9), 1306.

Kunz, C., Rudloff, S., Baier, W., Klein, N. & Strobel, S. (2000). Oligosaccharides in human milk: structural, functional, and metabolic aspects. *Annual review of nutrition*, *20*(1), 699-722.

Lamsal, B. P. (2012). Production, health aspects and potential food uses of dairy prebiotic galactooligosaccharides. *Journal of the Science of Food and Agriculture*, *92*(10), 2020-2028.

Le Blay, G., Michel, C., Blottière, H. M. & Cherbut, C. (1999). Prolonged intake of fructo-oligosaccharides induces a short-term elevation of lactic acid-producing bacteria and a persistent increase in cecal butyrate in rats. *The Journal of nutrition*, *129*(12), 2231-2235.

Lee, H. W., Park, Y. S., Jung, J. S. & Shin, W. S. (2002). Chitosan oligosaccharides, dp 2–8, have prebiotic effect on the< i> Bifidobacterium bifidium</i> and< i> Lactobacillus</i> sp. *Anaerobe*, *8*(6), 319-324.

Lee, Y. J., Kang, D. G., Kim, J. S. & Lee, H. S. (2008). < i> Lycopus lucidus</i> inhibits high glucose-induced vascular inflammation in human umbilical vein endothelial cells. *Vascular pharmacology*, *48*(1), 38-46.

Letellier, A., Messier, S., Lessard, L. & Quessy, S. (2000). Assessment of various treatments to reduce carriage of Salmonella in swine. *Canadian Journal of Veterinary Research*, *64*(1), 27.

Li, C. J. & Elsasser, T. H. (2005). Butyrate-induced apoptosis and cell cycle arrest in bovine kidney epithelial cells: involvement of caspase and proteasome pathways. *Journal of animal science*, *83*(1), 89-97.

Li, Q., Zhou, S., Jing, J., Yang, T., Duan, S., Wang, Z., ... & Liu, L. (2013). Oligosaccharide from apple induces apoptosis and cell cycle arrest in HT29 human colon cancer cells. *International journal of biological macromolecules*.

Lupton, J. R. & Kurtz, P. P. (1993). Relationship of colonic luminal short-chain fatty acids and pH to in vivo cell proliferation in rats. *The Journal of nutrition*, *123*(9), 1522.

Macfarlane, G. T., Macfarlane, S. & Gibson, G. R. (1998). Validation of a three-stage compound continuous culture system for investigating the effect of retention time on the ecology and metabolism of bacteria in the human colon. *Microbial Ecology*, *35*(2), 180-187.

Macfarlane, S. & Macfarlane, G. T. (2003, February). Regulation of short-chain fatty acid production. In *PROCEEDINGS-NUTRITION SOCIETY OF LONDON*(Vol. 62, No. 1, 67-72). CABI Publishing; 1999.

Macfarlane, S., Macfarlane, G. T. & Cummings, J. T. (2006). Review article: prebiotics in the gastrointestinal tract. *Alimentary pharmacology & therapeutics*, *24*(5), 701-714.

Maiyar, A. C. & Norman, A. W. (1992). Effects of sodium butyrate on 1, 25-dihydroxyvitamin D< sub> 3</sub> receptor activity in primary chick kidney cells. *Molecular and cellular endocrinology*, *84*(1), 99-107.

Malinow, M. R., Bostom, A. G. & Krauss, R. M. (1999). Homocyst (e) ine, diet, and cardiovascular diseases: a statement for healthcare professionals from the Nutrition Committee, American Heart Association. *Circulation*, *99*(1), 178-182.

Manderson, K., Pinart, M., Tuohy, K. M., Grace, W. E., Hotchkiss, A. T., Widmer, W., ... & Rastall, R. A. (2005). In vitro determination of prebiotic properties of oligosaccharides derived from an orange juice manufacturing by-product stream. *Applied and environmental microbiology*, *71*(12), 8383-8389.

Meijer, K., de Vos, P. & Priebe, M. G. (2010). Butyrate and other short-chain fatty acids as modulators of immunity: what relevance for health?. *Current Opinion in Clinical Nutrition & Metabolic Care*, *13*(6), 715-721.

Mihatsch, W. A., Hoegel, J. & Pohlandt, F. (2006). Prebiotic oligosaccharides reduce stool viscosity and accelerate gastrointestinal transport in preterm infants. *Acta Paediatrica*, *95*(7), 843-848.

Miller, T. L. & Wolin, M. J. (1996). Pathways of acetate, propionate, and butyrate formation by the human fecal microbial flora. *Applied and environmental microbiology*, *62*(5), 1589-1592.

Moro, G., Minoli, I., Mosca, M., Fanaro, S., Jelinek, J., Stahl, B. & Boehm, G. (2002). Dosage-related bifidogenic effects of galacto-and fructooligosaccharides in formula-fed term infants. *Journal of pediatric gastroenterology and nutrition*, *34*(3), 291-295.

Morrison, D. J., Mackay, W. G., Edwards, C. A., Preston, T., Dodson, B. & Weaver, L. T. (2006). Butyrate production from oligofructose fermentation by the human faecal flora: what is the contribution of extracellular acetate and lactate? *British Journal of Nutrition*, *96*(3), 570-577.

Mussatto, S. I. & Mancilha, I. M. (2007). Non-digestible oligosaccharides: A review. *Carbohydrate polymers*, *68*(3), 587-597.

Mustalahti, K., Lohiniemi, S., Collin, P., Vuolteenaho, N., Laippala, P. & Maki, M. (2002). Gluten-free diet and quality of life in patients with screen-detected celiac disease. *Effective clinical practice*, *5*(3), 105-113.

Nakamura, Y., Nosaka, S., Suzuki, M., Nagafuchi, S., Takahashi, T., Yajima, T., ... & Moro, I. (2004). Dietary fructooligosaccharides up-regulate immunoglobulin A response and polymeric immunoglobulin receptor expression in intestines of infant mice. *Clinical & Experimental Immunology*, *137*(1), 52-58.

Newburg, D. S. (2000). Oligosaccharides in human milk and bacterial colonization. *Journal of pediatric gastroenterology and nutrition*, *30*, S8-S17.

Ohtsuka, K., Benno, Y., Endo, K., Ueda, H., Ozawa, O., Uchida, T. & Mitsuoka, T. (1989). Effects of 4'-galactosyllactose intake on human fecal microflora. *Bifidus- Flores, Fructus et Semina*, *2*(2), 143-149.

O'Sullivan, D. J. (2000). Methods for analysis of the intestinal microflora. *Current issues in intestinal microbiology*, *1*(2), 39-50.

Ouwehand, A., Isolauri, E. & Salminen, S. (2002). The role of the intestinal microflora for the development of the immune system in early childhood. *European journal of nutrition*, *41*(1), i32-i37.

Paineau, D., Payen, F., Panserieu, S., Coulombier, G., Sobaszek, A., Lartigau, I., ... & Bornet, F. R. (2008).The effects of regular consumption of short-chain fructo-oligosaccharides on digestive comfort of subjects with minor functional bowel disorders. *British Journal of Nutrition*, *99*(02), 311-318.

Pan, X. D., Chen, F. Q., Wu, T. X., Tang, H. G. & Zhao, Z. Y. (2009). Prebiotic oligosaccharides change the concentrations of short-chain fatty acids and the microbial population of mouse bowel. *Journal of Zhejiang University SCIENCE B*, *10*(4), 258-263.

Paraman, I., Supriyadi, S., Wagner, M. E. & Rizvi, S. S. (2013). Prebiotic fibre-incorporated whey protein crisps processed by supercritical fluid extrusion. *International Journal of Food Science & Technology*.

Patel, S. & Goyal, A. (2011). Functional oligosaccharides: production, properties and applications. *World Journal of Microbiology and Biotechnology*, *27*(5), 1119-1128.

Pierre, F., Perrin, P., Champ, M., Bornet, F., Meflah, K. & Menanteau, J. (1997). Short-chain fructo-oligosaccharides reduce the occurrence of colon tumors and develop gut-associated lymphoid tissue in Min mice. *Cancer Research*, *57*(2), 225-228.

Pool-Zobel, B. L., Neudecker, C., Domizlaff, I., Ji, S., Schillinger, U., Rumney, C., ... & Rowland, I. (1996). *Lactobacillus-and bifidobacterium-mediated antigenotoxicity in the colon of rats.*

Pool-Zobel, B., Van Loo, J., Rowland, I. & Roberfroid, M. B. (2002). Experimental evidences on the potential of prebiotic fructans to reduce the risk of colon cancer. *British Journal of Nutrition*, *87*(2), 273-282.

Prasad, L. N., Sherkat, F. & Shah, N. P. (2013). Influence of Galactooligosaccharides and Modified Waxy Maize Starch on Some Attributes of Yogurt. *Journal of food science*, *78*(1), M77-M83.

Prentice, A., Ewing, G., Roberts, S. B., Lucas, A., MacCarthy, A., Jarjou, L. M. A. & Whitehead, R. G. (1987). The Nutritional Role of Breast-Milk IgA and Lactoferrin. *Acta Pædiatrica*, *76*(4), 592-598.

Priebe, M. G., Vonk, R. J., Sun, X., He, T., Harmsen, H. J. & Welling, G. W. (2002). The physiology of colonic metabolism. Possibilities for interventions with pre-and probiotics. *European journal of nutrition*, *41*(1), i2-i10.

Protonotariou, S. V., Karali, E., Evageliou, V., Yanniotis, S. & Mandala, I. (2013). Rheological and sensory attributes of cream caramel desserts containing fructooligosaccharides as substitute sweeteners. *International Journal of Food Science & Technology*, *48*(3), 663-669.

Pryde, S. E., Duncan, S. H., Hold, G. L., Stewart, C. S. & Flint, H. J. (2002). The microbiology of butyrate formation in the human colon. *FEMS microbiology letters*, *217*(2), 133-139.

Puupponen-Pimiä, R., Aura, A. M., Oksman-Caldentey, K. M., Myllärinen, P., Saarela, M., Mattila-Sandholm, T. & Poutanen, A. K. (2002). Development of functional ingredients for gut health. *Trends in Food Science & Technology*, *13*(1), 3-11.

Rafter, J. (2002). Lactic acid bacteria and cancer: mechanistic perspective. *British Journal of Nutrition*, *88*(1), 89-94.

Reddy, B. S., Narisawa, T. & Weisburger, J. H. (1976). Colon carcinogenesis in germ-free rats with intrarectal 1, 2-dimethylhydrazine and subcutaneous azoxymethane. *Cancer research*, *36*(8), 2874-2876.

Reddy, B. S., Sharma, C., Simi, B., Engle, A., Laakso, K., Puska, P. & Korpela, R. (1987). Metabolic epidemiology of colon cancer: effect of dietary fiber on fecal mutagens and bile acids in healthy subjects. *Cancer research*, *47*(2), 644-648.

Remesy, C., Levrat, M. A., Gamet, L. & Demigné, C. (1993). Cecal fermentations in rats fed oligosaccharides (inulin) are modulated by dietary calcium level. *American Journal of Physiology-Gastrointestinal and Liver Physiology*, *264*(5), G855-G862.

Roberfroid, M. (2002). Functional food concept and its application to prebiotics. *Digestive and Liver Disease*, *34*, S105-S110.

Roberfroid, M. B. (1998). Prebiotics and synbiotics: concepts and nutritional properties. *The British journal of nutrition*, *80*(4), S197-202.

Roberfroid, M. B. & Delzenne, N. M. (1998). Dietary fructans. *Annual review of nutrition*, *18*(1), 117-143.

Roberfroid, M. B., Cumps, J. & Devogelaer, J. P. (2002). Dietary chicory inulin increases whole-body bone mineral density in growing male rats. *The Journal of nutrition*, *132*(12), 3599-3602.

Roberfroid, M., Gibson, G. R., Hoyles, L., McCartney, A. L., Rastall, R., Rowland, I., ... & Meheust, A. (2010). Prebiotic effects: metabolic and health benefits. *British Journal of Nutrition*, *104*(S2), S1-S63.

Rodrigues, D., Rocha-Santos, T. A., Gomes, A. M., Goodfellow, B. J. & Freitas, A. C. (2012). Lipolysis in probiotic and synbiotic cheese: The influence of probiotic bacteria, prebiotic compounds and ripening time on free fatty acid profiles. *Food Chemistry*, *131*(4), 1414-1421.

Romeo, J., Nova, E., Wärnberg, J., Gómez-Martínez, S., Díaz Ligia, L. E. & Marcos, A. (2010). Immunomodulatory effect of fibres, probiotics and synbiotics in different life-stages. *Nutr Hosp*, *25*(3), 341-9.

Rößle, C., Brunton, N., Gormley, R. T., Ross, P. R. & Butler, F. (2010). Development of potentially synbiotic fresh-cut apple slices. *Journal of Functional Foods*, *2*(4), 245-254.

Rouet-Benzineb, P., Aparicio, T., Guilmeau, S., Pouzet, C., Descatoire, V., Buyse, M. & Bado, A. (2004). Leptin counteracts sodium butyrate-induced apoptosis in human colon cancer HT-29 cells via NF-κB signaling. *Journal of Biological Chemistry*, *279*(16), 16495-16502.

Saavedra, J. M., Bauman, N. A., Perman, J. A., Yolken, R. H. & Oung, I. (1994). Feeding of Bifidobacterium bifidum and Streptococcus thermophilus to infants in hospital for prevention of diarrhoea and shedding of rotavirus. *The Lancet*, *344*(8929), 1046-1049.

Sabater-Molina, M., Larqué, E., Torrella, F. & Zamora, S. (2009).Dietary fructooligosaccharides and potential benefits on health. *Journal of physiology and biochemistry*, *65*(3), 315-328.

Saito, Y., Takano, T. & Rowland, I. (1992). Effects of soybean oligosaccharides on the human gut microflora in *in vitro* culture. *Microbial Ecology in Health and Disease*, *5*(2), 105-110.

Sako, T., Matsumoto, K. & Tanaka, R. (1999). Recent progress on research and applications of non-digestible galacto-oligosaccharides. *International Dairy Journal*, *9*(1), 69-80.

Scheppach, W. & Weiler, F. (2004). The butyrate story: old wine in new bottles?. *Current Opinion in Clinical Nutrition & Metabolic Care*, *7*(5), 563-567.

Scheppach, W., Luehrs, H., Melcher, R., Gostner, A., Schauber, J., Kudlich, T., ... & Menzel, T. (2004). Antiinflammatory and anticarcinogenic effects of dietary fibre. *Clinical nutrition supplements*, *1*(2), 51-58.

Schneeman, B. O. (2002). Gastrointestinal physiology and functions. *British Journal of Nutrition*, *88*(S2), S159-S163.

Schneeman, B. O. (2003). Evolution of dietary guidelines. *Journal of the American Dietetic Association*, *103*(12), 5-9.

Scholz-Ahrens, K. E. & Schrezenmeir, J. (2002). Inulin, oligofructose and mineral metabolism—experimental data and mechanism. *British Journal of Nutrition*, *87*(S2), S179-S186.

Scholz-Ahrens, K. E., Schaafsma, G., van den Heuvel, E. G. & Schrezenmeir, J. (2001). Effects of prebiotics on mineral metabolism. *The American journal of clinical nutrition*, *73*(2), 459s-464s.

Schumacher, G., Bendas, G., Stahl, B. & Beermann, C. (2006). Human milk oligosaccharides affect P-selectin binding capacities: in vitro investigation. *Nutrition*, *22*(6), 620-627

Sheth, M. K. (2013). Feasibility and acceptability of fructooligosaccharides substituted popular Indian foods. *Nutrition & Food Science*, *43*(3), 3-3.

Singh, R., Sharma, P. K. & Malviya, R. (2012). Prebiotics: future trends in health care. *Mediterranean Journal of Nutrition and Metabolism*, *5*(2), 81-90.

Slavin, J. & Feirtag, J. (2011). Chicory inulin does not increase stool weight or speed up intestinal transit time in healthy male subjects. *Food & Function*, *2*(1), 72-77.

Ślusarczyk, S., Hajnos, M., Skalicka-Woźniak, K. & Matkowski, A. (2009). Antioxidant activity of polyphenols from<i>Lycopus lucidus</i> Turcz. *Food Chemistry*, *113*(1), 134-138.

Stahl, B., Thurl, S., Zeng, J. R., Karas, M., Hillenkamp, F., Steup, M. & Sawatzki, G. (1994). Oligosaccharides from human milk as revealed by matrix-assisted laser desorption/ionization mass spectrometry. *Analytical biochemistry*, *223*(2), 218-226.

Swennen, K., Courtin, C. M. & Delcour, J. A. (2006). Non-digestible oligosaccharides with prebiotic properties. *Critical reviews in food science and nutrition*, *46*(6), 459-471.

Tahiri, M., Tressol, J. C., Arnaud, J., Bornet, F., Bouteloup-Demange, C., Feillet-Coudray, C., ... & Coudray, C. (2001). Five-week intake of short-chain fructo-oligosaccharides increases intestinal absorption and status of magnesium in postmenopausal women. *Journal of bone and mineral research*, *16*(11), 2152-2160.

Takasaki, M., Inaba, H., Ohta, A., Motohashi, Y., Sakai, K., Morris, H. & Sakuma, K. (2000). Dietary short-chain fructooligosaccharides increase calbindin-D9k levels only in the large intestine in rats independent of dietary calcium deficiency or serum 1, 25 dihydroxy vitamin D levels. *International journal for vitamin and nutrition research*, *70*(5), 206-213.

Terrazas, L. I., Walsh, K. L., Piskorska, D., McGuire, E. & Harn, D. A. (2001). The schistosome oligosaccharide lacto-N-neotetraose expands Gr1+ cells that secrete anti-inflammatory cytokines and inhibit proliferation of naive CD4+ cells: a potential mechanism for immune polarization in helminth infections. *The Journal of Immunology*, *167*(9), 5294-5303.

Thurl, S., Henker, J., Siegel, M., Tovar, K. & Sawatzki, G. (1997). Detection of four human milk groups with respect to Lewis blood group dependent oligosaccharides. *Glycoconjugate journal*, *14*(7), 795-799.

Thurl, S., Müller-Werner, B. & Sawatzki, G. (1996). Quantification of individual oligosaccharide compounds from human milk using high-pH anion-exchange chromatography. *Analytical biochemistry*, *235*(2), 202-206.

Topping, D. L. (1996). Short-chain fatty acids produced by intestinal bacteria. *Asia Pacific Journal of Clinical Nutrition*, *5*, 15-19.

Trebichavsky, I., Rada, V., Splichalova, A. & Splichal, I. (2009). Cross-talk of human gut with bifidobacteria. *Nutrition reviews*, *67*(2), 77-82.

Tsai, G. J., Zhang, S. L. & Shieh, P. L. (2004). Antimicrobial activity of a low-molecular-weight chitosan obtained from cellulase digestion of chitosan. *Journal of Food Protection®*, *67*(2), 396-398.

Tungland, B. C. (2003, January). Fructooligosaccharides and other fructans: structures and occurrence, production, regulatory aspects, food applications, and nutritional health significance. In *ACS Symposium Series* (Vol. *849*, pp. 135-152). Washington, DC; American Chemical Society; 1999.

Tungland, B. C. & Meyer, D. (2002). Nondigestible Oligo-and Polysaccharides (Dietary Fiber): Their Physiology and Role in Human Health and Food. *Comprehensive Reviews in food science and food safety*, *1*(3), 90-109.

Tuohy, K. M., Rouzaud, G. C. M., Bruck, W. M. & Gibson, G. R. (2005). Modulation of the human gut microflora towards improved health using prebiotics-assessment of efficacy. *Current pharmaceutical design*, *11*(1), 75-90.

van den Broek, L., Struijs, K., Verdoes, J., Beldman, G. & Voragen, A. (2003). Cloning and characterization of two α-glucosidases from Bifidobacterium adolescentis DSM20083. *Applied microbiology and biotechnology*, *61*(1), 55-60.

van den Heuvel, E. G., Muys, T., van Dokkum, W. & Schaafsma, G. (1999). Oligofructose stimulates calcium absorption in adolescents. *The American journal of clinical nutrition*, *69*(3), 544-548.

Van Dokkum, W., Wezendonk, B., Srikumar, T. S. & Van den Heuvel, E. G. (1999). Effect of nondigestible oligosaccharides on large-bowel functions, blood lipid concentrations and glucose absorption in young healthy male subjects.*European journal of clinical nutrition*, *53*(1), 1.

Van Loo, J., Cummings, J., Delzenne, N., Englyst, H., Franck, A., Hopkins, M., ... & Van den Heuvel, E. G. H. M. (1999). Functional food properties of non-digestible oligosaccharides: a consensus report from the ENDO project (DGXII AIRII-CT94-1095). *British Journal of Nutrition*, *81*, 121-132.

Velázquez, M., Davies, C., Marett, R., Slavin, J. L. & Feirtag, J. M. (2000). Effect of oligosaccharides and fibre substitutes on short-chain fatty acid production by human faecal microflora. *Anaerobe*, *6*(2), 87-92.

Velupillai, P. & Harn, D. A. (1994). Oligosaccharide-specific induction of interleukin 10 production by B220+ cells from schistosome-infected mice: a mechanism for regulation of CD4+ T-cell subsets. *Proceedings of the National Academy of Sciences*, *91*(1), 18-22.

Vivatvakin, B., Mahayosnond, A., Theamboonlers, A., Steenhout, P. G. & Conus, N. J. (2010). Effect of a whey-predominant starter formula containing LCPUFAs and oligosaccharides (FOS/GOS) on gastrointestinal comfort in infants. *Asia Pac J Clin Nutr*, *19*(4), 473-480.

Vos, A. P., Van Esch, B. C., Stahl, B., M'Rabet, L., Folkerts, G., Nijkamp, F. P. & Garssen, J. (2007). Dietary supplementation with specific oligosaccharide mixtures decreases parameters of allergic asthma in mice. *International immunopharmacology*, *7*(12), 1582-1587.

Vulevic, J., Drakoularakou, A., Yaqoob, P., Tzortzis, G. & Gibson, G. R. (2008). Modulation of the fecal microflora profile and immune function by a novel trans-galactooligosaccharide mixture (B-GOS) in healthy elderly volunteers. *The American journal of clinical nutrition*, *88*(5), 1438-1446.

Walker, A. R. (1995). Nutrition-related diseases in Southern Africa: with special reference to urban African populations in transition. *Nutrition Research*, *15*(7), 1053-1094.

Walter, J., Hertel, C., Tannock, G. W., Lis, C. M., Munro, K. & Hammes, W. P. (2001). Detection of Lactobacillus, Pediococcus, Leuconostoc, and Weissella species in human feces by using group-specific PCR primers and denaturing gradient gel electrophoresis. *Applied and Environmental Microbiology*, *67*(6), 2578-2585.

Watzl, B., Girrbach, S. & Roller, M. (2005). Inulin, oligofructose and immunomodulation. *British Journal of Nutrition*, *93*(1), 49-56.

Weaver, L. T. (2003). Improving infant milk formulas: Near the end of the trail for the holy grail?. *Journal of pediatric gastroenterology and nutrition*, *36*(3), 307-310.

Westerbeek, E. A. M., Hensgens, R. L., Mihatsch, W. A., Boehm, G., Lafeber, H. N. & van Elburg, R. M. (2011). The effect of neutral and acidic oligosaccharides on stool viscosity, stool frequency and stool pH in preterm infants. *Acta Paediatrica*, *100*(11), 1426-1431.

Westerbeek, E. A., van Elburg, R. M., van den Berg, A., van den Berg, J., Twisk, J. W., Fetter, W. P. & Lafeber, H. N. (2008). Design of a randomised controlled trial on immune effects of acidic and neutral oligosaccharides in the nutrition of preterm infants: carrot study. *BMC pediatrics*, *8*(1), 46.

Wichienchot, S., Jatupornpipat, M. & Rastall, R. A. (2010). Oligosaccharides of pitaya (dragon fruit) flesh and their prebiotic properties. *Food chemistry*, *120*(3), 850-857.

Winkler, P., Ghadimi, D., Schrezenmeir, J. & Kraehenbuhl, J. P. (2007). Molecular and cellular basis of microflora-host interactions. *The Journal of nutrition*, *137*(3), 756S-772S.

Wollowski, I., Rechkemmer, G. & Pool-Zobel, B. L. (2001). Protective role of probiotics and prebiotics in colon cancer. *The American Journal of Clinical Nutrition*, *73*(2), 451s-455s.

Wong, J. M., Kendall, C. W., de Souza, R., Emam, A., Marchie, A., Vidgen, E., ... & Jenkins, D. J. (2010). The effect on the blood lipid profile of soy foods combined with a prebiotic: a randomized controlled trial. *Metabolism*, *59*(9), 1331-1340.

Yang, X., Zhao, Y., He, N. & Croft, K. D. (2010). Isolation, Characterization, and Immunological Effects of α-Galacto-oligosaccharides from a New Source, the Herb Lycopus lucidus Turcz. *Journal of agricultural and food chemistry*, *58*(14), 8253-8258.

Yates, A. A., Schlicker, S. A. & Suitor, C. W. (1998). Dietary reference intakes: the new basis for recommendations for calcium and related nutrients, B vitamins, and choline. *Journal of the American Dietetic Association*, *98*(6), 699-706.

Younes, H., Demigné, C. & Rémésy, C. (1996). Acidic fermentation in the caecum increases absorption of calcium and magnesium in the large intestine of the rat. *British Journal of Nutrition*, *75*(02), 301-314.

Zafar, T. A., Weaver, C. M., Zhao, Y., Martin, B. R. & Wastney, M. E. (2004). Nondigestible oligosaccharides increase calcium absorption and suppress bone resorption in ovariectomized rats. *The Journal of nutrition*, *134*(2), 399-402.

Zahn, S., Pepke, F. & Rohm, H. (2010).Effect of inulin as a fat replacer on texture and sensory properties of muffins. *International Journal of Food Science & Technology*, *45*(12), 2531-2537.

Zhang, J. P. & Normark, S. (1996). Induction of gene expression in Escherichia coli after pilus-mediated adherence. *Science*, *273*(5279), 1234-1236.

Ziemer, C. J. & Gibson, G. R. (1998). An overview of probiotics, prebiotics and synbiotics in the functional food concept: perspectives and future strategies. *International Dairy Journal*, *8*(5-6), 473-479.0

In: Oligosaccharides
Editors: L. S. Schweizer, S. J. Krebs
ISBN: 978-1-62948-328-3

Chapter II

Prebiotic Oligosaccharides, Resistant Starch, and Sugar Alcohols in Lentils: Implications for Obesity

***Casey Johnson*[1], *Dil Thavarajah*[1*], *Pushparajah Thavarajah*[1] *and Shiv Kumar*[2]**
[1]School of Food Systems, North Dakota State University, Fargo, ND, US
[2]ICARDA/Rabat Office, Rabat-Institutes, Rabat, Morocco

Abstract

Imbalanced energy intake, coupled with low concentrations of bioactive compounds in foods, has fueled the rising epidemic of obesity and related non-communicable diseases. Currently, over 35% of American adults are obese, including nearly 17% of children and adolescents. In relation, chronic non-communicable diseases result in an estimated 36 million deaths globally each year. To combat these diet-related disorders, support health, and contribute to overall sustainability of health care, it is important to develop foods that supply proper amounts of energy and nutrients. Lentil (*Lens culinaris* L.), a cool season food legume, is a good source of protein (20-30%), essential fatty acids, micronutrients, and prebiotic carbohydrates and elicits a low glycemic index. Carbohydrate profiles of lentil that contribute to healthful effects include prebiotics: raffinose-family oligosaccharides, fructooligosaccharides, sugar alcohols, and resistant starch. The book chapter will provide insights on prebiotic carbohydrates in lentil, including genetic diversity and growing environmental and processing effects. Additionally, the chapter will include a synopsis of health consequences associated with consumption of these important prebiotic carbohydrates, and, in closure, will discuss avenues of further research.

1. Introduction

While for centuries, the greatest disease threats facing humanity were infectious, now, chronic non-communicable diseases (obesity, type 2 diabetes, cardiovascular diseases, hypertension, etc.) account for an unprecedented 63% of global disease burden (United Nations, 2012). Dietary behaviors contribute to the etiology and prognosis of such disorders (Singh et al., 1992). Increased caloric intake and altered diet composition, e.g., refined sugar and vegetable oils, are associated with a drastic increase in the prevalence of obesity and related non-communicable diseases (Austin, Ogden, & Hill, 2011). Furthermore, along with the addition of refined, high-energy foods to the typical diet, traditional foods with natural protective agents have been displaced (Kearney, 2010). A variety of bioactive compounds that exist in traditional foods are now being realized for their capacity to reduce risk factors of obesity and its comorbidities. As the burden of disease escalates, demand for these traditional staple crops – previously a pillar of the food system – will increase.

Lentil (*Lens culinaris* L.), a cool-season food legume and a staple in many Eastern diets, is an important component of a sustainable food system. Lentil is a good source of protein (20-30%), carbohydrates (~60%), essential fatty acids, and a range of vitamins and minerals (Bhatty, 1988; Thavarajah et al., 2011). Contributing to its low glycemic index (Jenkins et al., 1981), lentil also has a unique profile of carbohydrates including several healthful prebiotic compounds: raffinose-family oligosaccharides (RFO), fructooligosaccharides (FOS), sugar alcohols, and resistant starch (RS) (Bhatty, 1988; Wang, Hatcher, Toews, & Gawalko, 2009).

A prebiotic is a component of food which is neither digested nor absorbed in the small intestine, is passed to the large intestine and fermented, and elicits its effects via interactions with the microbial flora (Gibson & Roberfroid, 1995). Diet rich in prebiotics contributes to human health and well-being through multiple facets, both physiological and pathophysiological, including reduction of risk factors for obesity and non-communicable diseases (Roberfroid et al., 2010). These attributes make prebiotics and prebiotic-rich foods such as lentil an interesting research topic for prevention of obesity.

The objectives of this book chapter are to (1) overview the current obesity epidemic, and (2) discuss those prebiotics which are present in lentil and their human health consequences. We will also review the need for producing appropriate crops for human nutrition and address lentil and its importance in healthy food systems. The remainder of the chapter will focus on prebiotic oligosaccharides, polysaccharides, and sugar alcohols in lentil, and the effects of genetics, growing environment, and cooking and processing on their concentration.

2. Obesity

Obesity is simply defined as having excess body fat (CDC). The most commonly used measurement of excess body fat is body mass index (BMI), a calculation from a person's height and weight (m^2/kg). In the US, BMI is used to characterize overweight (BMI >25) and obese (BMI >30) individuals (Expert Panel on the Identification, Evaluation, and Treatment of Overweight in Adults). Essentially, BMI is a tool to be used in epidemiological studies that lead to a better understanding of the accumulation of body fat and the development of obesity.

The accumulation of excess body fat is the result of a metabolic imbalance of energy, *i.e.*, more energy is consumed than is utilized (Horton et al., 1995). The human body naturally stores any available energy exceeding its requirements, most notably in the forms of glycogen and lipid, the latter being preferred for long-term storage (Horton et al., 1995). Lipids are energy-dense and require no water to store, making them ideal as an energy reserve. Fat storage in the body is a survival mechanism: when food is unavailable the body utilizes stored fat reserves to maintain function for extended periods of time (Cahill, 1970). Moreover, adipocytes, or lipid depot cells, are responsible for sequestering circulating glucose and triglycerides and for maintaining plasma insulin concentrations (Gavrilova et al., 2000; Seip & Trygstad, 1996). So, if our body fat is so important for metabolism and general well-being, why is obesity of so much concern?

The death toll associated with obesity is over 300,000 each year in the US (U.S. Department of Health & Human Services). Obesity greatly increases the risk of a long list of health consequences – heart disease, type-2 diabetes, and certain types of cancer, stroke, arthritis, breathing problems, and psychological disorders, such as depression (Popkin, Kim, Rusev, Du, & Zizza, 2006). Globally, over 500 million people are obese (Finucane et al., 2011); this includes about 36% of US adults (Flegal, Carroll, Kit, & Ogden, 2012). In recent years, the percentage of obese adults (over 20 years) in the US increased by 8% in men and 3% in women (Table 1). The combined prevalence of overweight and obesity has reached about 70% in the US (Flegal et al., 2012). In addition to morbidity and mortality concerns, the estimated economic cost of obesity in the US was $117 billion in the year 2000, seen in medical services and loss of worker income and productivity (U.S. Department of Health & Human Services). To put the consequences of obesity in perspective, in addition to causing a drastic reduction in quality of life and disability-free life years, obesity and related comorbidities account for 63% of global deaths (Lopez, Mathers, Ezzati, Jamison, & Murray, 2006; United Nations, 2012). The situation deserves global attention.

3. Obesity and the Current Food System: Are They Related?

The question of how obesity and non-communicable diseases came to be a global problem and how they progressed cannot be easily answered. However, it will be valuable to consider a few key elements in order to create a logical framework to solve this global health problem. Two factors that contributed to the high prevalence of obesity will be briefly discussed: (1) increased production and consumption of foods that are energy-dense and deficient in bioactive compounds, and (2) decreased production and consumption of traditional pulse crops.

A significant change in the global food system can be dated back to the start of the agricultural, or so-called 'green', revolution (Welch & Graham, 1999). In an effort to preclude impending famine and starvation, technologically-advanced agricultural practices were implemented in many regions to increase productivity (Evenson & Gollin, 2003). The program successfully increased land productivity and food availability per person. Since then, food availability per person increased by about 350 kcal per capita per day, a 15% increase in energy within 30 years (FAOSTAT).

Table 1. Trends in the age-adjusted and age-specific prevalence of obesity (BMI ≥30) in US adults aged 20 years or older for 1999-2008

	(%) of Adults							
	Age ≥20 y[a]	(%) Change Over 10 y	Ages 20-39 y	(%) Change Over 10 y	Ages 40-59 y	(%) Change Over 10 y	Age ≥60 y	(%) Change Over 10 y
Men[b]								
1999-2000	28		24		29		32	
2001-2002	28		22		32		30	
2003-2004	31		28		35		30	
2005-2006	33		28		40		32	
2007-2008	32		28		34		37	
2009-2010	36	29	33	38	37	28	37	16
Women								
1999-2000	33		28		38		35	
2001-2002	33		30		36		35	
2003-2004	33		29		39		32	
2005-2006	35		31		41		34	
2007-2008	36		34		38		34	
2009-2010	36	9	32	14	36	-5	42	20

Abbreviations: BMI, body mass index (calculated as weight in kilograms divided by height in meters squared). (Continued on next page)

Sources: Adapted from (1) Flegal, K; Carroll, M; Ogden, C; Curtin, L. (2010) Prevalence and trends in obesity among US adults, 1999-2008. JAMA, 303(3):235-241[20] and (2) Flegal, K; Carroll, M; Kit, B; Ogden, C. (2012) Prevalence of obesity and trends in the distribution of body mass index among US adults, 1999-2010. JAMA, 307(5):491-497.

[a] Age adjused by the direct method to the year 2000 Census population using the age groups 20-39 years, 40-59 years, and 60 years or older.

[b] Includes racial and ethnic groups not shown separately.

Table 2. Calories from major commodities (kcal per capita per day) in developing countries, North America, and the World

	Year	Pulses	% Change Three Decades	Starchy Roots	% Change Three Decades	Cereals	% Change Three Decades	Vegetable Oils	% Change Three Decades	Sugar & Sweeteners	% Change Three Decades
Developing Countries	1970	119		131		1288		104		147	
	1985	91		122		1316		137		173	
	2000	84	-29	145	11	1355	5	178	71	168	14
North America	1970	28		99		592		347		561	
	1985	29		100		682		523		539	
	2000	43	54	110	11	872	47	627	81	626	12
World	1970	72		178		1188		142		222	
	1985	60		131		1309		200		239	
	2000	56	-22	141	-21	1306	10	247	74	228	3

Data from: FAOSTAT (http://faostat.fao.org/site/368/Desktop.Default.aspx?PageID=368#ancor).

Cereal crops including wheat (*Triticum* spp.), rice (*Oryza sativa* L.), and corn (*Zea mays* L.) were primary contributors to this energy boost (Table 2). The green revolution is commended for preventing food shortages in many regions.

The next agricultural revolution, however, will need to address hidden consequences of the last revolution – malnutrition and obesity.

Swinburn et al. (2011) emphasize that a global energy overbalance increases the obesity epidemic. As opposed to previous generations where energy expenditure determined energy intake, currently, energy intake is driving energy expenditure (Figure 1). As food availability increased, the world prevalence of obesity surged. Globally, the average BMI has increased significantly since 1980 (Finucane et al., 2011), and the prevalence of overweight and obesity among children in many countries has more than doubled since the 1970s (Figure 2). While excess food availability is certainly a large contributor to obesity, the question remains, "Does all food contribute to obesity equally?"

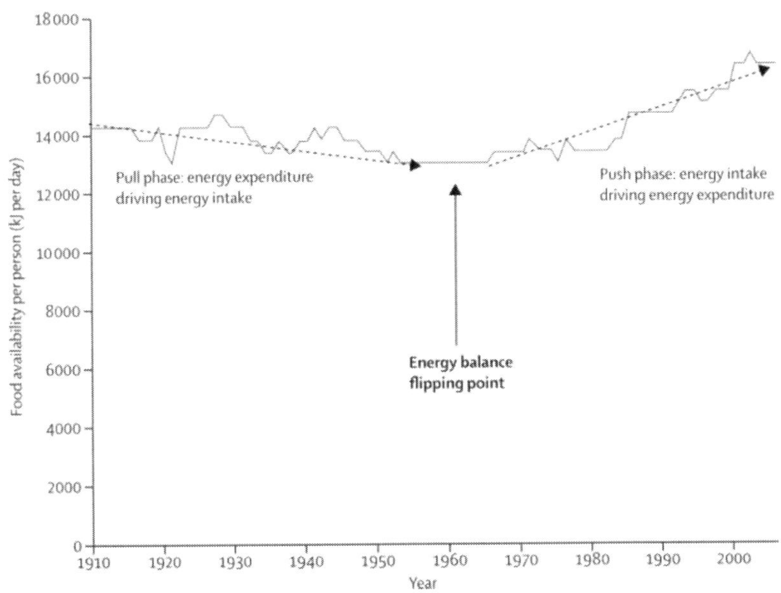

Figure 1. Food availability for the USA, 1910–2006. There are two distinct phases: a decrease in food energy supply (postulated to be pulled down by reduced energy expenditure requirements for daily living), followed by an increase in food energy supply (postulated to be pushed up by increasing food access). An energy balance flipping point is proposed, marking the change in how the US population generally achieved energy balance. Reproduced with permission from Swinburn et al. 2011.

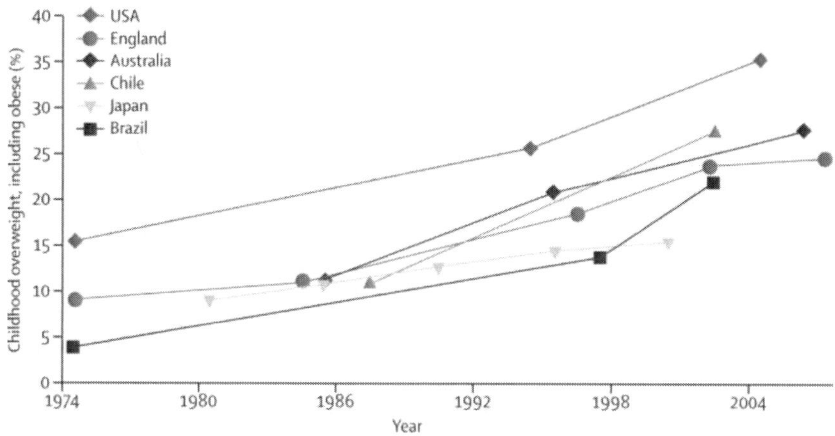

Figure 2. Estimates of percentage of childhood population overweight, including obese (with use of International Obesity Taskforce cutoffs) in a selection of countries. Reproduced with permission from (Swinburn et al., 2011).

During the green revolution, little attention was given to the nutritional quality of the food system as a whole.

Certain crops were produced disproportionately: an over-abundance of high yielding cereals and a displacement of micronutrient-rich crops, especially pulses. The world transitioned from traditional food staples – pulses, roots, and tubers – to processed cereal-based foods and foods rich in added fats, vegetable oils, and sugars (Table 2). Over the three decades between the 1970s and 2000s, availability of high-energy food products (cereals, vegetable oils, sugar, etc.) increased in North America and developing countries. Meanwhile, pulses, roots, and tubers availability decreased globally. In North America, although the amount of pulses, roots, and tubers per person increased, availability of these crops remained quite low as the major part of the pulses produced in North America is exported to developing countries.

The associations of increased prevalence of obesity with dietary patterns raised many questions. How much of the obesity epidemic can be attributed to the overall increase in calorie consumption? How much to the displacement of traditional foods with their diverse protective agents from the diet? Also, what caused the increase in calorie consumption? Did excess food availability lead to excess consumption, or did altered diet composition lead to excess consumption, or both? These are difficult questions which cannot be fully answered by any single group. Nevertheless, agricultural, nutrition, and food scientists can logically gather some helpful cues on which to focus attention:

(1) understanding of the chemistry of healthful bioactive compounds in foods and (2) production of foods that provide appropriate energy and nutrition. Thus, having reviewed the problems to be addressed, we will discuss an important element of the first of these objectives – prebiotics.

4. Gut Microbiota and Prebiotics

Prebiotics emerged in the literature in 1995 with the discovery that certain oligosaccharides could provide host benefits by altering the microbial ecology in the gut (Gibson & Roberfroid, 1995). Soon after, the gut microbiota and prebiotics were hot topics in the area of human nutrition including nutrient absorption, immunology, evolution, and epidemiology (Krajmalnik-Brown, Ilhan, Kang, & DiBaise, 2012; Backhed et al., 2004; Lee & Mazmanian, 2010; Turnbaugh et al., 2006; Roberfroid et al., 2010; Rowland, 2009). Prior to these advances, surprisingly little was known about the complex relationship between the gut microbiota, its substrates, and the gastrointestinal tract.

The large intestine was, for many years, thought to have only two main functions: (1) waste excretion and (2) water absorption (Welch, 1936). We now understand that without the cooperative role of the gut microbiota in the large intestine human hosts are incapable of performing several vital physiological, metabolic, and immunological functions (Turnbaugh et al., 2007; Gill et al., 2006; Backhed, Ley, Sonnenburg, Peterson, & Gordon, 2005). The gastrointestinal microbiota is also involved in the development of miscellaneous human pathophysiological conditions (Rowland, 2009; Turnbaugh et al., 2006; Rabot et al., 2010). The intestinal epithelium and the gut microbial community function interdependently, cooperatively forming an intricate organ system – a partially external and partially "inhuman" organ system (Backhed et al., 2005). Thus with the wealth of recently generated information, the microbiota is now recognized as a key player in health and well-being.

The human gastrointestinal tract hosts about 10^{12} - 10^{14} microorganisms, varying greatly in composition, function, and location of colonization between individuals (Savage, 1977). The concentration of live microorganisms in the stomach is about 10^{3} CFU/mL of contents, in the small intestine about 10^{4}-10^{6} CFU/mL of contents, and in the large intestine about 10^{12} CFU/g of contents (Holzapfel, Haberer, Snel, & Schillinger, 1998). Over 1000 commensal species in the human hindgut were identified in a cohort of 124 individuals, with each individual being host to approximately 160 different species (Qin et al., 2010). The dominant phyla present are the Firmicutes and Bacteroidetes, followed by Actinobacteria, and Proteobacteria (Table 3). These dominant groups are comprised of various genera, some potentially beneficial, some potentially harmful, and others have the potential to be either harmful or beneficial.

For example, many species within these phyla provide energy to the colonocytes in the form of short-chain fatty acids (SCFA) such as acetate, propionate, and butyrate (Table 3). Additionally, certain species can produce essential vitamins (e.g., vitamins K_2 and B_{12}) and other beneficial metabolites; other species, however, can produce toxic, genotoxic, or carcinogenic metabolites (Pandeya et al., 2012).

When the composition of commensal groups and their metabolites exist in the right balance, or 'normobiosis,' that is the potentially health-promoting microorganisms predominate over potentially harmful microorganisms, the human host is benefitted (Gibson & Roberfroid, 1995; Cummings & Kong, 2004). On the other hand, an unbalance, or 'dysbiosis,' in the gut microbiota results in a harmful relationship, causing inflammation and disease. The concept of prebiotics is based in the coexistence of these beneficial and harmful bacterial genera. A dietary prebiotic provides the right microbial 'food' to selectively alter the concentrations and functions of the microbial populations leading to 'normobiosis'. The most extensively researched genus that is stimulated by prebiotics and is an important part of the normobiotic phenomenon is *Bifidobacterium* (Roberfroid et al., 2010).

Lactobacillus is also recognized as a beneficial genus, and other genera will likely be included as more data accrue, e.g. *Eubacterium, Faecalibacterium,* and *Roseburia* (Roberfroid et al., 2010).

Prebiotics have attracted enormous attention (mostly for marketing purposes), and the need for strict criteria to define them became apparent, leading to the establishment of the following requirements to be classified as a prebiotic food ingredient:

- Resist degradation by processes in the upper gastrointestinal tract (acidity, pancreatic enzymes, brush boarder enzymes, etc.)
- Be fermented by intestinal microbiota
- Selectively alter the composition/activity of certain microbes resulting in health benefits to the host

Table 3. Composition and characteristics of dominant phyla of human gut microbiota and several subgroups of bacteria and their substrates and products

Phyla	Bacterial Subgroup	Approx. CFU/g of Feces	Approx. (%) of Microbiota	Mode of Action on Substrate(s)	Fermentation Product(s)
Firmicutes		$3\text{-}5.3 \times 10^{10}$	30-53%		
	Clostridia			saccharolytic, some aa-fermenting species	Ac, Pr, Bu, La, e
	Eubacteriaceae			saccharolytic, some aa-fermenting species	Ac, Bu, La
	Rumminococcus			saccharolytic	Ac
	Lactobacillus	1×10^{8}	1%	saccharolytic	La
	Streptococcus			carbohydrate and aa-fermentation	La, Ac
Bacteroidetes		$0.9\text{-}4.2 \times 10^{10}$	9-42%		
	Bacteroides			saccharolytic	Ac, Pr, Su
Actinobacteria		$0.2\text{-}2.5 \times 10^{10}$	2-25%		
	Bifidobacterium	$0.7\text{-}1.0 \times 10^{10}$	1-14%	saccharolytic	Ac, La, f, e
	Collinsella-Atopobium	$0.3\text{-}4.0 \times 10^{9}$	0.7-10%		
Proteobacteria		$0.7\text{-}4.0 \times 10^{9}$	1-10%		
	Escherichia			carbohydrate and aa-fermentation	Mixed acids
	Desulfovibrio			various	Ac

aa, amino acid; Ac, acetate; Pr, propionate; Su, succinate; Bu, butyrate; La, lactate; f, formate; e, ethanol.
Soures: (Pandeya et al., 2012; Roberfroid et al., 2010; Roberfroid, 2008).

Table 4. Main areas of pathophysiological interest in which prebiotics have been investigated

Effects	Primary model	References
Functional effects		
Intestinal/colonic functions (e.g., fecal bulking, stool production)	Human	Causey et al. 2000
	Human	Cummings et al. 2002
Resistance to intestinal infections	Human	*Gibson et al. 2005*
	Human	*Bosscher et al. 2006*
Immunostimulation	Dog	*Field et al. 1999*
	Human	*Guigoz et al. 2002*
Satiety and appetite	Human	*Cani et al. 2009*
	Rat	*Parnell et al. 2012*
Influence on gastrointestinal peptides (e.g., glucagon-like peptide 1 (GLP-1) and ghrelin)	Human	*Cani et al. 2009*
	Rat	*Parnell et al. 2012*
Influence on serum lipids and glucose	Human	*Delzenne et al. 2001*
	Rat	*Pereira et al. 2002*
Bioavailability of minerals, especially Ca and Mg	Human	*Bosscher et al. 2003*
	Human	*Franck 2006*
Disease risk reduction		
Infectious diarrhea	Human	*Chouraqui et al. 2008*
	Human	*Bosscher et al. 2006*
Inflammatory bowel diseases	Human	*Friedman et al. 2000*
	Human	*Furrie et al. 2005*
	Human	*Lindsay et al. 2006*
Obesity	Rat/human	*Daubioul et al. 2000*
	Rat	*Cani et al. 2007*
	Rat/human	*Delzenne et al. 2010*
Metabolic syndrome	Rat/human	*Delzenne et al. 2005*
	Rat	*Cani et al. 2007*
	Rat/human	*Delzenne et al. 2010*
Osteoporosis	Rat	*Roberfroid et al. 2002*
	Human	*Abrams et al. 2005*
Colon cancer	Rat	*Wollowski et al. 2001*
	Rat	*Le Leu et al. 2010*
	Rat	*Conlon et al. 2012*

The most recent and widely-accepted definition of a dietary prebiotic is a "selectively fermented ingredient that results in specific changes, in the composition and/or activity of the gastrointestinal microbiota, thus conferring benefit(s) upon host health" (ISAPP, 2008).

A key condition of this definition is 'selectively'. There are many fibers and dietary components that are fermented by the microbiota, but only those which are selectively fermented by certain beneficial microbes are prebiotic. Therefore, though it is likely that more carbohydrates will be considered prebiotic in the future, only several currently have sufficient experimental support to meet the necessary requirements (Roberfroid et al., 2010). Those that have prebiotic 'status' are fructooligosaccharides (FOS), galactooligosaccharides (GOS), and lactulose (Kolida & Gibson, 2008).

Prebiotics are included under the broad category of low-digestible carbohydrates (Grabitske & Slavin, 2009). Low-digestible carbohydrates (LDC) are fermentable and are comprised of three groups of compounds: non-starch polysaccharides, sugar alcohols, and resistant starch (RS). Some examples of non-starch polysaccharides are RFO, FOS, and inulin. Sugar alcohols are collectively known as hydrogenated mono-, di-, or polysaccharides. Naturally occurring sugar alcohols include sorbitol and mannitol. The final subgroup, RS, occurs naturally in foods in two forms (RS_1 and RS_2), though other forms exist synthetically. The above mentioned LDCs (discussed in detail under subheading 6) are poorly digested by human enzymes and fermented in the large intestine (Grabitske & Slavin, 2009).

Fermentation of prebiotics and certain LDCs elicits a variety of health effects which can be subdivided into two main groups: functional effects and disease risk reduction (Table 4). Functional effects are physiological effects that can be measured relatively easily, including induction of satiety (Parnell & Reimer, 2012), reduction of caloric intake (Cani et al., 2009), and reduction of serum cholesterol, triglycerides, and glucose concentrations (Pereira & Gibson, 2002). Disease risk reduction, as implied by the name, is the compounding effect over time of one or more functional effects to reduce the risk/severity of chronic diseases. For example, prebiotic-induced satiety, reduced caloric intake, and improved serum lipid profile contribute to reducing the risk and severity of obesity and metabolic syndrome (Delzenne & Kok, 2001; Delzenne, Neyrinck, Backhed, & Cani, 2011; Delzenne, Neyrinck, & Cani, 2013).

5. Lentil

Lentil is highly nutritious food crop, often consumed either as a whole food or dehulled and split. The proximate composition of lentil is as follows: moisture (c.a. 10-12%), carbohydrate (c.a. 60-65%), starch (c.a. 40-55%), protein (c.a. 20-30%), ash (c.a. 3%), and lipid (c.a. 1 – 3%) (Bhatty, 1988). The seed consists of three parts: the seed coat, cotyledons, and embryo which account for 8%, 90%, and 2% of the seed weight, respectively (Singh, Singh, & Sikka, 1968). Each of these components has a different chemical composition and nutritional quality. The seed coat is formed mostly of fibers – cellulose, hemicellulose, and lignin (Bhatty, 1988). Many of the seed's minerals and free and fiber-bound polyphenolics, flavonoids, and tannins are also present in the seed coat (Duenas, Sun, Hernandez, Estrella, & Spranger, 2003; Xu & Chang, 2010). The cotyledons are the main energy store of the seed, containing the starch fraction and about 90% of the total protein and lipids (Singh et al., 1968). Various sugar alcohols and mono-, di-, and oligosaccharides are also present in the cotyledons (c.a. 5-10% of dry matter (DM)) including glucose, sucrose, RFO (raffinose, stachyose, and verbascose), FOS (nystose), and various others in lesser concentrations (Tahir, Vandenberg, & Chibbar, 2011; Biesiekierski et al., 2011; Bhatty, 1988). Minerals and water-soluble vitamins such as ascorbic acid are also concentrated in the cotyledons (Ekinci & Kadakal, 2005). Lipids and fat-soluble vitamins are contained in both the cotyledons and the embryo. The bulk of the lipids in lentil are triacylglycerides: three fatty acid residues bound with ester linkages to a glycerol backbone (Bhatty, 1988). The fatty acid profile is as follows: linoleic acid (37%), oleic acid (16%), palmitic acid (13%), linolenic acid (9%), and less than 1% of stearic, arachadonic, and eicosenoic acids (Salunkhe, Kadam, & Chavan, 1985).

The unique food matrix of lentil leads to a number of desirable nutritional responses (Jenkins et al., 1981; Abeysekara, Chilibeck, Vatanparast, & Zello, 2012). Lentil has a low glycemic index (Jenkins et al., 1981). In other words, after a lentil meal the concentration of glucose in the serum does not increase greatly or rapidly. Jenkins et al. (1980) found that the glycemic responses to pulses in general was about 45% lower than to cereal grains, biscuits, pasta, and tubers. Moreover, consumption of lentil induces a higher degree of satiety after a meal than most foods (McCrory, Hamaker, Lovejoy, & Eichelsdoerfer, 2010). Although it has been suggested that this is accomplished via modulation of gastrointestinal hormones such as cholecystokinin and also through short-chain fatty acid production in the large intestine, a direct causal relationship to satiety has not been firmly established (Sufian, Hira, Asano, & Hara, 2007). Abeysekara et al. (2012) found that lentil diet reduced serum cholesterol in elderly by about 8% compared with regular diet. Comparing the physiological effects of lentil consumption to several effects of prebiotics (Table 4), directly or indirectly, prebiotic components likely play a role in low glycemic, satiating, and cholesterol-reducing responses in lentil (Cani et al., 2009).

Table 5. Nutrient concentration data in raw lentil and cereal grains

Proximates		Value per 100.0g			
		Lentil	Wheat, hard red spring	Brown rice, long-grain	White rice, long-grain, unenriched
Protein	g	26	15	8	7
Total lipid (fat)	g	1	2	3	1
Carbohydrate, by difference	g	60	68	77	80
Fiber, total dietary	g	31	12	4	1
Calcium, Ca	mg	56	25	23	28
Iron, Fe	mg	8	4	2	1
Potassium, K	mg	955	340	223	115
Zinc, Zn	mg	5	3	2	1
Vitamin C	mg	4	0	0	0
Riboflavin	mg	0.2	0.1	0.1	0.1
Niacin	mg	3	6	5	2
Folate, DFE	µg	479	43	20	8
Vitamin A, RAE	µg	2	0	0	0
Vitamin K (phylloquinone)	µg	5	2	2	0

Data obtained from the USDA Nutrient Database (USDA, 2012).

Compared with cereal grains, concentrations of vitamins, minerals, protein, and complex carbohydrates are comparatively greater in lentil (Table 5). Especially when consumed as a whole food, lentil is an excellent source of various nutrients; per 100 g, lentil contains ~31g dietary fiber, 8 mg iron, 5 mg zinc, 4 mg vitamin C, 479 µg folate, and 5 µg vitamin K. A ½-cup serving of cooked lentil can provide about one third of the recommended intake of dietary fiber (USDA, 2012). The large quantity of dietary fiber in lentil is a desirable trait for several reasons. First, dietary fiber is associated with reduced incidence of heart diseases and certain types of cancer (Fuchs et al., 1999; Pietinen et al., 1996), and second, some of the components of lentil dietary fiber are prebiotic carbohydrates (Brown, 2004; Martínez -Villaluenga, Frias, Vidal-Valverde, & Gomez, 2005).

6. Lentil Prebiotics

A number of prebiotic carbohydrates are widespread in plant-derived foods in varied concentrations; vegetables, roots, tubers, and legumes in particular often contain high concentrations of the one or more prebiotics (Table 6). Concentration of prebiotics in foods ranges from trace amounts, as is the case in white rice, to relatively high amounts in other foods, such as Jerusalem artichoke (van Loo, Coussement, De Leenheer, Hoebregs, & Smits, 1995). In lentil, several groups of prebiotic carbohydrates have been indicated, including certain non-digestible oligosaccharides (FOS and RFO), RS, and sugar alcohols (Wang et al., 2009; Tahir et al., 2011; Bhatty, 1988). There are gaps in our knowledge of these important compounds, however. What is the profile of prebiotic carbohydrates in lentil? How much variation in their concentration exists between lentil genoypes? between growing environments? How much prebiotic carbohydrates are found in commercially available lentils? How does dehulling, cooking, and cooling affect those concentrations? These issues, when pertainent, will be presented for various carbohydrates in the following sections. Additionally, chemical structure of these compounds, as well as their concentration in foods and respective health consequences, will be discussed.

6.1. FOS

Often synonymously called oligofructose, FOS are by far the most famous family of prebiotic oligosaccharides. FOS consist of small chains of β (2→1) D-fructose residues of varying length with a terminal α (1→2)-linked D-glucose (Lewis, 1993). The DP of FOS is between 3 and 10 (Kolida & Gibson, 2008). Polymers of β-D-fructofuranosyl units – having a DP greater than 10 – are known as inulin (Roberfroid, 2007). The few investigations of FOS in legumes have focused on the shortest chain length compounds, kestose (β -D-fructofuranosyl-(2→1)- β -D-fructofuranosyl α-D-glucopyranoside) and nystose (β-D- fructofuranosyl -(2→1)-β-D- fructofuranosyl -(2→1)-β-D-fructofuranosyl -(2→1)-α-D- glucopyranoside) (Biesiekierski et al., 2011).

The human small intestine lacks the necessary enzymes for degradation of FOS (Roberfroid, 1999). Case studies on patients with ileostomies have been the most important in confirming the non-digestibility of FOS (Kolida & Gibson, 2008). These non-digested compounds reach the large intestine intact, where they are fermented by the microbiota to produce SCFAs (Cummings, Macfarlane, & Englyst, 2001). The majority of this fermentation takes place in the cecum and ascending colon (Macfarlane, Gibson, & Cummings, 1992). The capacity of FOS to selectively stimulate microbial populations, especially bifidobacteria and lactobacilli, has been thoroughly demonstrated in recent years (Gibson & Roberfroid, 1995; Kruse, Kleessen, & Blaut, 1999). Supplementation of 15 g of FOS significantly increased counts of *Bifidobacterium* sp., while reducing counts of other prominent bacteria, including bacteroides, fusobacteria, and potentially pathogenic *Clostridium* sp. (Gibson & Roberfroid, 1995). The physiological and disease risk reduction effects of FOS have been widely examined, many of which are included in Table 4. In addition to maintenance of normal intestinal microbiota and prevention of pathogen colonization (Bouhnik et al., 1999), investigators demonstrated lower levels of circulating glucose and cholesterol in humans after ingestion of FOS (Pereira & Gibson, 2002).

Varying concentrations of FOS occur in over 36,000 plant species (Carpita, Kanabus, & Housley, 1989); high concentrations occur in chicory, Jerusalem artichoke, asparagus, garlic, and onion. Moderate concentrations of FOS have been observed in food legumes (Biesiekierski et al., 2011; Muir et al., 2009). In lentil, only small concentrations of FOS exist: ~100–200 mg/100g food weight (Biesiekierski et al., 2011). However, it may be possible to enhance this concentration, because FOS is already present in lentil seeds, suggesting that the genetic machinery leading to FOS accumulation, fructosyltransferase (Yun, 1996), is functional in lentil. To our knowledge, no studies have reported concentrations of FOS in lentil genotypes or taken into consideration changes in FOS concentration with growing environment.

6.2. RFO

The most well-known and studied prebiotics, such as FOS and GOS, are oligosaccarides. Not surprisingly, other oligosaccharides have attracted the attention of researchers for their potential to promote health. Oligosaccarides that are common in legumes include the members of the raffinose family: raffinose, stachyose, and verbascose (Guillon & Champ, 2002). The basic structure of RFO contains a sucrose backbone and one or more α (1→6)-linked galactose residues with a DP of les than 10. This structure differs from trans-GOS in that trans-GOS have a lactose backbone instead of sucrose and β (1→4)-linked galactose residues instead of α (1→6)-linkages (Barreteau, Delattre, & Michaud, 2006). Raffinose, stachyose, and verbascose have chain lengths of 3, 4, and 5 saccharide residues, respectively (Guillon & Champ, 2002).

Due to a lack of α galactosidase activity in the small intestine, RFO are non-digestible (Smiricky et al., 2002). Once RFO reach the large intestine intact, they are fermented by the hindgut microbiota (Desjardins, Roy, & Goulet, 1990). Studies suggest that fermentation of RFO results in the selective increase of bifidobacteria in the large intestine, which is commonly associated with prebiotic compounds (Benno, 1987; Saito, Takano, & Rowland, 1992; Hayakawa et al., 1990). Supplementation of 15g/day raffinose to healthy subjects resulted in increased counts of bifidobacteria (Benno, 1987). Moreover, total bacterial counts remained stable, and *Bacteroides* spp. and *Clostridium* spp. were significantly lesser after raffinose administration than before (Benno, 1987). These observations were confirmed by a double-blind, placebo-controlled study in which 2.5 to 10 g/day raffinose significantly stimulated bifidogenesis (Bouhnik et al., 2004).

Some of the first reports of RFO in lentil appeared in the late 1970s; total RFO concentrations ranged from 2.5 to 7.2% (Bhatty, 1988; Wang et al., 2009). The profile of individual RFO concentrations has also been reported in lentil [raffinose, 0.1–1.0 g ; stachyose, 1.1–4.0 g; and verbascose, not detectable–6.4 g per 100g DM] (Martinez-Villaluenga, Frias, & Vidal-Valverde, 2008). The majority of RFO are concentrated in the cotyledons of lentil; however, multiple investigations have observed significantly raffinose concentrations in the seed coat, but not stachyose or verbascose (Wang et al., 2009; Wang, Hatcher, & Gawalko, 2008). This factor results in raffinose concentration decrease with dehulling (Wang et al., 2009). Owing to the water soluble nature of RFO, boiling results in significant leaching into the cooking water. Discarding cooking water therefore results in significant decreases in RFO concentrations in food (Vidal-Valverde et al., 1994). Onigbinde & Akinyele (1983) observed another interesting effect of cooking – RFO in African legumes were partially hydrolyzed leading to lesser concentrations of higher degree of polymerization (DP) oligosaccharides and greater concentrations of short-chain oligosaccarides and sucrose. The authors attributed this to heat hydrolysis of the α (1→6)-linkages during cooking.

6.2.1. Sugar Alcohols

Sugar alcohols are low-digestible, hydrogenated monosaccharides, otherwise known as polyols (Grabitske & Slavin, 2009). They are neither sugars nor alcohols and have a representatively lower energy contribution compared with carbohydrates: sorbitol, 2.6 kcal/g; mannitol, 1.6 kcal/g; and carbohydrates, 4.0 kcal/g (Wolever, Piekarz, Hollands, & Younker, 2002). Sugar alcohols are found naturally in berries, mushrooms, and many higher plants (Makinen & Soderling, 1980), and are used extensively as artificial sweeteners for their low-calorie properties (Beards, Tuohy, & Gibson, 2010). In addition to the low glycemic index of sugar alcohols (Wolever et al., 2002), research suggests they may also have prebiotic action (de Vaux, Morrison, & Hutkins, 2002). In a mixed bacteria culture, addition of sorbitol to media resulted in the displacement of pathogenic bacteria, *Escherichia coli* O157:H7. Beards *et al.* (2010) assessed the prebiotic capacity of sugar alcohol and other confectionary sweeteners in a human trial and reported beneficial changes in the microflora, based on predominant prebiotic markers: bifidobacteria, lactobacilli, and SCFAs.

6.2.2. RS

Native starch is made up of two polysaccharides, amylose and amylopectin (Tester, Karkalas, & Qi, 2004). These polymers are acted upon within the upper gastrointestinal tract by a cohort of digestive enzymes (e.g., α-amylase, β-amylase, and amyloglucosidase) (Gray, 1992). Amylose is linear, consisting of α (1→4) linked glucose moieties, and is hydrolyzed by exo- and endo-enzymes (Tester et al., 2004). The average DP of amylose varies among food sources (Zobel, 1988). Amylopectin is highly branched and therefore additionally requires debranching enzymes such as amyloglucosidase for complete hydrolysis (Gray, 1992). The average molecular weight and DP of branches also varies among starch sources (Zobel, 1988).

Resistant starch, as its name suggests, is resistant to hydrolysis by human digestive enzymes. There are a myriad of factors that contribute to this non-digestibility (Hoover & Zhou, 2003). To name but a few, differences in *in vitro* starch digestibility have been attributed to the following: amylose/amylopectin ratio (Hoover & Sosulski, 1985), starch granule size (Snow & O'Dea, 1981), degree of starch crystallinity (Hoover & Sosulski, 1985), starch with B-type crystallinity (Englyst & Macfarlane, 1986), amylose-lipid complexes (Guraya, Kadan, & Champagne, 1997; Nebesny, Rosicka, & Tkaczyk, 2002), enzyme inhibitors (Lajolo, Finardi Filho, & Menezes, 1991), protein and dietary fiber matrix (Dreher, Dreher, Berry, & Fleming, 1984), physical entrapment in cell structures (Wursch, Del Vedovo, & Koellreutter, 1986), and interactions of starch molecules (Dreher et al., 1984). The high resistance to hydrolysis of legume starch is a cumulative effect of high concentration of amylose, extensive physical entrapment by fibers and other food matrix factors, antinutrients, and stong interactions between amylose chains (Tovar, Francisco, Bjorek, & Asp, 1991; Hoover & Zhou, 2003; Deshpande & Cheryan, 1984).

There are five main types of RS which vary in structure and source (Bird, Conlon, Christophersen, & Topping, 2010). Current categorizations of RS are based on its source or derivation (Cummings, Beatty, Kingman, Bingham, & Englyst, 1996). RS1 refers to starch that is physically encapsulated in food, for example, in a fiber mesh or thick cell wall, and is therefore unavailable to enzymes. RS2 is naturally resistant starch due to crystallinity or tightly-packed and unhydrated nature. RS3 is derived from heating and cooling of gelatinized starch. RS4 has been modified chemically, which may include the formation of cross-linkages and esterification. RS5 is resistant to hydrolysis because of complexation with lipids (Bird et al., 2010).

By definition, RS is "the sum of starch and products of starch degradation not absorbed in the small intestine of healthy individuals" (Asp, 1992). Thus RS can come from any food containing starch, limiting the presence of RS to any starchy food, but its concentration varies greatly (Murphy, Douglass, & Birkett, 2008). The estimated consumption of RS in the United States is 4.9 grams per person per day, on average (Murphy et al., 2008). Even though relative concentrations of RS are low, bread, pasta, and non-legume vegetables are the major contributors to RS consumption because they are widely eaten.

There has been enormous interest and research emphasis on RS in recent decades because of prebiotic responses and its putative therapeutic and preventative role in obesity and NCDs (Cummings et al., 2001; Johnston, Thomas, Bell, Frost, & Robertson, 2010; Conlon et al., 2012). Highlighted responses to administration of RS include reducing glycemic response, reducing caloric intake, improving bowel health (Brown, 2004), increasing absorption of micronutrients (Scholz-Ahrens et al., 2007), preventing colorectal cancer (Conlon et al., 2012), and improving insulin sensitivity (Johnston et al., 2010). These and other responses related to reducing risk factors of obesity and NCDs have marked it as a target for therapeutic and food applications (Brown, 2004).

Table 6. Various prebiotics in common foods

Food	Prebiotics[a]	g per 100 g Food			Reference
		Mean	Min.	Max.	
Lentil (boiled, drained)	RS	3.4	1.6	9.1	*Murphy et al. 2008**
	RFO	0.4	0.2	0.5	*Biesiekierski et al. 2011*
	FOS	0.2	0.1	0.2	*Biesiekierski et al. 2011*
	SA	TR			*Biesiekierski et al. 2011*
Common bean (pinto, boiled, drained)	RS	1.9	1.8	2	*Murphy et al. 2008*
(red kidney, boiled, drained)	RFO	1.4			*Biesiekierski et al. 2011*
(red kidney, boiled, drained)	FOS	0.5			*Biesiekierski et al. 2011*
Chickpea (cooked/canned)	RS	2.6	0.8	4.3	*Murphy et al. 2008*
	RFO	0.2			*Biesiekierski et al. 2011*
	FOS	0.2			Biesiekierski et al. 2011
White rice (long grain, cooked)	RS	1.2	0	3.7	*Murphy et al. 2008*
	RFO	ND			*Biesiekierski et al. 2011*
	FOS	ND			*Biesiekierski et al. 2011*
White bread	RS	1.2	0.1	4.4	*Murphy et al. 2008*
	RFO	0.2			*Biesiekierski et al. 2011*
	FOS	0.7			*Biesiekierski et al. 2011*
Potato (boiled)	RS	1.3	0.3	4.5	*Murphy et al. 2008*
(boiled, cooled 4°C 24h)		~3 × greater			*Englyst et al. 1987*
					Muir et al. 1992
Jerusalem artichoke	FOS	12.2			*Muir et al. 2007*

RS, resistant starch; RFO, raffinose-family oligosaccharides; FOS, fructooligosaccharides; and SA, sugar alcohols; TR, trace amounts detected only; ND, not detected.

[a]Additional prebiotics may be present in selected foods. *Indicates individual reference is a comprehensive review.

Reported concentrations of RS in lentil range widely (Table 6). This may be dependent upon lentil cultivars, growing location, and whether or not the material was processed or analyzed as eaten or freeze dried (Skrabanja, Liljeberg, Hedley, Kreft, & Björck, 1999; Wang et al., 2009). Literature reports of RS concentration in cooked lentils have ranged from 1.6 to 5.2% (dry weight) (Chung et al., 2008; Wang et al., 2009) and from 1.6 to 9.1% (food weight) (Murphy et al., 2008).

Only limited data is available however on the effect of genotype of RS concentration in the lentil seed. In field pea RS concentrations were found to vary with genotype (Skrabanja et al., 1999). Processing also plays a large role in RS concentration in food. Mishra et al. (2008) nicely demonstrated that cooling of cooked potato increases the RS concentration by a factor of 2 or 3. Yadav et al. (2009) also reported increased RS concentrations in legumes with heating and cooling cycles. To date, demonstrations of changes in RS concentrations in commecially available lentils from the United States have not been reported.

7. Measurement of Prebiotics

Identification and quantification of prebiotic carbohydrates require different approaches. Oligosaccharides can be easily analyzed with simple instrument procedures. High performance anion-exchange chromatography with pulsed electrochemical detection (HPAEC-PED), high performance liquid chromatography- refractive index (HPLC-RI), capillary zone electrophoresis (CZE), gas chromatography with flame ionization detection (GC-FID), Matrix assisted laser desorption/ionization time-of-flight mass spectrometry (MALDI-TOF-MS) and nuclear magnetic resonance (NMR) spectroscopy are the most used instrumental procedures to analyze any non-polysaccharide carbohydrate including RFO, FOS, and sugar alcohols. Identification and quantification of RS, on the other hand, require enzymatic and/or chemical/physical treatment prior to analysis of resulting carbohydrates. RS, by definition, is the starch fraction that escapes hydrolysis by human digestive enzymes, does require prior enzyme treatment.

The HPAEC-PED could be considered as the most widely used accurate method to quantify RFO. This is mainly due the excellent chromatographic resolution of target compounds from each other and from other compounds. Furthermore, greater detection sensitivity provides advantage over other methods. The GC-FID and CZE also provide similar advantages as HPAEC-

PED. Survey of literature indicates HPAEC-PED is widely used procedure due to greater analytical accuracy and versatility in oligosaccharide separation. The MALDI-TOF-MS and NMR techniques are also powerful analytical chemistry techniques to determine molecular masses and chemical structural details of the carbohydrates. However, these instruments are relatively expensive and require highly skilled personal to operate and interpretate data.

Many carbohydrates are weak acids. At high pH, hydroxyl groups of carbohydrates are partially or totally transformed to oxyanions depending on the pKa values of those hydroxyl groups. High pH resistant strong anion-exchange columns with sodium hydroxide and/or sodium acetate mobile phases provide selective elution of carbohydrates based on their number of hydroxyl groups, isomerism, and degree of polymerization (DP). The eluted carbohydrates are then detected by PED. Therefore, HPAEC-PED is a versatile technique to analyze large number of carbohydrates in a single run. The difficulties in analysis of prebiotic carbohydrates with high DP could be overcome by comparison to commerical standards. For those carbohydrate with no commerical standards, those ones could be isolated employing anion exchange chromatography, and then selective acid/enzyme hydrolysing to determine their monosaccharide compositions.

Regarding quantification of RS in foods, one of the greatest obstacles has been validation of data (Champ, Kozlowski, & Lecannu, 2001). This difficulty is largely attributed to the vast amount of factors that lead to resistance to starch hydrolysis mentioned in the previous section. Major advances in RS determination were made with the use of *in vivo* comparisons obtained from ileostomy patients (Muir & O'Dea, 1993). Researchers continued to improve the existing methods, even developing standard reference material of known RS concentrations (Megazyme, 2012).

There are many ways of preparing foods. For example, lentil may be sprouted, boiled, boiled and cooled, and ground into a flour. Furthermore, analysis can be done with fresh samples, oven-dried samples, or freeze-dried samples. Changes in RS concentration may accompany any of these changes (Mishra, Monro, & Hedderley, 2008). Therefore, different RS values can be achieved for the same lentil genotype depending on the preparation. To assess RS in lentil, understanding of both native starch resistance and resistance after cooking or processing is informative, but the two may or may not be related, stressing the need for further development of the RS analyical procedures.

8. Future Directions

Understanding the problem is hard, how much more so the solution? Einstein is quoted saying "We cannot solve our problems with the same thinking we used when we created them."

The challenges we face in food security, nutrition, and obesity and its comorbidities are indeed difficult to understand and approach. Taken as a whole, the situation is overwhelming. However, focused efforts from many cooperative disciplines will yield results.

For the food scientist working with lentil, this requires answering several important questions. What is profile of these various prebiotic and low-digestible carbohydrates in lentil genotypes? Are variations in those traits heritable? How much prebiotics are contained in different lentil market classes? Additionally, how are dehulling, cooking, and refrigeration going to affect those concentrations? Furthermore, while it has been demonstrated that both lentil as a whole food and individual carbohydrates found in lentil contribute to reducing risk factors for obesity and NCDs, a causal link still remains to be established using animal and human trials.

In conclusion, although much remains to be elucidated and understood, lentil is a prime candidate as a dietary source of prebiotics and as a potential functional food. Lentil is a popular food in many countries, circumventing the problem of social or cultural rejection associated with many foods. Also, it can also be grown successfully in many regions of the world, so availability (at least under present circumstances) will not be an issue. Finally, prebiotic carbohydrates that are found in lentil have been repeatedly shown *in vitro* and *in vivo* to have beneficial health effects (Brown, 2004; Johnston et al., 2010; Conlon et al., 2012; Koo & Rao, 1991; Benno, 1987). With the necessary questions answered, lentil may be a useful tool in reducing obesity and NCDs (Hermsdorff, Zulet, Abete, & Martínez, 2011).

References

Abeysekara, S., Chilibeck, P. D., Vatanparast, H., & Zello, G. A. (2012). A pulse-based diet is effective for reducing total and LDL-cholesterol in older adults. *British Journal of Nutrition, 108,* S103.

Asp, N. G. (1992). Preface: resistant starch, proceedings from the second plenary meeting of Euresta: European FLAIR Concerted Action N11 on physiological implications of the consumption of resistant starch in man. *European Journal of Clinical Nutrition, 46,* S1.

Austin, G. L., Ogden, L. G., & Hill, J. O. (2011). Trends in carbohydrate, fat, and protein intakes and association with energy intake in normal-weight, overweight, and obese individuals: 1971–2006. *The American Journal of Clinical Nutrition, 93,* 836-843.

Backhed, F., Ley, R. E., Sonnenburg, J. L., Peterson, D. A., & Gordon, J. I. (2005). Host-bacterial mutualism in the human intestine. *Science, 307,* 1915-1920.

Backhed, F., Ding, H., Wang, T., Hooper, L. V., Koh, G. Y., Nagy, A. et al. (2004). The gut microbiota as an environmental factor that regulates fat storage. *Proceedings of the National Academy of Sciences of the United States of America, 101,* 15718-15723.

Barreteau, H., Delattre, C., & Michaud, P. (2006). Production of oligosaccharides as promising new food additive generation. *Food Technology and Biotechnology, 44,* 323.

Beards, E., Tuohy, K., & Gibson, G. (2010). A human volunteer study to assess the impact of confectionery sweeteners on the gut microbiota composition. *104,* 701-708.

Benno, Y. (1987). Effect of raffinose intake on human fecal microflora. *Bifidobacteria Microflora, 6,* 59-63.

Bhatty, R. S. (1988). Composition and Quality of Lentil (Lens-Culinaris Medik) - A Review. *Canadian Institute of Food Science and Technology Journal, 21,* 144-160.

Biesiekierski, J. R., Rosella, O., Rose, R., Liels, K., Barrett, J. S., Shepherd, S. J. et al. (2011). Quantification of fructans, galacto-oligosacharides and other short-chain carbohydrates in processed grains and cereals. *Journal of Human Nutrition and Dietetics, 24,* 154-176.

Bird, A. R., Conlon, M. A., Christophersen, C. T., & Topping, D. L. (2010). Resistant starch, large bowel fermentation and a broader perspective of prebiotics and probiotics. *Beneficial Microbes, 1,* 423-431.

Bouhnik, Y., Attar, A., Joly, F. A., Riottot, M., Dyard, F., & Flourie, B. (2004). Lactulose ingestion increases faecal bifidobacterial counts: a randomised double-blind study in healthy humans. *European Journal of Clinical Nutrition, 58,* 462-466.

Bouhnik, Y., Vahedi, K., Achour, L., Attar, A., Salfati, J., Pochart, P. et al. (1999). Short-chain fructo-oligosaccharide administration dose-dependently increases fecal bifidobacteria. *The Journal of Nutrition, 129,* 2286.

Brown, I. L. (2004). Applications and uses of resistant starch. *Journal of AOAC International, 87,* 727-732.

Cahill, G. F. (1970). Starvation in Man. *New England Journal of Medicine, 282,* 668-675.

Cani, P. D., Lecourt, E., Dewulf, E. M., Sohet, F. M., Pachikian, B. D., Naslain, D. et al. (2009). Gut microbiota fermentation of prebiotics increases satietogenic and incretin gut peptide production with consequences for appetite sensation and glucose response after a meal. *American Journal of Clinical Nutrition, 90,* 1236-1243.

Carpita, N. C., Kanabus, J., & Housley, T. L. (1989). Linkage Structure of Fructans and Fructan Oligomers from Triticum aestivum and Festuca arundinacea Leaves. *Journal of Plant Physiology, 134,* 162-168.

Center for Disease Control and Prevention (CDC). (2010). Adult Obesity: Obesity Rises Among Adults. *Vital Signs*

Champ, M., Kozlowski, F., & Lecannu, G. (2001). In vivo and in vitro methods for resistant starch measurement. *Advanced Dietary Fibre Technology,* 106-120.

Chung, H. J., Liu, Q., Dormer, E., Hoover, R., Warkentin, T. D., & Vandenberg, B. (2008). Composition, molecular structure, properties, and in vitro digestibility of starches from newly released Canadian pulse cultivars. *Cereal Chemistry, 85,* 473-481.

Conlon, M. A., Kerr, C. A., McSweeney, C. S., Dunne, R. A., Shaw, J. M., Kang, S. et al. (2012). Resistant Starches Protect against Colonic DNA Damage and Alter Microbiota and Gene Expression in Rats Fed a Western Diet. *The Journal of Nutrition, 142,* 832-840.

Cummings, J. H., Beatty, E. R., Kingman, S. M., Bingham, S. A., & Englyst, H. N. (1996). Digestion and physiological properties of resistant starch in the human large bowel. *British Journal of Nutrition, 75,* 733-747.

Cummings, J. H., Macfarlane, G. T., & Englyst, H. N. (2001). Prebiotic digestion and fermentation. *The American Journal of Clinical Nutrition, 73,* 415S-420S.

Cummings, J. H. & Kong, S. C. (2004). Probiotics, prebiotics and antibiotics in inflammatory bowel disease. In D. Chadwick & J. Goode (Eds.), *Inflammatory Bowel Disease: Crossroads of Microbes, Epithelium and Immune Systems* (pp. 99-114). Wiley Online Library.

de Vaux, A., Morrison, M., & Hutkins, R. W. (2002). Displacement of Escherichia coli O157:H7 from Rumen Medium Containing Prebiotic Sugars. *68,* 519-524.

Delzenne, N. M. & Kok, N. (2001). Effects of fructans-type prebiotics on lipid metabolism. *The American Journal of Clinical Nutrition, 73,* 456s-458s.

Delzenne, N. M., Neyrinck, A. M., Backhed, F., & Cani, P. D. (2011). Targeting gut microbiota in obesity: effects of prebiotics and probiotics. *Nature Reviews Endocrinology, 7,* 639-646.

Delzenne, N. M., Neyrinck, A. M., & Cani, P. D. (2013). Gut microbiota and metabolic disorders: how prebiotic can work? *British Journal of Nutrition, 109,* S81-S85.

Deshpande, S. S. & Cheryan, M. (1984). Effects of phytic acid, divalent cations, and their interactions on α-amylase activity. *Journal of Food Science, 49,* 516-519.

Desjardins, M. L., Roy, D., & Goulet, J. (1990). Growth of bifidobacteria and their enzyme profiles. *Journal of Dairy Science, 73,* 299-307.

Dreher, M. L., Dreher, C. J., Berry, J. W., & Fleming, S. E. (1984). Starch digestibility of foods: a nutritional perspective. *Critical Reviews in Food Science & Nutrition, 20,* 47-71.

Duenas, M., Sun, B., Hernandez, T., Estrella, I., & Spranger, M. I. (2003). Proanthocyanidin Composition in the Seed Coat of Lentils (Lens culinaris L.). *Journal of Agricultural and Food Chemistry, 51,* 7999-8004.

Ekinci, R. & Kadakal, C. (2005). Determination of seven water-soluble vitamins in tarhana, a traditional Turkish cereal food, by high-performance liquid chromatography. *ACTA Chromatographica, 15,* 289.

Englyst, H. N., Trowell, H., Southgate, D. A., & Cummings, J. H. (1987). Dietary fiber and resistant starch. *The American Journal of Clinical Nutrition, 46,* 873-874.

Englyst, H. N. & Macfarlane, G. T. (1986). Breakdown of resistant and readily digestible starch by human gut bacteria. *Journal of the Science of Food and Agriculture, 37,* 699-706.

Evenson, R. E. & Gollin, D. (2003). Assessing the impact of the Green Revolution, 1960 to 2000. *Science, 300,* 758-762.

Expert Panel on the Identification, Evaluation, and Treatment of Overweight in Adults. Clinical guidelines on the identification, evaluation, and treatment of overweight and obesity in adults: executive summary. (1998). *The American Journal of Clinical Nutrition, 68,* 899-917.

FAOSTAT. (2011). Food and Agricultural Commodities Production. Accessed June 5, 2013 from: http://faostat.fao.org/site/368/Desktop.Default.aspx?PageID=368#ancor

Finucane, M. M., Stevens, G. A., Cowan, M. J., Danaei, G., Lin, J. K., Paciorek, C. J. et al. (2011). National, regional, and global trends in body-mass index since 1980: systematic analysis of health examination surveys and epidemiological studies with 960 country-years and 9.1 million participants. *The Lancet, 377,* 557-567.

Flegal K. M., Carroll, M.D., Ogden, C.L., Curtin, L.R. (2010). Prevalence and trends in obesity among us adults, 1999-2008. *JAMA, 303,* 235-241.

Flegal, K. M., Carroll, M. D., Kit, B. K., & Ogden, C. L. (2012). Prevalence of obesity and trends in the distribution of body mass index among US adults, 1999-2010. *JAMA, 307,* 491-497.

Fuchs, C. S., Giovannucci, E. L., Colditz, G. A., Hunter, D. J., Stampfer, M. J., Rosner, B. et al. (1999). Dietary fiber and the risk of colorectal cancer and adenoma in women. *New England Journal of Medicine, 340,* 169-176.

Gavrilova, O., Marcus-Samuels, B., Graham, D., Kim, J. K., Shulman, G. I., Castle, A. L. et al. (2000). Surgical implantation of adipose tissue reverses diabetes in lipoatrophic mice. *Journal of Clinical Investigation, 105,* 271-278.

Gibson, G. R. & Roberfroid, M. B. (1995). Dietary Modulation of the Human Colonic Microbiota - Introducing the Concept of Prebiotics. *Pharmaceutiques, 125,* 1401-1412.

Gill, S. R., Pop, M., DeBoy, R. T., Eckburg, P. B., Turnbaugh, P. J., Samuel, B. S. et al. (2006). Metagenomic analysis of the human distal gut microbiome. *Science, 312,* 1355-1359.

Grabitske, H. A. & Slavin, J. L. (2009). Gastrointestinal effects of low-digestible carbohydrates. *Critical Reviews in Food Science and Nutrition, 49,* 327-360.

Gray, G. M. (1992). Starch digestion and absorption in nonruminants. *The Journal of Nutrition, 122,* 172-177.

Guillon, F. & Champ, M. M. J. (2002). Carbohydrate fractions of legumes: uses in human nutrition and potential for health. *British Journal of Nutrition, 88,* S293-S306.

Guraya, H. S., Kadan, R. S., & Champagne, E. T. (1997). Effect of rice starch-lipid complexes on in vitro digestibility, complexing index, and viscosity. *Cereal Chemistry, 74,* 561-565.

Hayakawa, K., Mizutani, J., Wada, K., Masai, T., Yoshihara, I., & Mitsuoka, T. (1990). Effects of soybean oligosaccharides on human faecal flora. *Microbial Ecology in Health and Disease, 3,* 293-303.

Hermsdorff, H. H., Zulet, M. Á., Abete, I., & Martínez, J. A. (2011). A legume-based hypocaloric diet reduces proinflammatory status and improves metabolic features in overweight/obese subjects. *European Journal of Nutrition, 50,* 61-69.

Holzapfel, W. H., Haberer, P., Snel, J., & Schillinger, U. (1998). Overview of gut flora and probiotics. *International journal of food microbiology, 41,* 85-101.

Hoover, R. & Sosulski, F. (1985). Studies on the functional characteristics and digestibility of starches from Phaseolus vulgaris biotypes. *Starch - Stärke, 37,* 181-191.

Hoover, R. & Zhou, Y. (2003). In vitro and in vivo hydrolysis of legume starches by α-amylase and resistant starch formation in legumes - a review. *Carbohydrate Polymers, 54,* 401-417.

Horton, T. J., Drougas, H., Brachey, A., Reed, G. W., Peters, J. C., & Hill, J. O. (1995). Fat and carbohydrate overfeeding in humans: different effects on energy storage. *The American Journal of Clinical Nutrition, 62,* 19-29.

ISAPP (2008). 6th Meeting of the International Scientific Association of Probiotics and Prebiotics, London, Ontartio. In.

Jenkins, D. J., Wolever, T. M., Taylor, R. H., Barker, H., Fielden, H., Baldwin, J. M. et al. (1981). Glycemic index of foods: a physiological basis for carbohydrate exchange. *The American Journal of Clinical Nutrition, 34,* 362-366.

Jenkins, D. J., Wolever, T. M., Taylor, R. H., Barker, H. M., & Fielden, H. (1980). Exceptionally low blood glucose response to dried beans: comparison with other carbohydrate foods. *British Medical Journal, 281,* 578.

Johnston, K. L., Thomas, E. L., Bell, J. D., Frost, G. S., & Robertson, M. D. (2010). Resistant starch improves insulin sensitivity in metabolic syndrome. *Diabetic Medicine, 27,* 391-397.

Kearney, J. (2010). Food consumption trends and drivers. *Philosophical Transactions of the Royal Society B: Biological Sciences, 365,* 2793-2807.

Kolida, S. & Gibson, G. R. (2008). The Prebiotic Effect: Review of Experimental and Human Data. In G. R. Gibson & M. B. Roberfroid (Eds.), *Handbook of Prebiotics* (pp. 69-92). Boca Raton, Fl: CRC Press.

Koo, M. & Rao, A. V. (1991). Long–term effect of bifidobacteria and Neosugar on precursor lesions of colonic cancer in cf1 mice. *Nutrition and Cancer, 16,* 249-57.

Krajmalnik-Brown, R., Ilhan, Z. E., Kang, D. W., & DiBaise, J. K. (2012). Effects of gut microbes on nutrient absorption and energy regulation. *Nutrition in Clinical Practice, 27,* 201-214.

Kruse, H. P., Kleessen, B., & Blaut, M. (1999). Effects of inulin on faecal bifidobacteria in human subjects. *British Journal of Nutrition, 82,* 375-382.

Lajolo, F. M., Finardi Filho, F., & Menezes, E. W. (1991). Amylase inhibitors in Phaseolus vulgaris beans. *Food Technology, 45,* 119-121.

Lee, Y. K. & Mazmanian, S. K. (2010). Has the Microbiota Played a Critical Role in the Evolution of the Adaptive Immune System? *Science, 330,* 1768-1773.

Lewis, D. H. (1993). Nomenclature and diagrammatic representation of oligomeric fructans – a paper for discussion. *New Phytologist, 124,* 583-594.

Lopez, A. D., Mathers, C. D., Ezzati, M., Jamison, D. T., & Murray, C. J. (2006). Global and regional burden of disease and risk factors, 2001: systematic analysis of population health data. *The Lancet, 367,* 1747-1757.

Macfarlane, G. T., Gibson, G. R., & Cummings, J. H. (1992). Comparison of fermentation reactions in different regions of the human colon. *Journal of Applied Microbiology, 72,* 57-64.

Makinen, K. K. & Soderling, E. (1980). A quantitative study of mannitol, sorbitol, xylitol, and xylose in wild berries and commercial fruits. *Journal of Food Science, 45,* 367-71, 374.

Martínez-Villaluenga, C., Frias, J., & Vidal-Valverde, C. n. (2008). Alpha-galactosides: antinutritional factors or functional ingredients? *Critical Reviews in Food Science and Nutrition, 48,* 301-316.

Martínez-Villaluenga, C., Frias, J., Vidal-Valverde, C., & Gomez, R. (2005). Raffinose family of oligosaccharides from lupin seeds as prebiotics: Application in dairy products. *Journal of Food Protection, 68,* 1246-1252.

McCrory, M. A., Hamaker, B. R., Lovejoy, J. C., & Eichelsdoerfer, P. E. (2010). Pulse consumption, satiety, and weight management. *Advances in Nutrition, 1,* 17-30.

Megazyme. (2012). Resistant Starch Assay Procedure. RSTAR 11/02. Megazyme International Ltd.

Mishra, S., Monro, J., & Hedderley, D. (2008). Effect of processing on slowly digestible starch and resistant starch in potato. *Starch – Stärke, 60,* 500-507.

Muir, J. G. & O'dea, K. (1992). Measurement of resistant starch: factors affecting the amount of starch escaping digestion in vitro. *The American Journal of Clinical Nutrition, 56,* 123-127.

Muir, J. G. & O'Dea, K. (1993). Validation of an in vitro assay for predicting the amount of starch that escapes digestion in the small intestine of humans. *The American Journal of Clinical Nutrition, 57,* 540-546.

Muir, J. G., Rose, R., Rosella, O., Liels, K., Barrett, J. S., Shepherd, S. J. et al. (2009). Measurement of Short-Chain Carbohydrates in Common Australian Vegetables and Fruits by High-Performance Liquid Chromatography (HPLC). *Journal of Agriculture and Food Chemistry, 57,* 554-565.

Muir, J. G., Shepherd, S. J., Rosella, O., Rose, R., Barrett, J. S., & Gibson, P. R. (2007). Fructan and Free Fructose Content of Common Australian Vegetables and Fruit. *Journal of Agricultural and Food Chemistry, 55,* 6619-6627.

Murphy, M. M., Douglass, J. S., & Birkett, A. (2008). Resistant starch intakes in the United States. *Journal of the American Dietetic Association, 108,* 67-78.

Nebesny, E., Rosicka, J., & Tkaczyk, M. (2002). Effect of Enzymatic Hydrolysis of Wheat Starch on Amylose–Lipid Complexes Stability. *Starch - Stärke, 54,* 603-608.

Onigbinde, A. O. & Akinyele, I. O. (1983). Oligosaccharide content of 20 varieties of cowpeas in Nigeria. *Journal of Food Science, 48,* 1250-1251.

Pandeya, D. R., D'Souza, R., Rahman, M. M., Akhter, S., Kim, H. J., & Hong, S. T. (2012). Host-microbial interaction in the mammalian intestine and their metabolic role inside. *Biomedical Research, 23,* 9-21.

Parnell, J. A. & Reimer, R. A. (2012). Prebiotic fibres dose-dependently increase satiety hormones and alter Bacteroidetes and Firmicutes in lean and obese JCR:LA-cp rats. *British Journal of Nutrition, 107,* 601-613.

Pereira, D. I. A. & Gibson, G. R. (2002). Effects of consumption of probiotics and prebiotics on serum lipid levels in humans. *Critical Reviews in Biochemistry and Molecular Biology, 37,* 259-281.

Pietinen, P., Rimm, E. B., Korhonen, P., Hartman, A. M., Willett, W. C., Albanes, D. et al. (1996). Intake of Dietary Fiber and Risk of Coronary Heart Disease in a Cohort of Finnish Men The Alpha-Tocopherol, Beta-Carotene Cancer Prevention Study. *Circulation, 94,* 2720-2727.

Popkin, B. M., Kim, S., Rusev, E. R., Du, S., & Zizza, C. (2006). Measuring the full economic costs of diet, physical activity and obesity–related chronic diseases. *Obesity Reviews, 7,* 271-293.

Qin, J., Li, R., Raes, J., Arumugam, M., Burgdorf, K. S., Manichanh, C. et al. (2010). A human gut microbial gene catalogue established by metagenomic sequencing. *Nature, 464,* 59-65.

Rabot, S., Membrez, M., Bruneau, A., Gerard, P., Harach, T., Moser, M. et al. (2010). Germ-free C57BL/6J mice are resistant to high-fat-diet-induced insulin resistance and have altered cholesterol metabolism. *The FASEB Journal, 24,* 4948-4959, 10.

Roberfroid, M. B., Gibson, G. R., Hoyles, L., McCartney, A. L., Rastall, R., Rowland, I. et al. (2010). Prebiotic effects: metabolic and health benefits. *British Journal of Nutrition, 104,* S1-S63.

Roberfroid, M. B. (2007). Prebiotics: The Concept Revisited. *The Journal of Nutrition, 137,* 830S-837S.

Roberfroid, M. B. (1999). Caloric value of inulin and oligofructose. *The Journal of Nutrition, 129,* 1436S-1437s.

Roberfroid, M. B. (2008). General Introduction: Prebiotics in Nutrition. In Glenn R Gibson & Marcel B Roberfroid (Eds.), *Handbook of Prebiotics* (pp. 1-11). Boca Raton, Fl: CRC Press.

Rowland, I. R. (2009). The role of the gastrointestinal microbiota in colorectal cancer. *Current Pharmaceutical Design 15,* 1524-1527.

Saito, Y., Takano, T., & Rowland, I. (1992). Effects of soybean oligosaccharides on the human gut microflora in in vitro culture. *Microbial Ecology in Health and Disease, 5,* 105-110.

Salunkhe, D. K., Kadam, S. S., & Chavan, J. K. (1985). *Postharvest biotechnology of food legumes.* (pp. 1-176). CRC Press.

Savage, D. C. (1977). Microbial ecology of the gastrointestinal tract. *Annual Reviews in Microbiology, 31,* 107-133.

Scholz-Ahrens, K. E., Ade, P., Marten, B., Weber, P., Timm, W., Agil, Y. et al. (2007). Prebiotics, probiotics, and synbiotics affect mineral absorption, bone mineral content, and bone structure. *Journal of Nutrition, 137,* 838S-846S.

Seip, M. & Trygstad, O. (1996). Generalized lipodystrophy, congenital and acquired (lipoatrophy). *Acta Paediatrica, 85,* 2-28.

Singh, R. B., Rastogi, S. S., Verma, R., Laxmi, B., Singh, R., Ghosh, S. et al. (1992). Randomised controlled trial of cardioprotective diet in patients with recent acute myocardial infarction: results of one year follow up. *BMJ: British Medical Journal, 304,* 1015.

Singh, S., Singh, H. D., & Sikka, K. C. (1968). Distribution of nutrients in the anatomical parts of common Indian pulses. *Cereal Chemistry, 45,* 13-18.

Skrabanja, V., Liljeberg, H. G., Hedley, C. L., Kreft, I., & Björck, I. M. (1999). Influence of genotype and processing on the in vitro rate of starch hydrolysis and resistant starch formation in peas (Pisum sativum L.). *Journal of Agricultural and Food Chemistry, 47,* 2033-2039.

Smiricky, M. R., Grieshop, C. M., Albin, D. M., Wubben, J. E., Gabert, V. M., & Fahey, G. C. (2002). The influence of soy oligosaccharides on apparent and true ileal amino acid digestibilities and fecal consistency in growing pigs. *Journal of Animal Science, 80,* 2433-2441.

Snow, P. & O'Dea, K. (1981). Factors affecting the rate of hydrolysis of starch in food. *The American Journal of Clinical Nutrition, 34,* 2721-2727.

Sufian, M. K. N. B., Hira, T., Asano, K., & Hara, H. (2007). Peptides derived from dolicholin, a phaseolin-like protein in country beans (Dolichos lablab), potently stimulate cholecystokinin secretion from enteroendocrine STC-1 cells. *Journal of Agricultural and Food Chemistry, 55,* 8980-8986.

Swinburn, B. A., Sacks, G., Hall, K. D., McPherson, K., Finegood, D. T., Moodie, M. L. et al. (2011). The global obesity pandemic: shaped by global drivers and local environments. *The Lancet, 378,* 804-814.

Tahir, M., Vandenberg, A., & Chibbar, R. N. (2011). Influence of environment on seed soluble carbohydrates in selected lentil cultivars. *Journal of Food Composition and Analysis, 24,* 596-602.

Tester, R. F., Karkalas, J., & Qi, X. (2004). Starch–composition, fine structure and architecture. *Journal of Cereal Science, 39,* 151-165.

Thavarajah, D., Thavarajah, P., Wejesuriya, A., Rutzke, M., Glahn, R. P., Combs, G. F. et al. (2011). The potential of lentil (Lens culinaris L.) as a whole food for increased selenium, iron, and zinc intake: preliminary results from a 3 year study. *180,* 123-128.

Tovar, J., Francisco, A. d., Bjorek, I., & Asp, N. G. (1991). Relationship between microstructure and in vitro digestibility of starch in precooked leguminous seed flours. *Food Structure, 10.*

Turnbaugh, P. J., Ley, R. E., Hamady, M., Fraser-Liggett, C. M., Knight, R., & Gordon, J. I. (2007). The human microbiome project. *Nature, 449,* 804-810.

Turnbaugh, P. J., Ley, R. E., Mahowald, M. A., Magrini, V., Mardis, E. R., & Gordon, J. I. (2006). An obesity-associated gut microbiome with increased capacity for energy harvest. *Nature, 444,* 1027-1031.

United Nations. (2012). Prevention and control of non-communicable diseases. Report of the Secretary General.

U.S. Department of Health & Human Services. Overweight and Obesity: Health Consequences.

Accessed July 16, 2013 from: http://www.surgeongeneral.gov/library/calls /obesity/fact_consequences.html.

USDA (2012). Nutrient data for 16070, Lentils, mature seeds, cooked, boiled, without salt. http://ndb.nal.usda.gov/ndb/foods/show/4684?fg=&man= &lfacet=&format=&count=&max=25&offset=&sort=&qlookup=lentil

van Loo, J., Coussement, P., De Leenheer, L., Hoebregs, H., & Smits, G. (1995). On the presence of inulin and oligofructose as natural ingredients in the Western diet. *Critical Reviews in Food Science and Nutrition, 35,* 525-552.

Vidal-Valverde, C., Frias, J., Estrella, I., Gorospe, M. J., Ruiz, R., & Bacon, J. (1994). Effect of processing on some antinutritional factors of lentils. *Journal of Agriculture and Food Chemistry, 42,* 2291-2295.

Wang, N., Hatcher, D. W., Toews, R., & Gawalko, E. J. (2009). Influence of cooking and dehulling on nutritional composition of several varieties of lentils (Lens culinaris). *LWT- Food Science and Technology, 42,* 842-848.

Wang, N., Hatcher, D. W., & Gawalko, E. J. (2008). Effect of variety and processing on nutrients and certain anti-nutrients in field peas (Pisum sativum). *Food Chemistry, 111,* 132-138.

Welch, C. (1936). Function of the large intestine of man in absorption and excretion: Study of a subject with an ileostomy stoma and an isolated colon. *Archives of Internal Medicine, 58,* 1095-1110.

Welch, R. M. & Graham, R. D. (1999). A new paradigm for world agriculture: meeting human needs: productive, sustainable, nutritious. *Field Crops Research, 60,* 1-10.

Wolever, T. M. S., Piekarz, A., Hollands, M., & Younker, K. (2002). Sugar alcohols and diabetes: a review. *Canadian Journal of Diabetes, 26,* 356-362.

Wursch, P., Del Vedovo, S., & Koellreutter, B. (1986). Cell structure and starch nature as key determinants of the digestion rate of starch in legume. *American Journal of Clinical Nutrition, 43*.

Xu, B. & Chang, S. K. C. (2010). Phenolic Substance Characterization and Chemical and Cell-Based Antioxidant Activities of 11 Lentils Grown in the Northern United States. *Journal of Agriculture and Food Chemistry, 58,* 1509-1517.

Yadav, B. S., Sharma, A., & Yadav, R. B. (2009). Studies on effect of multiple heating/cooling cycles on the resistant starch formation in cereals, legumes and tubers. *International Journal of Food Science and Nutrition, 60,* 258-272.

Yun, J. W. (1996). Fructooligosaccharides–occurrence, preparation, and application. *Enzyme and Microbial Technology, 19,* 107-117.

Zobel, H. F. (1988). Molecules to granules: a comprehensive starch review. *Starch - Stärke, 40,* 44-50.

In: Oligosaccharides
Editors: L. S. Schweizer, S. J. Krebs
ISBN: 978-1-62948-328-3

Chapter III

Behaviour of Dietary Oligosaccharides in the Human Gastrointestinal Tract

Fraser L. Courts, Liz K. Wells, Stephanie K. Allen and Huw S. Jones
HONEI, Centre for Cardiovascular and Metabolic Research,
Hull York Medical School, University of Hull, Hull, UK

Abstract

Dietary oligosaccharides are classically assumed to remain inert during passage through the human gastrointestinal (GI) tract, being resistant to both hydrolysis and uptake into systemic circulation. Advances in the fields of paediatric nutrition, gut microbiology and animal nutrition have more recently provided intriguing evidence for diverse physiological and physicochemical behaviours of oligosaccharides in the mammalian GI tract however, including mucosal adhesive properties, prebiotic activity, and modulation of intestinal nutrient uptake. The GI tract is also now known to process some dietary oligosaccharides via acidic and enzymatic hydrolysis of specific glycosidic linkages, forming shorter and less-branched saccharides, the latter of which may be actively transported across the intestinal mucosa to participate in glycolytic metabolism. This evidence, fuelled by the advent of advanced analytical methodologies for carbohydrate analysis and the global rise in demand for functional food product innovation has resulted in interest surrounding the functional behaviour of dietary oligosaccharides through the human GI tract, and downstream effects on health.

In this chapter, current and future experimental strategies for modelling oligosaccharide behaviour in the mammalian GI tract are summarised, focusing on the biological models and methods of oligosaccharide analysis used in the literature. The application of these techniques towards elucidating dietary oligosaccharide pharmacokinetics is described, revealing often stark differences between oligosaccharide families in susceptibility to hydrolysis and transmucosal uptake into systemic circulation. Evidence for the ability of surviving oligosaccharide structures in the lumen to prevent pathogen adhesion to the intestinal mucosa, act as excipients for nutrient uptake and as prebiotic substrates for the colonic microflora, selectively encouraging the growth of particular microbial genera or species is also detailed, with implications for intestinal bacterial invasion and gut health discussed. Overall, we present evidence for the importance of oligosaccharides in maintaining a healthy GI tract and the modulatory role of the GI tract in determining the potential protective effects of oligosaccharides against disease and dysfunction.

1. Introduction

The human gastrointestinal (GI) tract is a complex and dynamic system designed to liberate, digest and selectively absorb a wide range of nutrients from their respective matrices. Classically, few classes of molecules from the diet have been considered more stable whilst traversing this gauntlet than the oligosaccharides, being assumed to be resistant to both hydrolysis and uptake into systemic circulation.

This traditional view is highly simplistic however, and we now understand far more about the complex behaviour of these seemingly inert structures in the human GI tract, largely due to information gleaned as the result of advances in the fields of paediatric nutrition, gut microbiology and animal nutrition. This information is assembled in this chapter to provide a better picture of the gastrointestinal fate and activities of dietary oligosaccharides, alongside information on current and future strategies for modelling oligosaccharide behaviour in the mammalian GI tract.

2. Oligosaccharide Hydrolysis and Stability in the Human Gastrointestinal Tract

From the moment the dietary oligosaccharide load enters the buccal cavity, it is susceptible to hydrolysis of glycosidic bonds by endogenous and microfloral enzymes, and changes in pH throughout the GI tract. Hydrolysis can result in the formation of monosaccharides from precursor dietary oligosaccharides that are often actively transported into circulation from the small intestine. These carbohydrates are classified as 'digestible' and play an important role in health as they trigger hormone secretion and serve as substrates and regulators in major metabolic pathways [1]. Conversely, 'non-digestible' oligosaccharides remain largely intact, partially owing to the ability of the anomeric carbon (*C*1 or *C*2) to confer resistance to mammalian hydrolytic enzymes [2]. These species therefore may have various distinct physiological and nutritional effects in humans [3].

By far the best studied of the non-digestible oligosaccharides are those reported to have prebiotic function. It is stated that many of these species, including human milk oligosaccharides, xylooligosaccharides and fructooligosaccharides, are largely resistant to degradation by both the acidic environment of the stomach and endogenous enzymes, however they may still be susceptible to low rates of hydrolysis. The current knowledge regarding oral and intestinal digestion of these oligosaccharides is outlined below, and strategies for assessing the digestion of oligosaccharides are also highlighted and critiqued.

2.1. Oral Hydrolysis

Digestion of oligosaccharides in the oral cavity is mediated not only via salivary enzymes but also by the oral microflora, of which *Streptococcus mutans* is a commonly detected species [4]. The presence of such bacteria allows the possible fermentation of sugars as an energy source and thus utilisation of different oligosaccharides. Indeed, these bacterial strains are known to secrete hydrolytic enzymes such as invertase (allowing the cleavage of sucrose), and there is also evidence suggesting the secretion of α-glucosidase into the extracellular environment [4]. The α-glucobiose oligosaccharides display different extents of utilisation by *S.mutans*, in a manner dependent on the position of the glycosidic linkage. For example, trehalose (1→1 α-glucobiose) and both 1→4 and 1→6 α-glucobioses are utilised by *S.mutans*, whereas the 1→2 and 1→3 α-glucobioses kojibiose and nigerose are not [4]. For β-linked glucobioses, *S.mutans* utilisation is also affected by location of the glycosidic linkage. Gentibiose (β-1→6) is resistant to bacterial utilisation, whereas the β-1→2, β-1→3 and β-1→4 structures (sophorose, laminaribose and cellobiose respectively) were all shown to be utilised by *S. mutans* [4]. In addition to this, cereal fructans with a DP less than 5 have been shown to potentially act as substrates for oral microflora [5, 6]. Fructooligosaccharides, hydrogenated isomaltooligosaccharides and xylooligosaccharides have been shown to be resistant to human salivary amylase [7-9]. The pectin-derived oligosaccharide homogalacturonan has also been reported to be stable when incubated with saliva [10], as have both neutral and acidic human milk oligosaccharides [11].

It should be noted that as the oral microflora consists of multiple different species it is important that the effects of different varieties of bacteria (e.g. *Actinomyces*, *Bifidobacterium* or *Lactobacillus*) are assessed for their ability to digest oligosaccharides. This information is currently missing from the literature. It is also worth noting that hydrolysis in the mouth is of particular relevance to dental health, as hydrolysis by the salivary enzymes releases monosaccharides that represent metabolic substrates for cariogenic bacteria [4]. Careful selection of poorly hydrolysed oligosaccharides enables their use as low cariogenic sugar substitutes in food products, and as a result are often added to confectionery, chewing gums, yoghurts and drinks [12].

2.2. Gastric Hydrolysis

Despite the low potential of enzymic hydrolysis of oligosaccharides in the human stomach, the hydrolytic potential of the acidic environment has been assessed for some discreet classes of oligosaccharides, particularly fructooligosaccharides. As a general rule, oligosaccharides generally exhibit good stability at low pH, however there is now some evidence that hydrolysis can occur in varying amounts between the different classes of oligosaccharides. In a study using human simulated gastric fluid (SGF) of modulated pH, up to 15% of incubated fructooligosaccharide fractions of D.P. 3 - 9 underwent hydrolysis below pH 1.8 over a 60-minute incubation. In the same study, hydrogenated isomaltooligosaccharides were shown to be stable in SGF at pH 1.0 - 2.0. Xylooligosaccharides have also been reported to be stable over a broad pH range (pH 2.5 - 8.0), and are not hydrolysed by SGF [9]. Similarly, homogalacturonan remains stable in SGF [10], and human milk oligosaccharides showed only minor loss of acidic oligosaccharides in phosphate buffered saline adjusted to pH 2.0 [11]. In addition to these SGF studies, Björck and Nilsson (1991) used an alternative approach to answering the question of gastric stability by employing a novel rat model to determine the specific role of low gastric pH in inulin hydrolysis. Antibiotic-treated rats fed with inulin and omeprazole (to prevent gastric acid secretion) showed significantly lower inulin hydrolysis when compared to a group of control rats fed inulin alone. The lack of confounding microfloral hydrolysis in this study due to the use of antibiotics therefore suggests a role for acidic hydrolysis in the stomach in the production of fructose and short chain fructans from inulin.

2.3. Upper-Intestinal Hydrolysis

As with oral digestion, the relative resistance of oligosaccharides to hydrolytic enzyme activity found in the small intestine varies not only with degree of polymerisation, but also with the class of oligosaccharide. For example, isomaltooligosaccharides with a D.P. > 3 have good resistance to hydrolysis, and fructans with hexose sugars contatining β-1→2 linkages are entirely resistant to hydrolysis, whereas maltooligosaccharides (α-D-glucose - linked with α-1→4 linkages) are hydrolysed and absorbed in the small intestine [12]. Using rat intestinal extracts (containing intestinal enzymes such as α-glucosidase), Hodonicky et al. (2012) showed that the α-1→2, α-1→3 and α-1→6 -linked glucobioses were either hydrolysed at a lower rate than maltose, or were resistant to hydrolysis [4].

Fructooligosaccharides have been reasonably well studied in terms of their intestinal digestion. The β-1→2 linkages of fructooligosaccharides are resistant to digestion due to the lack of specific enzymatic activity within the upper-intestinal lumen, and these oligosaccharides are also resistant to pancreatic amylases [8]. Other evidence corroborating the upper-intestinal stability of fructooligosaccharides includes a lack of hydrolysis in the small intestine of both normal human subjects and those with ileostomies, supported by similar stability of fructooligosaccharides exhibited in human duodenal mucosa preparations. In the same study, Molis et al. (1996) concluded that recovery from the small intestine fructooligosaccharides mimicked that of an unabsorbed marker compound [13].

Hydrogenated isomaltooligosaccharides are resistant to porcine pancreatic amylases [7]. When incubated with the acetone powder of rat intestinal mucosa, however, moderate successive hydrolysis of glucose from the reducing end of tri-, tetra- and penta-saccharide alcohols occurred, but disaccharides were resistant to cleavage in this study [7]. Both neutral and acidic human milk oligosaccharides have been shown to be resistant to digestion by both human and porcine brush-border preparations and pancreatic glycosidases, however the digestion of non-fucosylated (but not fucosylated) neutral oligosaccharides has been demonstrated using enriched porcine brush border preparations [11]. Conversely, xylooligosaccharides are not hydrolysed by intestinal homogenates [9]. Using enzyme preparations derived from the homogenates of both suckling and adult rat intestines, Morita et al. (2008) demonstrated that cellobiose was hydrolysed at approximately 50% the rate of lactose. Interestingly, the hydrolytic activity of intestinal preparations from suckling rats was approximately three times that of adult rat preparations, however gentiobiose remained stable.

Ileorectostomised rats were employed to demonstrate the significance of low gastric pH on cellobiose or fructooligosaccharide hydrolysis (65% and 35% of the total oral dose respectively). These quantities were unaffected by the presence of a broad-spectrum antibiotic, included to remove the potential for bacterial hydrolysis. Unsurprisingly, it was noted that these observations in rat models conflict with human data. Hydrogen breath test experiments suggested that the majority of administered cellobiose undergoing fermentation [14], and the detection of low quantities of ingested fructooligosaccharides undergoing hydrolysis in the terminal ileum or stomach [13]. The authors of the latter study point out that it is possible that this low level of digestion of cellobiose in the human intestine results from intestinal lactase activity, due to its ability to cleave β-1→4 linkages.

Incubation of homogalacturonan with ileostomy effluent from healthy human subjects resulted in almost complete digestion over a 10-hour period, although it should be stressed that there was some inter-individual variation between subjects in the rate of this digestion over shorter periods [10].

2.4. Methods of Assessing Gastrointestinal Oligosaccharide Hydrolysis

Due to the inherent complexity and compartmentalisation of the digestive system, modelling the various processes that occur throughout the entire GI tract is challenging. Current strategies range from using single compartment models that assess either one or a small number of individual processes, such as oral microbiota function, to more elaborate multi-compartment systems that assess not only enzymatic processes, but physical processing and induction or modulation over time as well, such as ileostomised human subjects. This section highlights examples of different models that can be used to assess the digestion of oligosaccharides, primarily focusing upon single compartment models. Although these are simplified versions of the biological system they represent, they can provide useful data to guide *in vivo* studies. It should also go without statement when developing and using such models that human-derived materials and relevant bacterial species should be used where possible in order to reduce the risk of drawing misleading conclusions due to inter-species variation in GI tract biology.

2.4.i. Modelling Oral Hydrolysis

The most commonly used approach for simulating the digestion of oligosaccharides in the oral cavity involves the incubation of the test compound with either saliva collected from human subjects, or the use of purified enzymes such as salivary amylase. Other enzymes, such as those secreted by oral bacteria, may also be used. For the assessment of the contribution of the oral microflora, culture of the selected strain and species (such as *S.mutans* – see Hodoniczky *et al.* 2012 for details), followed by re-suspension in Stephan's buffer to mimic the physiological conditions of the oral cavity and incubation with the test oligosaccharide allows the collection of samples over time, followed by appropriate analysis (discussed in section 3). The measurement of pH changes resulting from the oral bacterial metabolism of incubated oligosaccharides have also been used as a marker of oral digestion, as such changes in pH reflect the utilisation of hydrolysed sugars by the bacterial load.

2.4.ii. Modelling Gastric Hydrolysis

The use of simulated gastric fluid (SGF) to mimic digestion in the stomach is a simplistic but commonly used approach. SGF can be readily purchased or produced in the laboratory using published composition tables, although it should be noted that some of the commercially-sourced versions do not contain pepsin or other digestive enzymes such as gastric lipase which must therefore be supplemented prior to use. Incubation of test carbohydrates with SGF may then be followed by a suitable analytical technique (section 3) to estimate hydrolysis of the oligosaccharides. An alternative to SGF is to collect gastric fluid from human subjects or animal models with or without stimulation using subcutaneous administration of pentagastrin. The obvious limitation of *ex vivo* or synthetic fluids is that they do not reproduce either the mechanical stresses or the different distinct regions of the stomach observed *in vivo*. The use of antibiotic treatments (to eliminate gut microflora and thus exclude the fermentation from the experiment) and treatment with omeprazole to eliminate gastric acid secretion in animal models can provide some evidence of hydrolysis in the stomach, however such results should be interpreted with caution due to potential inter-species variation.

2.4.iii. Modelling Upper-Intestinal Hydrolysis

There are several different strategies for assessing the digestion of oligosaccharides in the small intestine. For *in vitro* systems, purified enzymes such as pancreatic amylase and intestinal lactase can be incubated with the test oligosaccharides and assessed for degradation. At best, this simplified approach provides maximal rates of hydrolysis in single enzyme systems. The use of intestinal brush border preparations (homogenates prepared from intestinal biopsies) from human subjects provides a more realistic mixed enzyme system than isolated purified enzymes, but still does not reflect the modulation of these hydrolytic systems over time.

The most realistic strategy to determine the extent of small intestine digestion of oligosaccharides is via the use of ileostomised human subjects as this approach allows collection of the digested milieu prior to colonic microfloral utilisation, accurately represents the digestive processes of the small intestine and stomach in the complete absence of microfloral artefacts. As availability of such subjects can limit study scope, this approach can also be used in animal models. High inter-individual variability in the extent of hydrolysis is another potential limitation of the use of ileostomists, resulting in variable data from the usually low numbers of individuals used in this type of study [10].

Overall, the ideal strategy for assessing the hydrolysis and digestion of oligosaccharides combines several of the different types of system outlined above. The use of ileostomised subjects provides an unrivalled assessment of the potential for digestion in humans, but access can be limiting. Also, it can be difficult to identify mechanistic detail regarding the processes and enzymatic systems involved in oligosaccharide digestion. The use of *in vitro* experiments to determine the relative contributions of salivary enzymes, oral microflora, gastric juice, and intestinal secretions and extracellular enzymes can provide not only useful data for predicting such processes *in vivo*, but can also add to the mechanistic data on these processes. With regards to the use of rodent models for *in vivo* experiments, and non-human derived enzymes and homogenates, caution should be applied when interpreting such data due to species differences, and the use of human subjects or human-derived systems is usually preferred.

3. Oligosaccharide Absorption and Pharmacokinetics

Fractional hydrolysis of the total oral oligosaccharide load in the upper GI tract inevitably results in a significant proportion of intact or cleaved oligosaccharide entities reaching the lumen of the small intestine; a site that is specialised for nutrient uptake. An increase in studies detailing biological responses to oral oligosaccharides calls the traditional view that oligosaccharides are not bioavailable into question. Colleagues in the field have begun to investigate if, to some greater or lesser extent, oligosaccharides may be taken-up from the GI tract into systemic circulation, thus allowing their presence to exert pharmacological effects. Pharmacokinetic principals may be applied to better understand this process. Pharmacokinetic studies are a well established branch of pharmacology, tasked with determining the physical and chemical fate of molecules *in vivo*. Whilst most commonly applied to synthetic organic molecules and toxins, the principals of oral pharmacokinetic studies have more recently been applied to small organic xenobiotic molecules such as polyphenols, usually arising in the human diet from consumed plant foods [15]. Despite the value of such studies in predicting dose, efficacy and fate of potentially biologically active xenobiotics, and recent interest surrounding the biological activities of dietary oligosaccharides, studies concerning the oral pharmacokinetics of oligosaccharides are far less common in the scientific literature, presumably due to a number of confounding factors.

Firstly, it is only rather recently that it has been accepted that oral oligosaccharides may be found in the central compartment, rather than simply traversing the luminal course of the gastrointestinal tract. The small intestinal epithelium has perhaps too often been regarded as an impenetrable barrier to all but selected essential substrates, for which specific transporters exist to allow facilitated or active transcellular diffusion into systemic circulation. In fact, the dynamic epithelium may permit the passive paracellular uptake of a vast array of xenobiotic structures, controlled largely by physicochemical parameters such as molecular size, polarity and pKa, but also via energy-dependant mechanisms such as trancytosis [16, 17]. Use of in *vitro* models of the small intestinal epithelium to study parameters of oligosaccharide uptake is now emerging and is discussed in this section. It is worth noting that the potential for hydrolytic structural modulation of oligosaccharides in the GI tract lumen adds to the complexity, providing a greater array of possible oligosaccharide structures available for intestinal uptake (as detailed in section 2). Finally, analytical difficulties in the tracing and measurement of xenobiotic oligosaccharides make traditional methods of pharmacokinetic assessment less applicable to this area of investigation. The wider availability and use of alternative, complimentary methods are therefore required in order to fully understand the fate of oligosaccharide structures following oral consumption. It is therefore useful to outline approaches to tackling the oligosaccharide pharmacokinetic conundrum.

3.1. Analytical Method Development for Pharmacokinetic Tracing of Dietary Oligosaccharides

Several complimentary approaches must be applied when designing oligosaccharide-tracing experiments, considering both sample matrix and analytical equipment availability. Matrix interference presents the greatest challenge in the biological setting, owing to the large number of similar xenobiotic oligosaccharide structures that each may be present at low concentration, the transient nature of carbohydrates in metabolically active systems, and the potential for high concentrations of interfering analogous structures within the matrix. Additionally, methods for carbohydrate analysis are notorious for their requirement for compromise in terms of selectivity or sensitivity, principally due to high structural polarity, the vast array of similar structures, and lack of a useful chromaphore or fluorophore. Combined, the analytical challenge of quantifying trace oligosaccharides in complex biological matrices is great, but by no means insurmountable. Techniques can be split into labelled and native approaches, both of which may be employed in a complimentary fashion.

3.1.i. Oligosaccharide Labelling Approaches

The most commonly employed technique for tracing of crude oligosaccharide mixtures is via molecular labelling or tagging approaches. The principal behind this approach is usually to covalently link a synthetic fluorophore or chromophore to the oligosaccharide structure to endow differentiation from the matrix at the analysis stage [18]. If used with care, labelling techniques for oligosaccharide tracing and analysis may be used interchangeably; applying similar reactions to derivitise the oligosaccharide-rich sample either prior to biological application or during sample preparation for analysis.

Generally, oligosaccharides may be derivitised at carboxyl, carbonyl, or hydroxyl groups, however the choice of reactive group for tagging should be governed by the requirement for quantification versus semi-quantification, specificity of detection to acidic or neutral oligosaccharides and proposed method(s) of detection.

Reactions at the carbonyl group are often favoured for labelling broad oligosaccharide mixtures, presumably due to the ability of the reaction to react stoichiometrically with all reducing sugars, regardless of structure, as each molecule contains a single reducing end. This allows quantification of individual oligosaccharide structures within crude mixtures when coupled to a separation-based analytical technology such as capillary or column chromatography [19]. Reductive amination of the carbonyl group is most commonly selected, using an array of cheap and easily accessible cyclic amine-containing reagents, such as 2-aminopyridine (2-AP) or 8-aminonaphthalene-1,3,6-trisulfonic acid (ANTS) [20]. A comprehensive review of these compounds and their application is given by Lamari *et al.* (2003) [18]. A significant problem in the selection of reductive methodologies to introduce a biological tracer to the oligosaccharide structure is the resulting open-ringed terminus saccharide residue. This represents a significant structural modification to the saccharide moiety besides the presence of the tracer moiety itself, and should be used with care. Maness and Mort (1989) point out that one major advantage of this technique is the ability to chemically hydrolyse the tag from the terminal saccharide group, allowing for preparative approaches to be used downstream of analysis [21].

Uronic acid containing oligosaccharides may be selectively tagged by condensation of the carboxyl group with sulfanilic acid or 7-aminonaphthalene-1,3-disulfonic acid in the presence of carbonised imine [22]. Stoichiometric derivitisation is not a feature of this approach, as a variable number of fluorescent or UV absorbing moieties may react with each oligosaccharide structure dependant on the number of uronic acid residues (and therefore number of reactive carboxyl groups) present in the structure. At best, this can therefore only be considered to provide semi-quantification, however provides the advantage of good reactive specificity to acidic oligosaccharides. Another significant limitation of condensation reactions at the carboxyl group is the relative change in oligosaccharide pKa, representing a major physicochemical difference if substitution is frequent throughout the structure.

Ubiquitous decoration of oligosaccharide structures with hydroxyl groups makes this a useful target for a derivitisation approach, especially where sensitive detection of oligosaccharides of greater DP is required. Classically, DeBelder and Granath describe a flexible labelling approach using the commonly applied fluorescent reagent, fluorescein isothiocyanate (FITC) [23]. It is worth noting that 5-(4,6-dichlorotriazinyl) aminofluorescein (DTAF) is sometimes used as an alternative to FITC for use in biological systems due to reported improved stability [24]. High oligosaccharide purity is required during hydroxyl derivitisation to prevent extensive non-specific reactions however, and the number of hydroxyl groups available for reaction makes the derivitisation unpredictable and non-stoichiometric. Thus, this is not a recommended technique for quantitative tagging of individual oligosaccharide structures, rather a useful pre-experimental labelling technique to monitor bulk localisation of crude oligosaccharide mixtures.

Regardless of the group chosen for derivitisation, utmost care must be taken to eliminate unreacted label from the oligosaccharide mixture to ensure good specificity of detection in the biological setting, especially if the mode of detection is not preceded by a separation technology. Usually, precipitation of total oligosaccharides by washing with excess organic solvent and recovery by centrifugation or filtration as a final stage of preparation is sufficient, however dialysis and gel filtration chromatography may also be used where more thorough clean-up is demanded by the down-stream application [25]. The degree to which unreacted label is removed can be readily assessed by fluorimetric measurement in the washing supernatant, however losses of oligosaccharide are less easy to monitor. Care should therefore be taken to ensure that the correct solvent washing conditions are chosen to ensure maximum oligosaccharide yield, either by mass balance or total carbohydrate analysis. The relatively inexpensive and simple p-hydroxybenzoic acid hydrazide (PAHBAH) assay for total reducing saccharides may be useful for the optimisation of yield in this case [26].

Careful choice of a tag allows not only spatial localisation in tissues via fluorescent or confocal microscopy, but partial if not full quantification of total oligosaccharides extracted from biological fluids or structures. The addition of any tag to an oligosaccharide target prior to tracing within a biological system invariably involves significant alterations to the innate oligosaccharide structure however, and should therefore be treated with caution. Whilst a thorough understanding of the structural change imparted during labelling is vital, it remains impossible to fully predict the change in distribution that a given label may impart. Thus, wherever possible, the use of a complimentary, alternative method of detection is advantageous to confirm findings.

3.1.ii. Native Oligosaccharide Detection

One such approach is to use native (underivitised) detection technologies. These use the native properties of the oligosaccharide itself for detection, but often require specialist equipment to perform. Separation from the biological matrix is key to the success of these techniques, and therefore such methods are usually specialised forms of liquid chromatography. Refractive index (RI), pulsed-ampometric detection (PAD), and electrospray-coupled mass spectrometry (LC-ESI-MS) may all be used following separation of the native oligosaccharide mixture using specific stationary and/or mobile phases.

LC-ESI-MS methods are beginning to emerge as the most likely viable alternative to PAD-based techniques for the separation, identification and quantification of oligosaccharides. Despite their commercial availability for over 10 years, the use of graphitic carbon stationary phases for the separation of oligosaccharides prior to mass spectrometry is only now becoming more commonplace. This stationary phase has the advantage of being about to separate oligosaccharides at low pH in the presence of organic solvent gradients and without using buffer salts that would otherwise rapidly foul the ESI source. Mass spectrometric identification of separated oligosaccharides as sodium adducts is usual, with cross-ring cleavage being capable of yielding detailed information about glycosidic linkages [27]. Use of these columns to fractionate target oligosaccharides can also yield further structural information via offline matrix assisted laser desorption ionisation time-of-flight (MALDI-TOF) analysis [28].

3.2. Evidence for Intestinal Uptake

Limited evidence exists in the literature for the ability of oral oligosaccharides to be taken-up into systemic circulation by small intestinal absorption. This evidence arises from work using either the classic pharmacological insert filter-grown Caco-2 cell line model of the small intestinal epithelium or from rodent models. Caco-2 cells are a line of epithelial colorectal carcinoma cells that, once grown under particular culture conditions, differentiate and polarise to functionally and morphologically resemble the small intestinal epithelium [29]. When grown on a permeable filter, the mature monolayer may be suspended between two spatially independent buffer-filled reservoirs, therefore acting as a predictive physical and biological barrier between which solute transport may be observed. Coupled to carefully selected analytical techniques, some of which have been discussed previously in this section, the authors of several studies have thus far used this model to study transepithelial transport of neutral and acidic oligosaccharides from human milk (HMOs), maltodextrin and modified citrus pectin (MCP).

Transport mechanisms have putatively been shown to vary by oligosaccharide type, ranging from passive paracellular diffusion to transcytosis. Unsurprisingly, the small intestinal epithelium appears most specialised to transport small, neutral *N*-linked oligosaccharides from human milk. Evidence exists for energy-dependant, preferential flux of lacto-*N*-fucopentaose 1 (LNFP1) and lacto-*N*-tetraose (LNT) across Caco-2 monolayers in an apical to basal direction. This movement was not completely inhibited by the trancytosis inhibitors brefeldin A, bafilomycin A and was only partially attenuated by reduced temperature. The directionality of transport was also mild, indicating that passive diffusion may also contribute to the flux of these compounds. Several other *N*-linked neutral oligosaccharides from human milk were also found to move in an apical to basal fashion, although the degree of energy dependence has not been established for these structures [30, 31]. Other studies have also shown the potential of neutral oligosaccharide structures to traverse the intestinal epithelium. Data from our laboratory showed that hexose-based rhamnogalacturonan I fragments from MCP are able to cross Caco-2 monolayers in a TEER-dependent manner. Flux was almost entirely ablated at the highest epithelial cell junction tightness (measured by transepithelial electrical resistance) used in these experiments, confirming the paracellular pore as an alternative route for neutral oligosaccharide absorption [32]. Preferential transfer of low D.P. maltodextrin oligosaccharides across Caco-2 monolayers in a similar model provides further evidence for this phenomenon [33]. Such selectivity by D.P. is indicative of a size exclusion process, linked most often to passage via the paracellular pore [16].

Less evidence exists for the transepithelial permeability of acidic oligosaccharides, putatively due to physiochemical parameters. Several studies show that acidic HMOs do not participate in the transcytotic events demonstrated for neutral oligosaccharides from the same milk fraction. Gnoth *et al.* (2001) conclude that the transport of the major acidic HMO oligosaccharide 6'-NeuAc-Lac does not occur in a directional or temperature dependant manner, and is not reduced by inhibitors of transcytosis, pointing towards a paracellular mode of flux [30]. Both studies from our laboratory and the work of Eiwegger *et al.* (2010) attempted to quantify transepithelial flux of acidic homogalacturonan from modified citrus pectin [32, 33]. Using post-experimental 2-AP derivitisation followed by HPAEC-FLD, our work strongly suggested that galacturonic acid polymers of D.P. 2 – 26 were not capable of transepithelial flux, confirmed by the use of the m-hydroxydiphenyl (m-HDP) method for detection of total uronic acids. This method revealed no significant increase in total uronic acids of a Caco-2 transport experiment loaded with an apical (lumenal) concentration of 10 mg/ml modified citrus pectin. The work of Eiwegger *et al.* (2010) disagrees with this somewhat, demonstrating modest transepithelial permeability of Caco-2 monolayers to galacturonic acid polymers between D.P. 2 – 4 and some unsaturated galacturonic acid polymers between D.P. 2 – 6, representing around a 4% mean mass transfer over the 90 minute experiment. The predictive utility of Caco-2 cell models is highly useful and perhaps under-utilised, however *in vivo* data is also required to support evidence for passive paracellular uptake by identification in or downstream of systemic circulation (i.e. in plasma or urine respectively). The work of Nishikawa *et al.* (2008) provided such evidence, revealing the presence of low D.P. alginate oligosaccharides in mouse plasma and urine following a single oral dose of 10 mg extensively hydrolysed alginate solution with a D.P. of 2 – 4. This dose resulted in a plasma concentration of 25 μg/ml only 5 minutes following administration. The pharmacokinetic profile of oligosaccharides in plasma exhibited an early maxima (5 minutes), followed by rapid decay to baseline at 2 hours. Urinary excretion was more prolonged however, reaching the maximum concentration at 30 minutes followed by a gradual decay to 6 hours, although this did not reach baseline in this study. Significantly, little hydrolysis was observed (either lumenally or systemically), indicating that small concentrations of alginate oligosaccharides are absorbed in the upper GI tract, likely via passive paracellular diffusion [34].

4. Direct Influence of Oligosaccharides on Gastrointestinal Tract Function

Despite the contributions of both hydrolysis and intestinal uptake towards the modulation of the total oral oligosaccharide load, surviving and non-digestible oligosaccharide components are suggested to have numerous downstream effects on human GI tract function, mainly related to the colonic microflora. The food industry has recently shown increased interest in the use of oligosaccharides as functional ingredients for this reason, spurred-on by the interest of modern consumers in the link between their diet and personal health [35].

4.1. Pathogen Adhesion Suppression

Non-digestible oligosaccharides appear to assist in the reduction of diarrhoea symptoms in adults, especially when the condition is associated with intestinal infections [2]. These observations are supported by clinical studies having demonstrated the importance of HMOs in protecting against infant diarrhoea. In both cases, it is reasonable to suggest that prevention of diarrhoea-related pathogen adhesion to the gut epithelium is responsible. More specifically, clinical observations suggest that HMOs reduce the incidence of gastroenteritis in breast-fed infants in comparison to those fed commercial formula milk, a finding neatly supported by *in vitro* studies demonstrating reduced binding of bacterial strains such as *Escherichia coli*, *Salmonella fyris*, and *Vibrio cholera* to Caco-2 cells when challenged with various HMO fractions [36]. Interestingly, this study demonstrated that fractions containing low molecular weight neutral oligosaccharides, high molecular weight oligosaccharides, and acidic oligosaccharides had differing yet overlapping inhibitory function towards the adhesion of the various strains of diarrheal bacteria. In particular, it has been reported that moderate to severe diarrhoea, of all causes, occurred less often in infants whose milk contained high relative concentrations of total fucosyloligosaccharides [37]. Whilst these findings appear clear-cut, it must be considered that evidence also exists to suggest that excessive consumption of non-digestible oligosaccharides may cause intestinal discomfort, flatulence or even diarrhoea. For example, galactooligosaccharide consumption greater than 20g per day and fructooligosaccharide consumption greater than 40g per day are both reported to cause diarrhoea [2]. Clearly a balance must therefore be struck when using oral oligosaccharides clinically to suppress diarrhoea risk.

A wide range of oligosaccharides, such as fucosyloligosaccharides, neutral oligosaccharides and sialylated oligosaccharides present in human milk and bovine colostrum have also been reported to prevent bacterial adherence of strains such as *E. coli*, *Campylobacter jejuni*, *Listeria monocytogenes*,

Streptococcus pneumonia, *Haemophilus influenza*, *Helicobacter pylori* and *Mycoplasma pneumonia* [38, 39]. Similarly, oligosaccharides derived from bovine colostrom, and acidic oligosaccharides in particular, have been observed to prevent the adhesion of pathogens such as *Neisseria meningitides* and *E. coli* [39]. In addition to milk-derived oligosaccharides, prevention of bacterial adhesion has been shown both in guinea pigs and *in vitro* by both galactooligosaccharides and xylooligosaccharides [40].

It is well documented that many of the oligosaccharide structures found in human milk have similarities to the glycoproteins and glycolipids present upon the apical surfaces of intestinal epithelial cells. It has been proposed that the inhibitory mechanism of pathogen adhesion to epithelial cells relies upon the direct interaction between free oligosaccharides (such as HMOs) and adhesion molecules present upon the pathogen surface, used to recognise epithelial cell surface glycoproteins and glycolipids. The sequestration of these adhesion sites by free, luminal oligosaccharides thus renders the binding of these sites to the corresponding epithelial cell target impossible. Due to the large variation in the glycosylated domains recognised by such adhesion molecules (which also varies between bacterial species), it is reasonable to assume that complex mixtures of oligosaccharides (such as the total oligosaccharide load from in human milk) are more effective at preventing pathogen adhesion than single isolated oligosaccharides, with supporting evidence provided in the observations of overlapping, but differing, efficacies of different oligosaccharide fractions from human milk towards the prevention of different adherence by different bacterial strains. Importantly, the ability to provide decoy adhesion sites also extends to other oligosaccharides, such as galacto- and xylooligosaccharides [40]. It is also worth mentioning that there is early evidence that in addition to mimicking cell surface adhesion sites, xylooligosaccharides are capable of modulating the expression of genes involved in regulating bacterial adhesion within the bacteria [40].

It should be noted that such prevention of adhesion is not exclusive to bacterial species, but is also relevant for some viruses, fungi, and toxins such as *V.cholera* toxin and the heat-stable toxin of *E. coli*. Fucosylated oligosaccharides isolated from human milk have been shown to prevent the activation of guanylate cyclase by the heat-stable toxin of *E. coli*, and also to significantly reduce the binding of the toxin to cells *in vitro* [41]. Also, pectin-derived oligosaccharides have been shown to prevent the toxic effects of *E. coli* shiga toxin upon human colonic cells cultured *in vitro*, with a concentration of 10 mg/ml of pectin-derived oligosaccharides required to completely prevent cell death, assessed as a function of HT29 cell adherence after incubation with shiga toxin in the absence and presence of the pectin-derived oligosaccharides [42]. When lactose or pectin polymers were used in place of the pectin-derived oligosaccharides no protective effect against shiga toxin toxicity was observed. Overall, oral oligosaccharides of many varieties may be considered protective against intestinal pathogen invasion, yet clinically they remain underused beyond advised provision of human milk to infants.

4.2. Prebiotic Function

The stability of oral oligosaccharides as they traverse the GI tract has been previously discussed in this chapter, highlighting in particular the role of the gut microflora in the luminal hydrolysis of oligosaccharides. Downstream effects on gut health as a result of this process were not defined however. From a microbial standpoint, the process of oligosaccharide hydrolysis and metabolism by the anaerobic colonic microflora is known as 'fermentation', and results in the production of metabolic waste-products such as short chain fatty acids and various gases (H_2, CO_2, CH_4). Rather usefully, these products now assist clinically in quantifying the degree of microfloral metabolism via the method of gaseous breath testing [1, 2]. On the basis that many oligosaccharides resistant to upper-GI tract hydrolysis are metabolic substrates for colonic bacteria, they are therefore touted to function as prebiotics; a concept that was first introduced by Japanese researchers in the 1980s [43, 44].

A wealth of information has since been gathered on prebiotics, and despite traditional preference for the use of *pro*biotics in the context of human health this information has contributed to a surge in prebiotic interest [45]. Most recently, prebiotics have been defined as molecular entities possessing the ability to selectively stimulate the growth and activity of one or a limited number of microbial genera in the total gut microbiome that results in health benefits to the host. This microbiome comprises an estimated 160 bacterial strains in humans, with high inter-individual variation influenced by age, diet, socioeconomic status and antibiotic use. Within this microfloral array, *Bifidobacteria* and *Lactobacilli* are of most relevance to the prevention of infection, lowering gut pH and producing bacteriocins, therefore blockading culture of pathogens such as *Clostridium difficile* [46].

The list of prebiotic effects in the human body is growing as more is discovered about their mechanism and a deeper understanding is gained. The main oligosaccharides with an established prebiotic effect include inulins and fructans, and galactooligosaccharides. It is postulated that prebiotics can influence the immune system, aid blood lipid control, aid mineral absorption, prevent the development of colon cancer, and reduce risk factors for cardiovascular diseases [47, 48].

Bacteria-mediated fermentation of prebiotic oligosaccharides results in the production of short-chain fatty acids, acetate, propionate, butyrate, lactate, pyruvate and succinate [45]. These products are utilised by specific bacterial strains (such as *Bifidobacteria* and *Lactobacteria*) as metabolic substrates, preferentially stimulating the growth of these species. It is via this mechanism that the anti-infective properties of prebiotic oligosaccharides are mediated.

There is also evidence that some of the systemic effects of oligosaccharides are also mediated via the products of bacterial fermentation that result from oligosaccharide catabolism. The absorption of acetate from the intestine into systemic circulation is associated with a decrease in cholesterol synthesis and altered regulation of adipocytes, resulting in altered lipid profiles associated with decreased cardiovascular risk [49].

Colon cancer is the second most common cancer in the United Kingdom. The aetiology of this disease is not fully understood, with genetics, smoking, diet and inactivity all identified as risk factors. Butyrate (a product of bacterial oligosaccharide fermentation) has been shown to stimulate apoptosis in colon cancer cell lines [50]. Butyrate is largely metabolised by the colonic epithelium serving as a major energy substrate for healthy colonocytes as well as a regulator of cell growth and differentiation [45] . The resulting subversion of colonic metabolism from protein and lipid metabolism reduces protein degradation in the colon, thus reducing the toxic substances such as ammonia and amines associated with protein degradation, which are linked to colon cancer [45].

4.3. Modulation of Nutrient Absorption

Non-digestible oligosaccharides have been shown to increase the absorption of different metal ions such as iron, calcium, magnesium and zinc [51]. It is believed that these metal ions, in mineral form, bind to non-digestible oligosaccharides and therefore resist digestion until they reach the colon, where they are released from the carbohydrate matrix and absorbed [2]. The hypothesis most frequently used to explain this increased absorption of minerals is that the stimulation of absorption seems to be caused by the increased solubility of the minerals in the short chain fatty acid-rich conditions

[3]. In a study conducted on mice it was found that fructooligosaccharide supplementation resulted in an increase in absorption rates of calcium (7 – 9%), magnesium (19 – 26%) and iron (17 – 22%) [52]. This theory is further supported by human studies, as Van den Heuvel *et al.* (2009) found that short chain fructooligosaccharides increased magnesium absorption by 18% in adolescent girls. This absorption enhancing effect could be of value for those with a high mineral requirement and/or a relatively low oral intake of these minerals [53]. Osteoporosis is a major public health issue, with increased risk of bone fractures being associated with increased mortality and morbidity; as a result calcium intake should meet the recommended nutrient intake to maximise peak bone mass. The primary site for calcium absorption is via the small intestine, but it is thought that secondary absorption occurs throughout the length of the gut. It is hypothesised that prebiotics enhance calcium absorption by decreasing the pH of the gut lumen and by increasing the bioavailability of calcium by extending the site of mineral absorption towards the large intestines. The most active product for enhancing calcium and magnesium absorption is oligofructose-enriched inulin.

4.4. Negative Clinical Impacts of Oligosaccharide Consumption and The Gastro-Intestinal Tract

In addition to the effects described in the previous sections, a small but significant percentage of industrialised populations suffer adverse consequences of oligosaccharide consumption, collectively known as irritable bowel syndrome (IBS) [54]. Symptons such as diarrhoea, constipation, abdominal bloating and distension, pain, wind, changes in bowel habit, incomplete evacuation of the bowel and mucous being passed via the rectum result in susbstantially-reduced quality of life are associated with gastro-intestinal dysfunction with no structural or pathological cause [54]. Key symptoms of IBS can be explained through the indigestibility of oligosaccharides in the human gut, and upon reaching the colon, the colonic microbiota ferment the oligosaccharide leading to gas production, luminal distension and associated pain to individuals who experience visceral hypersensitivity [55]. Altered bowel habit and diarrhoea occur due to the osmotic effect of non-digestible oligosaccharides in the colon. Increased water is delivered to the proximal colon occurs in the presence of fermenting oligosaccharides [55]. Irritable bowel syndrome has also been linked to decreased *Bifidobacteria* composition within the gut microflora [45], and current treatment includes the reduction of fermentable oligosaccharides, disaccharides, monosaccharides and polyols (FODMAPs) from the diet. It has been shown that >75% of IBS patients will show improvement in symptoms on a low-FODMAP diet [56]. Symptom relief is normally achieved after 8 weeks on the diet however, it must be noted that a low FODMAP diet also has the negative impact of decreased luminal *Bifidobacteria*, as would be expected by reducing the prebiotic content of the diet [57]. The patient can begin to re-introduce some elements of the diet whilst monitoring for symptom recurrence to redress this deficiency.

5. Human Milk Oligosaccharides (HMOs)

During the first months of life, infants rely on milk as their sole source of nutrition, whether this is from infant formula or human milk. Breast feeding is considered the best nutrition for the growth and development of an infant, as human milk consists of a unique combination of lipids, proteins, carbohydrates, vitamins and minerals which changes over time as an infant grows. Breast feeding has been shown to improve the development of the immune system of the newborn, resulting in protection against enteric and respiratory infections [58, 59]. The beneficial effects of human milk cannot be solely attributed to a single ingredient but it is generally accepted that HMOs play a key role. [58].

HMOs are the third largest component of human breast milk and are produced in the mammary gland where several monosaccharides are added to a lactose core by action of specific transferases [59]. It is generally accepted that HMOs are comprised of five monosaccharides: D-glucose, D-galactose, *N*-acetyl-glucosamine, L-fucose and sialic acid [60]. HMO can be divided into two groups: neutral and acidic. Neutral HMOs are composed of glucose and galactose but they also contain several units of *N*-acetyl-glucosamine and are further decorated by fucose. Acidic HMO can contain all these monosaccharides in addition to one or more molecules of *N*-Acetyl-neuraminic acid (sialic acid) [61].

The intestinal microbiota of breast fed infants is generally dominated by *Bifidobacteria* and *Lactobacteria*, whereas the intestinal microbiota of formula-fed infants is more similar to that of adults, in that it contains greater loads of *Bacteroides*, *Clostridia* and *Enterobacteriaceae* [62]. In addition to this prebiotic affect, HMOs can share common structural motifs with glycans on the infant's intestinal epithelia, known to be receptors for pathogens. As these structures are known to act as pathogen receptors, their presence in milk may provide a defensive strategy by acting as decoys to prevent binding of pathogens to epithelial cells (section 4.1) [61].

Human colostrum contains approximately 20-23g/L oligosaccharides, whereas mature human milk is reported to contain 12-15g/L oligosaccharides. The structure of HMOs is very diverse showing more than 200 compositional and structural isomers in pooled milk [60]. In a review by Barile and Rastall (2013) it was stated that 80 of these have now been fully characterised from a structural point of view [61]. It has also been hypothesized that the large structural variety is to combat the diversity of pathogens that humans may encounter [63]. As a comparison only trace amounts of oligosaccharides are present in mature milk of domestic animals such as cows, goats, pigs and sheep and as a result infant formula, made from cow's milk, is also naturally low in oligosaccharides [60]. But interestingly marsupial and montreme milks also contain significant amounts of oligosaccharides [64].

Due to their complexity, oligosaccharides with structures identical to human milk, are not available as dietary ingredients [58]. In a recent review it was noted that even though great progress has been made in synthesizing kilograms of HMO, the technology has not yet reached its full potential to consistently and efficiently produce large amounts of HMO that could therefore be added to infant formulas [65]. Instead of HMOs, galactooligosaccharide (GOS) and/or fructooligosaccharides (FOS) are often added to infant formulas with the expectation that they produce the prebiotic effect which will help promote bacterial microflora [60].

Rijnierse and Jeurink *et al.* (2011) report that several clinical trials have demonstrated that feeding formula which is supplemented with galacto-oligosaccharide and/or fructooligosaccharides results in intestinal flora similar to that found in breast fed babies [58]. This is reinforced in Nauta and Garssen's (2013) research which suggests that this mixture of oligosaccharides is so far the best studied non-milk derived oligosaccharides, for which the prebiotic and linked immune effects have been proven in clinical settings [63]. Bruzzese *et al.* (2009) conducted a prospective, randomised, placebo-controlled open trial to determine the effect of an infant formula containing galactooligosaccharide/fructooligosaccharides compared to a standard infant formula, involving 342 infants over 12 months. The results showed that compared with the controls, the use of prebiotic-supplemented formula was associated with a significant reduction in the incidence of gastroenteritis and in the rate of children with ≥ 1 episode of acute diarrhoea [62].

There is also some speculation that HMOs may have a beneficial effect on atopic eczema. The intestinal flora of atopic children has been found to differ from that of controls, with atopic children having more *Clostridia* and tending to have fewer *Bifidobacteria* than non-atopic children. In one study cited in Roberfroid *et al.* (2010) they investigated the effect of the prebiotic mixture (GOS/FOS) on the intestinal flora and the incidence of atopic dermatitis during the first 6 months of life in 259 infants at risk of allergy. They found that the frequency of atopic eczema in the experimental group was significantly reduced compared with the placebo group. Follow up of the present study showed that at 2 years the cumulative incidences of atopic dermatitis, recurrent wheezing and allergic urticaria were higher in the placebo group than in the intervention group. Unfortunately, this study had a large dropout rate (48% at 2 years of age). Therefore, further investigation into the effect of prebiotic supplementation on the prevention of allergic disease and food hypersensitivity is required [50].

Conclusion

Oligosaccharides are far less inert than classically thought, with selective hydrolysis occurring throughout the GI tract. Some of the products of such digestion are absorbed across the small intestine and into circulation. Those that remain in the gastro-intestinal tract can have a variety of functions including the modulation of bacterial composition, metal ion uptake, and pathogen adhesion. Despite all these positive benefits of oligosaccharide consumption, high intake is associated with gastro-intestinal dysfunction, and intolerances such as irritable bowel syndrome have significant effects upon gastro-intestinal health and quality of life for patients. More data is required to improve understanding of the behavior of these diverse species in the GI tract, and the experimental strategies outlined in this chapter are anticipated to aid the generation of such information.

References

[1] Delzenne NM, Roberfroid MR. Physiological Effects of Non-Digestible Oligosaccharides. *Food Science and Technology*, 1994 27(1), 1-6.

[2] Mussatto SI, Mancilha IM. Non-digestible oligosaccharides: A review. *Carbohydrate Polymers*, 2007 68(3), 587-597.

[3] Sako T, Matsumoto K, Tanaka R. Recent progress on research and applications of non-digestible galacto-oligosaccharides. *International Dairy Journal*, 1999 9(1), 69-80.

[4] Hodoniczky J, Morris CA, Rae AL. Oral and intestinal digestion of oligosaccharides as potential sweeteners: A systematic evaluation. *Food Chemistry*, 2012 132(4), 1951-1958.

[5] Nilsson U, Oste R, Jagerstad M, Birkhed D. Cereal fructans: in vitro and in vivo studies on availability in rats and humans. *The Journal of Nutrition*, 1988 118, 1325-1330.

[6] Holliday EG, Maguire JM, Evans TJ, Koblar SA, Jannes J, Sturm JW, Hankey GJ, Baker R, Golledge J, Parsons MW. Common variants at 6p21.1 are associated with large artery atherosclerotic stroke. *Nature Genetics*, 2012 44(10), 1147-1151.

[7] Tsunehiro J, Okamoto K, Furuyama Y, Yatake T, Kaneko T. Digestibility of the hydrogenated derivative of an isomaltooligo-saccharide mixture by rats. *Bioscience, Biotechnology, and Biochemistry*, 1999 63(9), 1515-1521.

[8] Roberfroid MB. Chicory fructooligosaccharides and the gastrointestinal tract. *Nutrition*, 2000 16(7/8), 667-669.

[9] Vasquez MJ, Alonso JL, Dominguez H, Parajo JC. Xylooligosaccharides: manufacture and applications. *Trends in Food Science & Technology*, 2000 11, 387-393.

[10] Knaup B, Kempf M, Fuchs J, Valotis A, Kahle K, Oehme A, Richling E, Schreier P. Model experiments mimicking the human intestinal transit and metabolism of D-galacturonic acid and amidated pectin. *Molecular Nutrition and Food Research*, 2008 52, 840-848.

[11] Gnoth MJ, Kunz C, Kinne-Saffran E, Rudloff S. Human milk oligosaccharides are minimally digested in vitro. *The Journal of Nutrition*, 2000 130, 3014-3020.

[12] Crittenden RG, Playne MJ. Production, properties and applications of food-grade oligosaccharides. *Trends in Food Science & Technology*, 1996 7(11), 353-361.

[13] Molis C, Flourie B, Ouarne F, Gailing M-F, Lartigue S, Guibert A, Bornet F, Galmiche J-P. Digestion excretion and energy value of fructooligosaccharides in healthy humans. *American Journal of Clinical Nutrition*, 1996 64, 324-328.

[14] Nakamura S, Oku T, Ichinose M. Bioavailability of cellobiose by tolerance test and breath hydrogen excretion in humans. *Nutrition*, 2004 20(11-12), 979-983.

[15] Scalbert A, Williamson G. Dietary Intake and Bioavailability of Polyphenols. *The Journal of Nutrition*, 2000 130(8), 2073S-2085S.

[16] Lipinski CA, Lombardo F, Dominy BW, Feeney PJ. Experimental and computational approaches to estimate solubility and permeability in drug discovery and development settings. *Advanced Drug Delivery Reviews*, 2001 46(13), 3-26.

[17] Artursson P, Ungell A-L, Löfroth J-E. Selective Paracellular Permeability in Two Models of Intestinal Absorption: Cultured Monolayers of Human Intestinal Epithelial Cells and Rat Intestinal Segments. *Pharmaceutical Research*, 1993 10(8), 1123-1129.

[18] Lamari FN, Kuhn R, Karamanos NK. Derivatization of carbohydrates for chromatographic, electrophoretic and mass spectrometric structure analysis. *Journal of Chromatography B*, 2003, 793(1), 15-36.

[19] Hase S. Precolumn derivatization for chromatographic and electrophoretic analyses of carbohydrates. *Jo*

[20] *urnal of Chromatography* A, 1996 720(12), 173-182.

[21] Hase S, Hara S, Matsushima Y. Tagging of Sugars with a Fluorescent Compound, 2-Aminopyridine. *Journal of Biochemistry*, 1979 85(1), 217-220.

[22] Maness NO, Mort AJ. Separation and quantitation of galacturonic acid oligomers from 3 to over 25 residues in length by anion-exchange high-performance liquid chromatography. *Analytical Biochemistry*, 1989 178(2), 248-254.

[23] Mechref Y, El Rassi Z. Capillary zone electrophoresis of derivatized acidic monosaccharides. *Electrophoresis*, 1994 15(1), 627-634.

[24] De Belder A, Granath K. Preparation and properties of fluorescein-labelled dextrans. *Carbohydrate Research*, 1973 30(2), 375-378.

[25] Prigent-Richard S, Cansell M, Vassy J, Viron A, Puvion E, Jozefonvicz J, Letourneur D. Fluorescent and radiolabeling of polysaccharides: Binding and internalization experiments on vascular cells. *Journal of Biomedical Materials Research*, 1998 40(2), 275-281.

[26] Gunning AP, Bongaerts RJM, Morris VJ. Recognition of galactan components of pectin by galectin-3. *The FASEB Journal*, 2009 23(2), 415-424.

[27] Lever M. A new reaction for colorimetric determination of carbohydrates. *Analytical Biochemistry*, 1972 47(1), 273-279.

[28] Ruhaak LR, Deelder AM, Wuhrer M. Oligosaccharide analysis by graphitized carbon liquid chromatography-mass spectrometry. *Analytical and Bioanalytical Chemistry*, 2009 394(1), 163-174.
[29] Mechref Y, Novotny MV, Krishnan C. Structural Characterization of Oligosaccharides Using Maldi-TOF/TOF Tandem Mass Spectrometry. *Analytical Chemistry*, 2003 75(18), 4895-4903.
[30] Pinto M. Enterocyte-like differentiation and polarization of the human colon carcinoma cell line Caco-2 in culture. *Biology of the Cell*, 1983 47, 323-330.
[31] Gnoth MJ, Rudloff S, Kunz C, Kinne RKH. Investigations of the in Vitro Transport of Human Milk Oligosaccharides by a Caco-2 Monolayer Using a Novel High Performance Liquid Chromatography-Mass Spectrometry Technique. *Journal of Biological Chemistry*, 2001 276(37), 34363-34370.
[32] Gnoth MJ, Rudloff S, Kunz C, Kinne RKH. Studies on the intestinal transport of human milk oligosaccharides (HMO) using Caco-2 cells. *Food Research International*, 2002 35(23), 145-149.
[33] Courts FL. Profiling of modified citrus pectin oligosaccharide transport across Caco-2 cell monolayers. *PharmaNutrition*, 2013 1(1), 22-31.
[34] Eiwegger T, Stahl B, Haidl P, Schmitt J, Boehm G, Dehlink E, Urbanek R, Szépfalusi Z. Prebiotic oligosaccharides: *In vitro* evidence for gastrointestinal epithelial transfer and immunomodulatory properties. *Pediatric Allergy and Immunology*, 2010 21(8), 1179-1188.
[35] Nishikawa T, Yokose T, Yamamoto Y, Yamaguchi K, Oda T. Detection and Pharmacokinetics of Alginate Oligosaccharides in Mouse Plasma and Urine after Oral Administration by a Liquid Chromatography/ Tandem Mass Spectrometry (LC-MS/MS) Method. *Bioscience, Biotechnology, and Biochemistry,* 2008 72(8), 2184-2190.
[36] Qiang X, YongLie C, QianBing W. Health benefit application of functional oligosaccharides. *Carbohydrate Polymers*, 2009 77(3), 435-441.
[37] Coppa GV, Zampini L, Galeazzi T, Facinelli B, Ferrante L, Capretti R, Orazio G. Human Milk Oligosaccharides Inhibit the Adhesion to Caco-2 Cells of Diarrheal Pathogens: *Escherichia coli*, *Vibrio cholerae*, and *Salmonella fyris*. *Pediatric Research*, 2006 59(3), 377-382.
[38] Morrow AL, Ruiz-Palacios GM, Altaye M, Jiang X, Lourdes Guerrero M, Meinzen-Derr JK, Farkas T, Chaturvedi P, Pickering LK, Newburg DS. Human milk oligosaccharides are associated with protection against

diarrhea in breast-fed infants. *The Journal of Pediatrics*, 2004 145(3), 297-303.

[39] Hickey RM. The role of oligosaccharides from human milk and other sources in prevention of pathogen adhesion. *International Dairy Journal*, 2012 22(2), 141-146.

[40] Lane JA, Marino K, Naughton J, Kavanaugh D, Clyne M, Carrington SD, Hickey RM. Anti-infective bovine colostrum oligosaccharides: *Campylobacter jejuni* as a case study. *International Journal of Food Microbiology*, 2012 157(2), 182-188.

[41] Ebersbach T, Andersen JB, Bergstrom A, Hutkins RW, Licht TR. Xylo-oligosaccharides inhibit pathogen adhesion to enterocytes in vitro. *Research in Microbiology*, 2012 163(1), 22-27.

[42] Crane JK, Azar SS, Stam A, Newburg DS. Oligosaccharides from human milk block binding and activity of *Escherichia coli* heat stable enterotoxin (STa) in T84 intestinal cells. *The Journal of Nutrition*, 1994 124, 2358-2364.

[43] Olano-Martin E, Williams MR, Gibson GR, Rastall RA. Pectins and pectic oligosaccharides inhibit *E coli* o157H7 shiga toxin as directed towards the human colonic cell line HT29. *FEMS Microbiology Letters*, 2003 218, 101-105.

[44] Yazawa K, Imai K, Tamura Z. Oligosaccharides and polysaccharides specifically utilizable by *bifidobacteria*. Chemical & Pharmaceutical Bulletin, 1978 26(11), 3306-3311.

[45] Mitsuoka T, Hidaka H, Eida T. Effect of fructo-oligosaccharides on intestinal microflora. *Die Nahrung*, 1987 31(5-6), 427-436.

[46] Roberfroid M, Gibson G, Hoyles L, McCartney A, Rastall R, Rowland I, Wolvers D, Watzl B, Szajewska H, Stahl B. Prebiotic effects: metabolic and health benefits. *The British Journal of Nutrition*, 2010 104(S2), 63.

[47] Quigley EMM. Prebiotics and Probiotics: Their Role in the Management of Gastrointestinal Disorders in Adults. *Nutrition in Clinical Practice*, 2012 27(2), 195-200.

[48] Gibson GR, Roberfroid MB. Dietary Modulation of the Human Colonic Microbiota - Introducing the Concept of Prebiotics. *Journal of Nutrition*, 1995 125(6), 1401-1412.

[49] Slavin J. Fiber and prebiotics: mechanisms and health benefits. Nutrients, 2013 5(4), 1417-1435.

[50] Roberfroid MB. Introducing inulin-type fructans. *The British Journal of Nutrition*, 2005 93(S1), S13S-S25.

[51] Roberfroid M, Gibson GR, Hoyles L, McCartney AL, Rastall R, Rowland I, Wolvers D, Watzl B, Szajewska H, Stahl B. Prebiotic effects: metabolic and health benefits. *British Journal of Nutrition*, 2010 104, S1-S63.

[52] Delgado GTC, Tamashiro WMDSC, Marostica MR, Moreno YMF, Pastore GM. The putative effects of prebiotics as immunomodulatory agents. *Food Research International*, 2011 44(10), 3167-3173.

[53] Wang Y, Zeng T, Wang S-e, Wang W, Wang Q, Yu H-X. Fructo-oligosaccharides enhance the mineral absorption and counteract the adverse effects of phytic acid in mice. *Nutrition*, 2010 26(3), 305-311.

[54] van den Heuvel EGHM, Muijs T, Brouns F, Hendriks HFJ. Short-chain fructo-oligosaccharides improve magnesium absorption in adolescent girls with a low calcium intake. *Nutrition Research*, 2009 29(4), 229-237.

[55] Maxion-Bergemann S, Thielecke F, Abel F, Bergemann R. Costs of irritable bowel syndrome in the UK and US. *PharmacoEconomics*, 2006, 24(1), 21-37.

[56] Ong DK, Mitchell SB, Barrett JS, Shepherd SJ, Irving PM, Biesiekierski JR, Smith S, Gibson PR, Muir JG. Manipulation of dietary short chain carbohydrates alters the pattern of gas production and genesis of symptoms in irritable bowel syndrome. *Journal of Gastroenterology and Hepatology*, 2010 25(8), 1366-1373.

[57] Shepherd SJ, Parker FC, Muir JG, Gibson PR. Dietary triggers of abdominal symptoms in patients with irritable bowel syndrome: randomized placebo-controlled evidence. *Clinical Gastroenterology and Hepatology,* 2008 6(7), 765-771.

[58] Staudacher HM, Lomer MCE, Anderson JL, Barrett JS, Muir JG, Irving PM, Whelan K. Fermentable Carbohydrate Restriction Reduces Luminal *Bifidobacteria* and Gastrointestinal Symptoms in Patients with Irritable Bowel Syndrome. *Journal of Nutrition*, 2012 142(8), 1510-1518.

[59] Rijnierse A, Jeurink PV, van Esch BCAM, Garssen J, Knippels LMJ. Food-derived oligosaccharides exhibit pharmaceutical properties. *European Journal of Pharmacology*, 2011 668(S1), S117-S123.

[60] Wainwright L. Does the addition of prebiotics to infant formula have beneficial effects for the baby? *Journal of Neonatal Nursing*, 2006 12(4), 130-137.

[61] Han NS, Kim T-J, Park Y-C, Kim J, Seo J-H. Biotechnological production of human milk oligosaccharides. *Biotechnology Advances*, 2012 30(6), 1268-1278.

[62] Barile D, Rastall RA. Human milk and related oligosaccharides as prebiotics. *Current Opinion in Biotechnology*, 2013 24(2), 214-219.

[63] Bruzzese E, Volpicelli M, Squeglia V, Bruzzese D, Salvini F, Bisceglia M, Lionetti P, Cinquetti M, Iacono G, Amarri S. A formula containing galacto- and fructo-oligosaccharides prevents intestinal and extra-intestinal infections: An observational study. *Clinical Nutrition*, 2009 28(2), 156-161.

[64] Nauta AJ, Garssen J. Evidence-based benefits of specific mixtures of non-digestible oligosaccharides on the immune system. *Carbohydrate Polymers*, 2013 93(1):263-265.

[65] Brand-Miller JC, McVeagh P, McNeil Y, Messer M. Digestion of human milk oligosaccharides by healthy infants evaluated by the lactulose hydrogen breath test. *The Journal of Pediatrics*, 1998 133(1), 95-98.

[66] Casado B, Affolter M, Kussmann M. OMICS-rooted studies of milk proteins, oligosaccharides and lipids. *Journal of Proteomics* 2009 73(2), 196-208.

In: Oligosaccharides
Editors: L. S. Schweizer, S. J. Krebs
ISBN: 978-1-62948-328-3

Chapter IV

The Role of Human Milk Oligosaccharides in Preventing Respiratory Infections in Infants

***Mariarosaria Marotta*[1] *and Rita M. Hickey*[1,2*]
[1]Food for Health Ireland, Teagasc Food Research Centre, Moorepark, Fermoy, Co. Cork, Ireland
[2]Food Biosciences Department, Teagasc Food Research Centre, Moorepark, Fermoy, Co. Cork, Ireland

Abstract

Breast milk plays a major role in protecting infants from respiratory and gastrointestinal infections. In addition to IgA, molecules such as free oligosaccharides present in human milk have been implicated in this protective role. When considering respiratory tract infection, pathogen adhesion and colonization is often a prelude to infection, and intervention at this early stage can help prevent disease. Bacteria have evolved a multitude of colonisation mechanisms commonly targeting surface carbohydrate structures of the host. Many human milk oligosaccharides are considered to be soluble receptor analogs of epithelial cell surface

* Corresponding author: Food Biosciences Department, Teagasc Food Research Centre, Moorepark, Fermoy, Co. Cork, Ireland. Phone: +353 (0) 25 42227, Fax: +353 (0) 25 42340, E-mail: rita.hickey@teagasc.ie.

carbohydrates, displaying structural homology to host cell receptors. Thus, they function as receptor decoys, which pathogens can bind to instead of the host. Here, we review the ability of human milk oligosaccharides and related structures to prevent colonisation of pathogens to host respiratory cells. In particular, we examine the use of *in vitro* assays to study milk oligosaccharides anti-infective effect. We also access the limited data, which is available from infant studies involving the use of human milk oligosaccharides. Finally, intervention studies assessing the effects of galactooligosaccharides and fructooligosaccharides when incorporated into infant formula on respiratory infections in developed countries are also reviewed.

Keywords: human milk oligosaccharides, respiratory tract infections, bacterial adhesion, infants

1. Introduction

Breast milk is the food that has been naturally selected for the offspring of humans. Its role is linked to functions beyond basic nutrition. The World Health Organisation recommends to exclusively breastfeed infants for the first six months of life, given the accumulating evidence that exists on the benefits of breastfeeding for growth, development and health (World Health Organisation, 2001). Breast milk also plays a major role in protecting infants from respiratory and gastrointestinal infections in both developed and developing countries (Duijts, Ramadhani & Moll, 2009; Roth, Caulfield, Ezzati & Black, 2008; Bachrach, Schwarz & Bachrach, 2003). This is an important consideration given that almost 50% of all paediatric consultations in industrialized countries are associated with respiratory tract infections (RTIs). Moreover, in developed countries, such as the US, 6% of infants younger than 1 year are hospitalised with lower respiratory tract diseases on an annual basis (Bachrach et al., 2003). Acute otitis media (etiological agent *Streptococcus pneumoniae*) is one of the most frequent childhood diseases and prevention of infection is generally difficult. Indeed, epidemiological studies on otitis media have reported that 62% of infants will have at least 1 episode of acute otitis media during the first year of life, while 17% will have at least 3 episodes (Duncan, Ey, Holberg, Wright, Martinez & Taussig, 1993; Teele, Klein & Rosner, 1989).

A recent review (Duijts et al., 2009), which assessed the effect of breastfeeding on infections during infancy in industrialised countries, reported that 13 out of 16 studies concluded that breastfeeding protects infants against RTIs. However, the protective effect of breastfeeding seems to differ when comparing upper and lower RTIs. In general, breastfeeding was associated with fewer cases of hospitalisation due to respiratory disease in the first two years of life. McNiel, Labbok & Abrahams (2010) reviewed the risks associated with formula feeding when compared to "exclusively breastfed" infants in order to promote the use of exclusive breastfeeding. When exclusive breastfeeding is set as the normative standard, "any formula use" is associated with increased incidence of otitis media and hospitalization, which is secondary to lower RTIs in infants in multiple studies. In particular, these studies demonstrated a statistical significance when examining the association between formula use and increased occurrence of otitis media. The mechanisms involved in the protection of breast-fed infants from infectious diseases when compared to fed formula infants have received much interest in recent years. Human milk contains secretory IgA antibodies, which, having transited the infant gut, bind to enteric pathogens protecting the breastfed infant. Indeed, IgA has been considered the major protective component in human milk for some time (Hanson, Jalil, Ashraf, Bernini, Carlsson, Cruz, et al., 1991). More recently, other molecules present in human milk have been implicated in this protective role, such as ferritin and oligosaccharides (Newburg, 2009). Oligosaccharides are the third largest solid component of human milk following lactose and lipids, with concentrations reaching up to 50 g/L or more in colostrum to an average of 10-15 g/L in mature milk (Kunz & Rudloff, 2008; Kunz, Rudloff, Schad & Braun, 1999). Interestingly, although human milk contains such a large amount of oligosaccharides, infants are unable to digest them (Gnoth, Rudloff, Kunz & Kinne, 2001; Engfer, Stahl, Finke, Sawatzki & Daniel, 2000). Given their abundance, it became clear that the evolutionary persistence of oligosaccharides is likely due to their biological functions. One of the first functions ascribed to human milk oligosaccharides (HMO) was their prebiotic or bifidogenic effect, which results in modulation of the infant's intestinal microbiota composition (LoCascio, Ninonuevo, Freeman, Sela, Grimm, Lebrilla, et al., 2007; Coppa, Bruni, Morelli, Soldi & Gabrielli, 2004; Gyorgy, Norris & Rose, 1954). Furthermore, *in vitro* studies have demonstrated that HMO may modulate the expression of glycans present on epithelial cell surfaces, which can act as attachment sites for commensal and pathogenic bacteria (Angeloni, Ridet, Kusy, Gao, Crevoisier, Guinchard, et al., 2005). Other *in vitro* studies have also indicated that specific oligosaccharides in human milk may have effect on the immune system (Eiwegger, Stahl, Haidl, Schmitt, Boehm, Dehlink, et al.,

2010; Bode, Kunz, Muhly-Reinholz, Seeger & Rudloff, 2004a; Bode, Rudloff, Kunz, Strobel & Klein, 2004b). Moreover, certain HMO structures are considered to be soluble receptor analogs of many receptors on pathogenic bacteria and host cells (Lane, Mehra, Carrington & Hickey, 2010; Kunz, Rudloff, Baier, Klein & Strobel, 2000). This observation has urged researchers to investigate interactions between bacteria and milk oligosaccharides. Studies on *Helicobacter pylori, Campylobacter jejuni* and *Clostridium difficile* have shown that HMO can influence bacterial-host interactions in the infant gut (reviewed by Lane et al., 2010). While a large number of mainly *in vitro* studies have focused on oligosaccharide activity in the gut, few studies have investigated the ability of milk oligosaccharides in preventing colonisation of bacteria to urinary (Martín-Sosa, Martín & Hueso, 2002; Coppa, Gabrielli, Giorgi, Catassi, Montanari, Varaldo, et al., 1990) or respiratory epithelia (reviewed in this chapter). Colonisation of the respiratory epithelia begins, as with all the other epithelia, from the first hours after the delivery and can occur through inhalation of air. The upper respiratory tract (nose and nasopharynx) is colonised by streptococci, staphylococci, *Corynebacterium* and several Gram-negative cocci (Tlaskalová-Hogenová, Stepánková, Hudcovic, Tucková, Cukrowska, Lodinová-Zádníková, et al., 2004). However, the nasopharynx and the oropharynx can in some cases be colonised also by opportunistic pathogens, which can cause life-threatening infections (such as pneumonia, meningitis) or infections, which are less severe (such as otitis, pharyngitis, sinusites). The lower respiratory tract (trachea, bronchi and lungs) is physiologically sterile because of anatomical organisation and effective cleansing mechanisms, such the action of ciliated epithelia and the trapping of bacteria in mucus (Tlaskalová-Hogenová et al., 2004).

The initial event in bacterial infection of a mucosal surface is attachment. Most commonly, this involves the specific recognition of a host cell surface glycoconjugate by a bacterial protein. It has long been suggested that interference with attachment, particularly by carbohydrates or homologues, could be an effective means of preventing the course of infection, and potentially, of transmission (Idänpään-Heikkilä, Simon, Zopf, Vullo, Cahill, Sokol, et al., 1997). However, demonstrating that milk oligosaccharides can act as anti-infectives *in vivo*, particularly in clinical studies, has been hampered because of their limited availability, which in turn prohibits their addition to infant formula. To overcome these issues, in recent years other oligosaccharides such as galactooligosaccharides (GOS, enzymatically synthesised from lactose) and fructooligosaccharides (FOS, of plant origin) have been produced and are currently used to supplement infant formula with the aim of emulating the beneficial effects of HMO. Indeed, GOS and FOS have been shown to modulate the numbers of beneficial microflora in the gastrointestinal tract. However, it is unlikely that these molecules exert the same effects as HMO given that their structure and composition is very different from their milk derived counterparts. For instance, certain biological properties such as prevention of pathogen adhesion, seem ascribed mainly to HMO given that a single group of oligosaccharides (GOS or other) invariably cannot match the anti-adhesive properties of the extremely diverse HMO structures (Fanaro, Boehm, Garssen, Knol, Mosca, Stahl, et al., 2005). It should be noted, however, that a small number of clinical studies have investigated the effect of GOS and/or FOS when added to infant formula and, in some cases in combination with probiotics, they appear to influence respiratory diseases, as will be discussed below (Picaud, Chapalain, Paineau, Zourabichvili, Borner & Duhamel, 2010; Bruzzese, Volpicelli, Squeglia, Bruzzese, Salvini, Bisceglia, et al., 2009; Kukkonen, Savilahti, Haahtela, Juntunen-Backman, Korpela, Poussa, et al., 2008; Arslanoglu, Moro & Boehm, 2007).

This chapter will focus on the link between HMO and related structures and prevention/reduction of RTIs. In particular, studies on human and bovine milk oligosaccharides in preventing/reducing adhesion of respiratory pathogenic bacteria using *in vitro* assays will be discussed. The limited data that is available from infants studies involving HMO will also be summarised. Finally, intervention studies assessing the effects of GOS and FOS in formula on respiratory infections in developed countries will also be reviewed.

2. Chemical Composition of HMO and Their Role in Preventing Infection

The biological activity of HMO is dependent on their individual chemical structures. Indeed, in relation to human milk, oligosaccharide composition is very complex, generally ranging in size from three to eight sugars, although highly complex oligosaccharides may contain up to 32 sugars (Stahl, Thurl, Zeng, Karas, Hillenkamp, Steup, et al., 1994). Currently, more than 90 structures have been elucidated and approximately 200 species have been identified (Ninonuevo, Ward, LoCascio, German, Freeman, Barboza, et al., 2007; Kunz et al., 2000; Finke, Stahl, Pfenninger, Karas, Daniel & Sawatzki, 1999). The monomers of milk oligosaccharides are D-glucose (Glc), D-galactose (Gal), N-acetylglucosamine (GlcNAc), L-fucose (Fuc) and the form of sialic acid called N-acetylneuraminic acid (NeuAc). These molecules are synthesized in the breast starting with lactose at the reducing end (Kunz et al., 2000). The simplest HMO are trisaccharides, such as the fucosylated 2'-fucosyllactose and 3'-fucosyllactose and the sialylated 3'-sialyllactose and 6'-sialyllactose (Bode, 2009). More complex oligosaccharides contain a core molecule, which is characterised by repetitive attachment of Gal and GlcNAc in β-glycosidic linkage to lactose (Boehm & Stahl, 2003). In particular, type 1 oligosaccharides are characterised by a Galβ1-3GlcNAc unit and type 2 oligosaccharides by a Galβ1-4GlcNAc unit (Newburg, Ruiz-Palacios, Altaye, Chaturvedi, Meinzen-Derr, Guerrero Mde, et al., 2004). Thus, a large number of core structures can be formed with further variations occurring due to the attachment of sialic acid and/or fucose to the respective core molecules (Stahl et al., 1994; Egge, 1993) (Figure 1). In this respect, human milk is considered to be unique in its high content of complex fucosylated and sialylated oligosaccharides (Kunz et al., 1999).

HMO are synthesised by the same glycosytransferases responsible for the synthesis of human cell surface glycans (Newburg, Ruiz-Palacios & Morrow, 2005). This common lineage is important when considering pathogenic colonization of the respiratory tract. For instance, the receptor for *S. pneumoniae* on human nasopharyngeal epithelial cells, Galβ1-4GlcNAcβ1-3Gal, shares similarity with lacto-N-neoTetraose, Galβ1-4GlcNAcβ1-3Galβ1-4Glc (LNnT), one of the major components of HMO. The structures characterising the neutral HMO Le^a (fucose α1-4 bound to GlcNAc of the type 1 chain) and Le^x (fucose α1-3 bound to GlcNAc of the type 2 chain), which are part of many carbohydrate chain receptors present in human airway mucins, are recognised by *Pseudomonas aeruginosa* (Scharfman, Degroote, Beau, Lamblin, Roussel & Mazurier, 1999). Also, the disaccharide moieties of type 1 and type 2 milk oligosaccharides have been identified as receptors for *P. aeruginosa* (Ramphal, Carnoy, Fievre & Michalski, 1991). As pathogenic bacteria utilise cell surface glycans to identify and bind to their target cells as the critical first step in pathogenesis (Newburg et al., 2005), soluble glycans, such as HMO, could act as decoys, thus inhibiting the ability of pathogens to bind to receptors on epithelial surfaces (Figure 2). This may explain why breastfed infants display increased protection from gastrointestinal, respiratory and urinary infections.

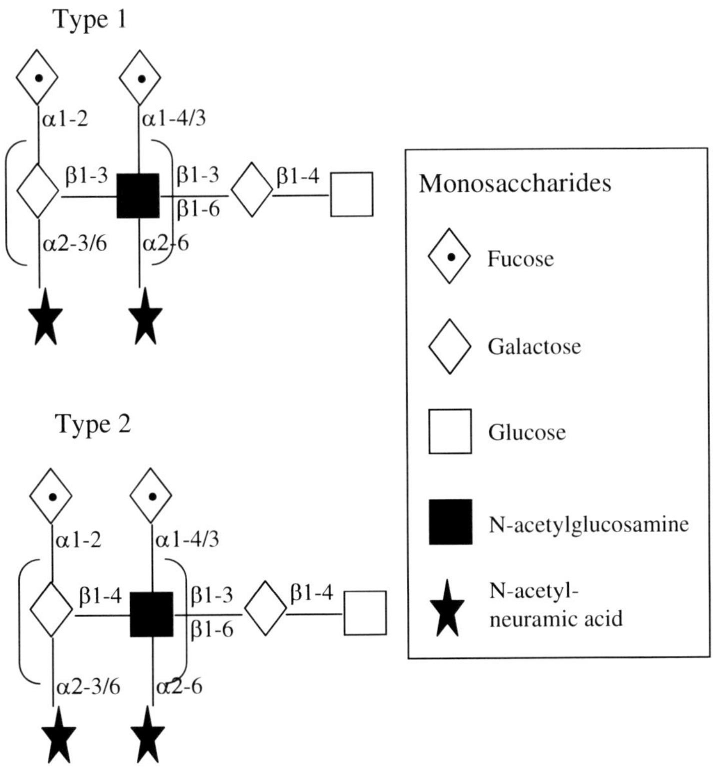

Figure 1. Structural composition of complex HMO. HMO are characterised by repetitive units of type 1 or type 2 chain (in brackets), with lactose at the reducing end. Fucose and/or N-acetylneuraminic acid may occur at the terminal end. Not all the linkages are present in individual oligosaccharides.

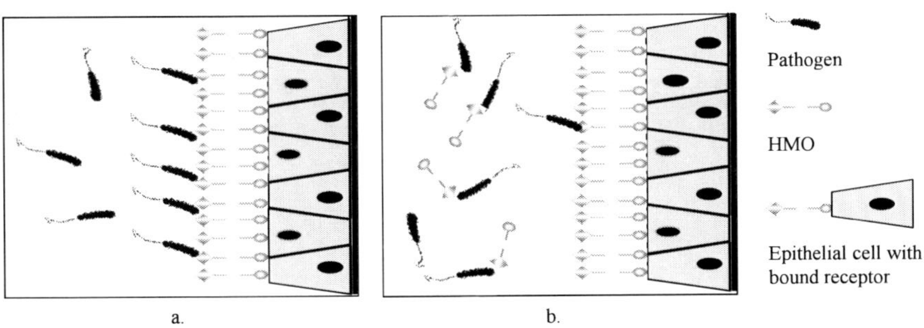

Figure 2. Anti-adhesive effects of HMO. (a) Pathogenic bacteria bind to sugar receptors on epithelial cells and cause infection. (b) HMO mimic the sugar receptors on the epithelial cells and block the bacteria from binding.

3. *In vitro* and Animal Studies Demonstrating Interactions between Milk Oligosaccharides and Respiratory Pathogens

A summary of the *in vitro* and animal studies investigating the interactions between milk oligosaccharides and respiratory pathogens is presented in Table 1. One of the first studies, which demonstrated the potential of HMO as anti-adhesive molecules for respiratory pathogens was reported by Andersson, Porras, Hanson, Lagergård & Svanborg-Edén (1986). The researchers focused on *S. pneumoniae* and *Haemophilus influenzae*, which are major etiological agents of RTIs in infants and children. The study demonstrated that the attachment of *S. pneumoniae* to human oropharyngeal epithelial cells and of *H. influenzae* to human buccal epithelial cells was inhibited by human milk. The inhibition of attachment did not correlate with the total immunoglobulin content or specific antibody activity. When the milk was fractionated the inhibition of *H. influenzae* was related to the high molecular weight fraction (HMWF), whereas *S. pneumoniae* attachment was blocked by both HMWF and low molecular weight fraction (LMWF) and by the free oligosaccharides purified from the LMWF. In particular, LNnT and lactotetraose were proposed as receptors for pneumococcal attachment. Based on these results, Idänpään-Heikkilä et al. (1997) tested oligosaccharides such as LNnT and its α2-3 and α2-6-sialylated derivatives for their ability to influence the course of pneumonia and to prevent colonisation of the nasopharinx in animal models. When streptococci exposed to LNnT and its α2-6-sialylated derivative Neu5Acα2-6Galβ1-4 GlcNAcβ1-3Galβ1-4Glc (LSTc), which is also present in human milk, were administered intratracheally at a challenge dose of $1x10^5$ cfu/mL into the right mainstem bronchus of rabbits, a dramatic decrease in pneumococcal load in the lungs was observed. Furthermore, the oligosaccharides conferred protection from bacteremia. When the challenge dose was increased to 1×10^7 cfu/mL, the oligosaccharides reduced the number of bacteria in the lungs and blood, but they did not prevent bacteremia or pneumonia. The authors also tested the ability of LNnT to interrupt the course of established pneumonia. When LNnT was administered 6 h after the pneumococcal challenge, they found that LNnT was protective at a threshold between 10 and 50 μM, reducing the bacterial load after 48 h, which persisted until 96 h. In the same study, neoglycoconiugates of the active oligosaccharides, when administered intranasally, prevented colonisation of the nasopharinx in infant rats. To understand the mechanism of action of the oligosaccharides, the authors performed *in vitro* assays in which the oligosaccharides were investigated for inhibition of streptococcal attachment to human lung cells, A549. The oligosaccharides LNnT, LSTc and 3'GL (GalNAcβ1-3Lac) were the most effective in blocking the attachment of streptococci to the lung cells, suggesting that oligosaccharides may act as direct competitive inhibitors of pneumococcal adherence to the host cells. Further evidence supporting this conclusion was documented by Barthelson, Mobasseri, Zopf & Simon (1998). In this study, oligosaccharides were investigated as inhibitors of *S. pneumoniae* binding to monolayers of primary normal human bronchial and tracheal cells (NHBE cells). The oligosaccharides were chosen based on their previous identification as targets for *S. pneumoniae* (Andersson et al., 1986). In addition, several other compounds comprising of sugar sequences known to occur on the surfaces of many eukaryotic cells were included in the study. The group concluded that *S. pneumoniae* relies to a significant extent upon sialylated oligosaccharide ligands terminating in NeuAcα2-3(or 6)Galβ1, such as 3'SLNnT and 6'SLNnT (Neu5Ac α2-6Gal β1-4 GlcNAc β1-3 Gal β1-4Glc), which naturally occur in human milk, for adherence to epithelial cells. Furthermore, the group also demonstrated that the HMO, 6'-sialyllactosamine (NeuAcα2-6Galβ1-4GlcNAc), which is also one of the most abundant acidic oligosaccharides in bovine milk (Barile, Marotta, Chu, Mehra, Grimm, Lebrilla, et al., 2010; McJarrow & van Amelsfort-Schoonbbeek, 2004; Martín-Sosa, Martín, García-Pardo & Hueso, 2003; Nakamura, Kawase, Kimura, Watanabe, Ohtani, Arai, et al., 2003; Gopal & Gill, 2000), inhibited adherence of *S. pneumoniae* R6 to Chang conjunctival cells at all stages during the bacterial growth cycle. A further study by Tong, McIver, Fisher & DeMaria (1999), demonstrated that adherence of *S. pneumoniae* to Galβ1-4GlcNAcβ1-3Gal, one of the receptors on human nasopharyngeal epithelial cells, is inhibited by LNnT. Moreover, the oligosaccharide was effective in a concentration dependent manner in reducing adhesion of *S. pneumoniae* to the chinchilla tracheal epithelium. The chinchilla trachea organ perfusion culture system used in the study models the Eustachian tube conditions, therefore the results suggest that LNnT may be useful in preventing or reducing streptococcal colonisation of the middle ear.

Pseudomonas aeruginosa is a ubiquitous, gram-negative, opportunistic pathogen, which is often associated to community- and hospital-acquired respiratory and urinary infections, such as those commonly found in preterm infants. *P. aeruginosa* also plays a major role in pneumonia, which develops in cystic fibrosis patients. The antibiotic resistance associated with this pathogen makes it difficult to treat and cure successfully (Strateva & Yordanov, 2009). Therefore, much interest has focused on finding molecules with the ability to reduce the morbidity and mortality associated with this bacterial infection. Thomas and Brooks (2004a) tested the effect of oligosaccharides moieties on inhibition of *P. aeruginosa* adherence to lung cells (A549). Among the oligosaccharides tested, some were of milk origin. For instance, 3'-sialyllactose and 6'-sialyllactose (0.05 mM) resulted in inhibition of *P. aeruginosa* PAK adhesion to lung cells by approximately 62 and 55%, respectively. In contrast, 2'-fucosyllactose and 3'-fucosyllactose (0.15 mM), also from milk, did not block the adhesion of *P. aeruginosa* to the A549 cells. The researchers also tested the type 1 and type 2 disaccharide chains. Galβ1-3GlcNAc and Galβ1-4GlcNAc, which resulted in greater than 60% inhibition of adhesion. The results from this study suggest that milk oligosaccharides containing type 1 and type 2 chains may inhibit the adhesion of *P. aeruginosa* to lung cells. Hemagglutination inhibition tests revealed that human milk resulted in inhibition of certain lectins of *P. aeruginosa* namely, PA-IIL and to a lesser extent, PA-IL (Lesman-Movshovich & Gilboa-Garber, 2003; Lesman-Movshovich, Lerrer & Gilboa-Garber, 2003). Since PA-IIL is a fucose-binding lectin, its interactions with fucosylated HMO, which represent the major components of HMO (Kunz et al., 2000), were investigated (Perret, Sabin, Dumon, Pokorná, Gautier, Galanina, et al., 2005). The study demonstrated that the most potent inhibitors of PA-IIL are those oligosaccharides with fucose linked to position 4 of GlcNAc, which belong to the Le^a series. Also, Le^x glucose analogues, in which GlcNAc is replaced by Glc and are characterized by fucose linked on position 3, were found to be good inhibitors.

Neisseria meningitidis, is a human-specific pathogen causing life-threatening septicemia and meningitis in newborns, children and young adults. The bacterium may colonize the nasopharynx of asymptomatic carriers and transmit person-to-person or translocate into the blood stream causing septicemia or through the blood-brain barrier causing meningitis (Nassif, 2000; Hardy, Christodoulides, Weller & Heckels, 2000; Nassif, 1999). Adhesion of *N. meningitidis* to epithelia is mediated by type IV pili (Nassif, 1999; Pujol, Eugène, de Saint Martin, & Nassif, 1997). Using a binding assay based on isolated meningococcal pili, Hakkarainen, Toivanen, Leinonen, Frängsmyr, Strömberg, Lapinjoki, et al. (2005) demonstrated that the binding of *N. meningitidis* to human salivary agglutinin was inhibited up to 50% in the presence of acidic bovine milk oligosaccharides. Given these results, the authors speculated that *N. meningitidis* receptors may be carbohydrates, even though no structures were identified.

Table 1. *In vitro* and animal studies investigating the interaction between milk oligosaccharides and respiratory pathogens

Pathogen	Milk Oligosaccharide Tested	Kind of Study	Reference
Streptococcus pneumoniae	Human milk oligosaccharides	*in vitro*	Andersson et al. (1986)
	Galβ1-4GlcNAcβ1-3Galβ1-4Glc (LNnT) and its α2-3 and α2-6-sialylated derivatives; GalNAc β1-3Lac (3'GL)	*in vitro*, animal	Idänpään-Heikkilä et al. (1997)
	α2-3 and α2-6-sialylated oligosaccharide, such as Neu5Ac α2-3Gal β1-4 GlcNAc β1-3 Gal β1-4Glc (3'SLNnT) and Neu5Ac α2-6Gal β1-4 GlcNAc β1-3 Gal β1-4Glc (6'SLNnT)	*in vitro*	Barthelson et al. (1998)
	LNnT	*ex vivo*	Tong et al. (1999)
Pseudomonas aeruginosa	3'-sialyllactose and 6'-sialyllactose; 2'-fucosyllactose and 3'-fucosyllactose; Galβ1-3GlcNAc, Galβ1-4GlcNAc	*in vitro*	Thomas and Brooks (2004a)
	fucosylated human milk oligosaccharides	*in vitro*	Perret et al. (2005)
Neisseria meningitidis	sialylated bovine milk oligosaccharides	*in vitro*	Hakkarainen et al. (2005)
Legionella pneumophila	3'-sialyllactose and 6'-sialyllactose; 2'-fucosyllactose and 3'-fucosyllactose; Galβ1-3GlcNAc, Galβ1-4GlcNAc	*in vitro*	Thomas and Brooks (2004b)

Legionella pneumophila is an opportunistic pathogen, which causes Legionnaires' disease. Legionnaires' disease presents as an acute pneumonia acquired from the inhalation of aerosols containing the microorganism (Fields & Benson, 2002; Winn, 1988). Attachment of *L. pneumophila* is likely mediated by interactions of multiple adhesins with the host cell surface, in particular with the oligosaccharides chains of glycolipids and glycoproteins. An extensive study was performed to assess the potential of certain oligosaccharides to inhibit legionella attachment to respiratory cells (Thomas & Brooks, 2004b). Among the oligosaccharides tested, some are present in human and bovine milk. The sialylated saccharides 3'-sialyllactose and 6'-sialyllactose (0.05 mM) inhibited the adhesion of *L. pneumophila* to human pneumocytes (A549) by 57 and 65%, respectively, while 2'-fucosyllactose and 3'-fucosyllactose (0.15 mM) resulted in only low levels of inhibition (28 and 35%, respectively). Inhibition of adhesion was also associated with type 1 and type 2 chains, such as Galβ1-3GlcNAc (60%) and Galβ1-4GlcNAc (70%) (1.5 mM). These disaccharides were also tested for their ability to inhibit *L. pneumophila* attachment to other cell lines, which were used to create an *in vitro* adhesion model descending the respiratory tract. In particular, Galβ1-4GlcNAc resulted in significant inhibition of *L. pneumophila* attachment to human nasal epithelial cells (RPMI-2650) (61%), to bronchial epithelial cells (BEAS2-B) (79%) and to human monocytes (U937) (76%). Galβ1-3GlcNAc significantly inhibited *L. pneumophila* attachment to human nasal and bronchial epithelial cells (70 and 68%, respectively) and moderately inhibited adhesion of the pathogen to human monocytes (56%).

The data presented above suggest that milk oligosaccharides have a direct effect on adhesion and can act as decoys for pathogens. In the upper respiratory tract, the frequent bathing in milk and its oligosaccharides might interrupt adherence of bacteria to epithelial cells, thereby reducing the load of colonising organisms and diminishing the risk of infection (Barthelson et al., 1998). In the lower respiratory tract, if HMO cross the gut epithelium, they may reach the respiratory epithelia through the systemic circulation, where they could modulate bacterial-host interactions as observed in the gut (Stepans, Wilhelm, Hertzog, Rodehorst, Blaney, Clemens, et al., 2006). In fact, HMO have been found in the urine of breast-fed infants indicating that they are absorbed intact in the infant gut (De Leoz, Shuai, Strum, Niñonuevo, Gaerlan, Mirmiran, et al., 2013; Rudloff, Pohlentz, Borsch, Lentze & Kunz, 2012; Obermeier, Rudloff, Pohlentz, Lentze & Kunz, 1999; Rudloff, Pohlentz, Diekmann, Egge & Kunz, 1996). Moreover, *in vitro* studies using human Caco-2 cells also suggest that HMO are able to cross the intestinal epithelium

(Gnoth et al., 2001). These observations suggest that HMO are present in the blood circulation of infants and they may exert activities at the systemic level.

Table 2. Human intervention studies on prevention or reduction of respiratory infections in children

Compound Studied	Administration Source	Main Outcome	Reference
LNnT	Supplemented formula	No difference in colonisation of mucosal cell surfaces with *S. pneumoniae* between the intervention and the control group	Prieto (2005)
3'SLNnT	Intranasally	No effect on the occurrence of acute otitis media or the nasopharyngeal carriage of bacterial pathogens in the intervention group	Ukkonen et al. (2000)
galactose and fucose	Inhalation in combination with tobramycin	No detection of *P. aeruginosa* in the respiratory tract of the patient within 2 weeks	von Bismarck et al. (2001)
LNF-II	Breast milk	LNF-II concentration inversely related with occurrences of respiratory illness at 6 weeks of age	Stepans et al. (2006)
FOS	Commercial synbiotic follow-on formula	Incidence of infectious diseases significantly lower in the intervention than in the control group	Picaud et al. (2010)
GOS	Synbiotic powder dissolved in liquid form	During intervention time, no difference in morbidity between the placebo and the intervention group. During follow-up, respiratory infections significant lower in the intervention when compared with the placebo group	Kukkonen et al. (2008)
GOS/FOS	Supplemented formula	No significant difference in the number of episodes of upper respiratory tract infections between the intervention and the control group. Significantly lower administration of antibiotics required in the intervention group when compared with the control group	Bruzzese et al. (2009)
scGOS/lcF OS	Supplemented hydrolysed formula	During the first 6 months of life, fewer episodes of all types of infections, particularly respiratory, in the intervention group when compared with the placebo group	Arslanoglu et al. (2007)

Abbreviations: LNnT, lacto-N-neoTetraose; 3'SLNnT, Neu5Acα2-3Galβ1-4GlcNAcβ1-3Gal β1-4Glc; LNF-II, lacto-N-fucopentaose II; FOS, fructooligosaccharides; GOS, galactooligosaccharides.

4. Human Studies Using HMO in Preventing or Reducing Respiratory Infections in Children

Despite the promising *in vitro*, *ex vivo* and *in vivo* animal studies described to date, which highlight the potential of milk oligosaccharides in the inhibition of infection by respiratory pathogens, human intervention studies are also required. Indeed, the lack of large amounts of milk oligosaccharides, either of human or animal origin, makes the feasibility of tailored (targeted, human) studies very challenging. Few human studies using HMO in preventing or reducing respiratory infections in children have been performed (Table 2). To best of our knowledge, only one clinical study (Prieto, 2005) was undertaken where a single human milk oligosaccharide, LNnT, was added to an infant formula. LNnT was chosen based on its anti-adhesive or anti-pathogenic activity against *S. pneumoniae* (Andersson et al., 1986). It was hypothesised that if infants aged 6-24 months of age were fed formulas with or without added LNnT (used at a concentration present in human milk i.e. 200 mg/L), different colonisation rates of *S. pneumoniae* would be evident in the throat. Children were monitored over a 16-week period which encompassed winter and spring, during which oropharyngeal swabs were taken every 2 weeks to ascertain colonisation of mucosal cell surfaces with *S. pneumoniae* (first variable). The results demonstrated that the colonisation in infants fed formula containing LNnT was higher than that observed in infants fed standard formula; however, the difference was not significant. When the second variable "abnormal ears" was analysed, the LNnT fed group had statistically significant less "abnormal ears" than the group fed the control formula. When the data were analysed taking into account the use of antibiotics, this difference was no longer significant. A different approach was used in a double-blind, placebo-controlled, outpatient study (Ukkonen, Varis, Jernfors, Herva, Joniken, Ruokokoski, et al., 2000). 3'SLNnT was given intranasally (daily dose: 8 mg) to children of an average age of 15.6 months to test the efficacy of the oligosaccharide for prophylaxis of acute otitis media and for its effect on nasopharyngeal carriage of *S. pneumoniae*, *H. influenzae* and *Moraxella catarrhalis*. 3'SLNnT was chosen because it occurs naturally in breast milk, inhibits binding of *S. pneumoniae* to monolayers of respiratory epithelial cells (Barthelson et al., 1998) and promotes clearance of *S. pneumoniae* from the airway of rabbits (Idänpään-Heikkilä et al., 1997). The results demonstrated that intranasally administration of the oligosaccharide did not have any beneficial effect on the occurrence of acute otitis media or the nasopharyngeal carriage of bacterial pathogens. The authors speculated that the concentration given was too low and possibly a preparation using multivalent oligosaccharides may give better results. A clinical case study by von Bismarck, Schneppenheim & Schumacher (2001) reported the use of galactose and fucose (which are monosaccharides present in many human milk oligosaccharides) in treating an infant, suffering from *P. aeruginosa* airway infection who was not responding to antibiotic treatment. The infant was treated with inhalations of tobramycin in combination with 0.01 M galactose and 0.01 M fucose. Within 2 weeks of the treatment, *P. aeruginosa* was not detected in the patient's respiratory tract. The rationale behind this treatment was related to the previously characterised interaction of the two lectins of *P. aeruginosa,* PA-IL and PA-IIL, with galactose and fucose, respectively (Gilboa-Gaber, 1982). These lectins act as *P. aeruginosa* attachment factors, but are also able to stop human ciliary beating, thus reducing an important non-specific airway defence mechanism. The authors considered that simple sugars such as galactose and fucose could be used to compete with a bound complex carbohydrate. Consequently, the monosaccharides were used to interfere with the lectin mediated attachment of bacteria to the epithelia. Yet another approach was used by Stepans et al. (2006) where the group instead of assessing the effects of milk oligosaccharides (which were administered via infant formula or intranasally), studied the association between the presence of a single oligosaccharide (lacto-N-fucopentaose II, LNF-II) in breast milk and maternal reports of respiratory and gastrointestinal problems in their breastfed infants. The researchers also studied the association between a single measure of HMO in faeces of breastfed infants and the same maternal reports of respiratory and gastrointestinal problems in their infants. Levels of HMO consumed by infants at 2 weeks of age, whether measured as concentrations of LNF-II in milk or fecal, were inversely related with occurrences of respiratory illness at 6 weeks of age. In addition, higher milk LNF-II levels at 2 weeks predict lower risk of respiratory illness at 12 weeks of age. The results from this study suggest that there is a link between LNF-II levels in mother's milk and lower rates of infant respiratory illness. However, the study cannot explain the mechanism whereby the HMO might inhibit respiratory disease. Direct effects as presented in the previous paragraph may be suggested in addition to indirect effects. For instance, *in vitro* studies have indicated that specific oligosaccharides in human milk may have effect on the immune system, by influencing leukocyte rolling (Bode et al., 2004a), serving as anti-inflammatory components (Bode et al., 2004b) or by imparting immunomodulatory effects (Eiwegger et al., 2010). Furthermore, milk oligosaccharides may protect from infectious diseases through their prebiotic effect. Prebiotics can positively influence the balance of gut microbiota, which in turn can influence the development of the gut mucosal immune system. It has been speculated that the gut microbiota may also affect the systemic immune system in remote organs such as the respiratory tract, however, this has yet to be determined (Coppa et al., 2004; Purchiaroni, Tortora, Gabrielli, Bertucci, Gigante, Iarino, et al., 2013). Indeed, the human studies outlined above suggest that the relationship between colonisation and pathogenesis is not as simple as *in vitro* or animal studies would suggest. Furthermore, it is noteworthy that only one oligosaccharide was tested in each study, where in fact the pool of oligosaccharides in human milk given its complexity may prove more effective. The use of a single oligosaccharide may explain the negative results reported by Prieto (2005) and Ukkonen et al. (2000).

5. Intervention Studies Using GOS and FOS in Preventing or Reducing Respiratory Infections in Children

Although it is widely accepted that breast feeding and HMO have beneficial effects on infant health, the large amounts of these molecules, which are required for intervention trials make their addition to infant formula currently challenging. As mentioned, oligosaccharides from other sources have been developed and added to infant formula with the aim of mimicking the beneficial effects of human milk on gut microflora. Among these, the most used commercial oligosaccharide ingredients are GOS and FOS. GOS are oligosaccharides characterised by D-galactose residues linked by β1-6 or β1-4 bonds with a terminal β1-4 linked D-glucose (Lamsal, 2012) (Figure 3). These molecules are synthesised from lactose via an enzymatic transgalactosylation using a β-galactosidase mainly of bacterial origin (Boehm & Moro, 2008). FOS can be prepared by the degradation of inulin or by the transfructosylation of sucrose using a β-fructosidase. FOS are characterised by D-fructose residues linked by β2-1 bonds with a terminal α1-2 linked D-glucose (Panesar, Kumari & Panesar, 2012) as illustrated in Figure 3.

Recently, studies have been performed to evaluate the effect of prebiotics, sometimes in combination with probiotics (synbiotics), on the incidence of infectious diseases (summarised in Table 2). Some of these studies have not only considered their effects on gut infections but also on respiratory infections. In a recent open study performed by Picaud et al. (2010), the authors compared the incidence and type of infectious diseases in healthy infants aged between 4 and 6 months, who were moving on to follow-on formula. The authors selected this age group as the incidence of infectious disease may increase as the infants are switching from exclusive breast feeding to follow-on formula. The infants were fed a commercial synbiotic-enriched follow-on formula (EF) or a standard follow-on formula (SF). The EF contained FOS (28 mg/g of powder) and two probiotic strains (*Bifidobacterium longum* at 10^7 cfu/g of powder and *Streptococcus thermophilus* at 10^6 cfu/g of powder). Neither formula contained nucleotides or other immunological factors. Among the infections observed, upper respiratory tract infections (URTIs) were the most common. Moreover the incidence of infectious diseases was significantly lower in EF fed infants when compared with SF fed infants. When the causative agents of the infections were evaluated, only the incidence of gastrointestinal infections differed between the groups, being significantly lower in the EF group when compared to the SF group ($p = 0.02$). The authors hypothesised that a synergistic effect (prebiotic and probiotic) was observed in the EF group and may have resulted in reducing pathogen numbers in the gut, given that the probiotics present may produce anti-microbial substances, block adhesion sites, compete for nutrients, degrade toxin receptors and stimulate immunity.

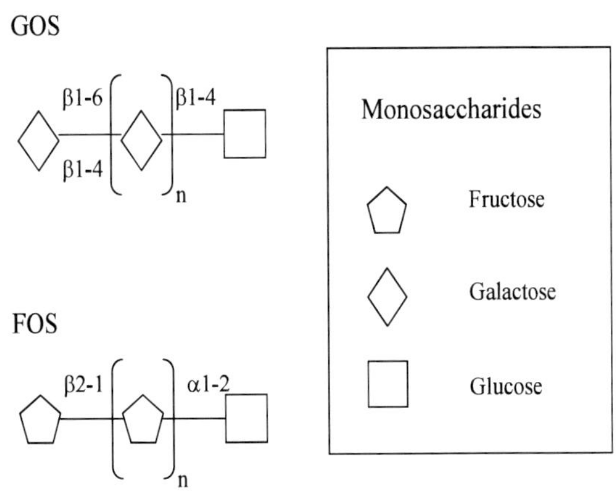

Figure 3. Structural composition of oligosaccharides currently used in infant formula. Galactooligosaccharides (GOS) are characterised by galactose units with glucose at the reducing end. Fructooligosaccharides (FOS) are characterised by fructose units with glucose at the reducing end.

A similar study by Kukkonen et al. (2008) aimed to investigate the safety and long-term effects of feeding synbiotics to newborn infants. Pregnant women carrying infants at high risk of allergy were recruited before delivery and randomly assigned to a group receiving 4 probiotic strains or to the placebo group. For 6 months after birth, their infants received the same probiotic and 0.8 g of GOS or placebo. The synbiotic powder was dissolved in liquid form (water, breast milk, formula). The follow-up continued up to 24 months and overall demonstrated that feeding synbiotics was safe. During the 6-month intervention, no difference in morbidity was observed between synbiotic and placebo groups. However, antibiotics were prescribed less often in the synbiotic group when compared to the placebo group. Throughout the follow-up period, respiratory infections were significant lower in the synbiotic group, while no differences were observed between groups when considering enteric infections. The authors inferred that feeding synbiotics promotes maturation of the immune system, resulting in a reduction of incidence of respiratory infections from 6 to 24 months of age.

A study by Bruzzese et al. (2009) aimed to test the hypothesis that prebiotics reduce the incidence of intestinal and respiratory infections in healthy infants. In this open-study, infants received either a standard formula or a prebiotic-supplemented formula for 12 months. The prebiotic-supplemented formula contained 0.4 g/100 mL of GOS/FOS in a ratio of 9:1. With regards to respiratory infections, the number of episodes of URTIs was lower in the GOS/FOS group when compared to the control group, but the difference was not significant ($p = 0.4$). There was no difference between the two groups in the number of infants with at least 1 episode of URTI. However, among the infants with at least 1 episode of URTI, the number of infants with recurrent URTI was lower in the group fed with the GOS/FOS formula when compared with the control and the difference was close to significance ($p = 0.06$). However, administration of antibiotics was significantly lower in the GOS/FOS group when compared with the control group ($p = 0.038$). The authors suggest that the data presented points toward a decreased incidence of URTI in infants fed GOS/FOS, as supported by the significant reduction in antibiotic usage.

Another study by Arslanoglu et al. (2007) involved a randomised, double-blind, placebo-controlled trial on healthy babies with a parental history of atopic eczema, allergic rhinitis or asthma. In the intervention group, the infants were fed a hydrolysed formula supplemented with 8g/L of a mixture of scGOS/lcFOS (Immunofortis, Numico). The placebo group was fed the same formula supplemented with 8g/L of maltodextrin. The study formula was fed to the infants for 6 months, during which weaning with food containing probiotics was prohibited. Mixed breast and formula feeding was accepted until week 6 of life. If breastfeeding was continued after 6 weeks of age, the infants were excluded from the study. The study demonstrated that during the first 6 months of life, infants in the intervention group displayed fewer episodes of all types of infections, particularly respiratory ($p = 0.07$), when compared with the placebo group. The authors concluded that prebiotics in early life could have a protective effect against infections through modification of the intestinal microbiota, which may in turn modulate the immune response. Taken together these studies suggest that GOS and FOS may protect infants from infectious diseases through indirect effects by influencing the stimulation of the immune system which in turn influences the incidence of respiratory infection. However, further studies are required to clarify the role of GOS and FOS in influencing the incidence of respiratory infections in infants, as 2 out of the 4 studies mentioned investigated the effects of synbiotics.

Conclusion

In summary, the data presented above suggest that HMO and related structures can have both direct and indirect effects in preventing or reducing RTIs. These molecules may exert direct effects by acting as receptor decoys and indirect effects through their prebiotic activity resulting in immunomodulation or stimulation of the immune system. In this respect, milk oligosaccharides have potential use in the development of functional foods and drugs to combat infectious diseases. However, further human intervention studies are required not only to demonstrate the efficacy of milk oligosaccharides in preventing or reducing RTIs, but also to determine the concentration necessary to induce a positive effect and if a cocktail of oligosaccharides would be more beneficial than an individual oligosaccharide. Before this scenario can be realized, economical and rapid enrichment and purification procedures must be put in place to exploit bovine or any suitable milk from domestic animals as a source of oligosaccharides. Recently, technologies have been developed to produce HMO in large scale, using chemical and enzymatic synthesis. Also, processes aiming to produce concentrates enriched in sialyllactose from milk streams that employ ultrafiltration and anion exchange chromatography have been developed. However, sufficient quantities of milk oligosaccharides at competitive costs for high volume commercial applications are as yet unavailable. The possibility of using non milk oligosaccharides, such as GOS and FOS, which may have similar functionality to those found in human milk despite their different chemical structures, should not be excluded from future studies on potential RTI anti-infectives.

Acknowledgments

The work described herein is supported by Enterprise Ireland under Grant Number CC20080001. The funding body had no involvement in the preparation of the article.

The authors have no relevant interest to declare.

References

Andersson, B., Porras, O., Hanson, L. A., Lagergård, T. & Svanborg-Edén, C. (1986). Inhibition of attachment of *Streptococcus pneumoniae* and *Haemophilus influenzae* by human milk and receptor oligosaccharides. *Journal of Infectious Diseases*, *153*, 232-237.

Angeloni, S., Ridet, J. L., Kusy, N., Gao, H., Crevoisier, F., Guinchard, S., et al. (2005). Glycoprofiling with micro-arrays of glycoconjugates and lectins. *Glycobiology*, *15*, 31-41.

Arslanoglu, S., Moro, G. E. & Boehm, G. (2007). Early supplementation of prebiotic oligosaccharides protects formula-fed infants against infections during the first 6 months of life. *Journal of Nutrition*, *137*, 2420–2424.

Bachrach, V. R. G., Schwarz, E. & Bachrach, L. R. (2003). Breastfeeding and the risk of hospitalization for respiratory disease in infancy. *Archives of Pediatrics and Adolescent Medicine*, *157*, 237-243.

Barile, D., Marotta, M., Chu, C., Mehra, R., Grimm, R., Lebrilla C. B., et al. (2010). Neutral and acidic oligosaccharides in Holstein-Friesian colostrum during the first 3 days of lactation measured by high performance liquid chromatography on a microfluidic chip and time-of-flight mass spectrometry. *Journal of Dairy Science*, *93*, 3940-3949.

Barthelson, R., Mobasseri, A., Zopf, D. & Simon, P. (1998). Adherence of *Streptococcus pneumoniae* to respiratory epithelial cells is inhibited by sialylated oligosaccharides. *Infection and Immunity*, *66*, 439-1444.

Bode, L. (2009). Human milk oligosaccharides: prebiotics and beyond. *Nutrition Reviews, 67(S2)*, S183-S191.

Bode, L., Kunz, C., Muhly-Reinholz, M., Seeger, W. & Rudloff, S. (2004a). Inhibition of monocyte, lymphocyte, and neutrophil adhesion to endothelial cells by human milk oligosacchrides. *Thrombosis and Haemostasis, 92*, 1402-1410.

Bode, L., Rudloff, S., Kunz, C., Strobel, S. & Klein, N. (2004b). Human milk oligosaccharides reduce platelet-neutrophil complex formation leading to a decrease in neutrophil b 2 integrin expression. *Journal of Leukocyte Biology, 76*, 820-826.

Boehm, G. & Moro, G. (2008). Structural and functional aspects of prebiotics used in infant nutrition. *Journal of Nutrition, 138*, S1818-S1828.

Boehm, G. & Stahl, B. (2003). Oligosaccharides. In T. Mattila-Sandholm (Ed.), *Functional Dairy Products* (vol. *1*, 203-243). Cambridge: Woodhead Publishing Limited.

Bruzzese, E., Volpicelli, M., Squeglia, V., Bruzzese, D., Salvini, F., Bisceglia, M., et al. (2009). A formula containing galacto- and fructo-oligosaccharides prevents intestinal and extra-intestinal infections: an observational study. *Clinical Nutrition, 28*, 156-161.

Coppa, G. V., Bruni, S., Morelli, L., Soldi, S. & Gabrielli, O. (2004). The first prebiotics in humans. *Journal of Clinical Gastroenterology, 38(S2)*, S80-S83.

Coppa, G. V., Gabrielli, O., Giorgi, P., Catassi, C., Montanari, M. P., Varaldo, P. E., et al. (1990). Preliminary study of breastfeeding and bacterial adhesion to uroepithelial cells. *Lancet, 335*, 569-571.

De Leoz, M. L., Shuai, W., Strum, J. S., Niñonuevo, M. R., Gaerlan, S. C., Mirmiran, M., et al. (2013). A quantitative and comprehensive method to analyze human milk oligosaccharide structures in the urine and feces of infants. *Analytical and Bioanalytical Chemistry, 405*, 4089-4105.

Duijts, L., Ramadhani, M. K. & Moll, H. A. (2009). Breastfeeding protects against infectious diseases during infancy in industrialized countries. A systematic review. *Maternal and Child Nutrition, 5*, 199-210.

Duncan, B., Ey, J., Holberg, C. J., Wright, A. L., Martinez, F. D., & Taussig, L. M. (1993). Exclusive breast-feeding for at least 4 months protects against otitis media. *Pediatrics, 91*, 867-872.

Egge, H. (1993). The diversity of oligosaccharides in human milk. In B. Renner & G. Sawatzki (Eds.), *New Perspectives in Infant Nutrition* (12–26). New York: Thieme, Stuttgart.

Eiwegger, T., Stahl, B., Haidl, P., Schmitt, J., Boehm, G., Dehlink, E., et al. (2010). Prebiotic oligosaccharides: *In vitro* evidence for gastrointestinal epithelial transfer and immunomodulatory properties. *Pediatric Allergy and Immunology*, *21*, 1179–1188.

Engfer, M. B., Stahl, B., Finke, B., Sawatzki, G. & Daniel, H. (2000). Human milk oligosaccharides are resistant to enzymatic hydrolysis in the upper gastrointestinal tract. *American Journal of Clinical Nutrition*, *71*, 1589-1596.

Fanaro, S., Boehm, G., Garssen, J., Knol, J., Mosca, F., Stahl, B., et al. (2005). Galactooligosaccharides and long-chain fructooligosaccharides as prebiotics in infant formulas: a review. *Acta Paediatrica*, *94*, 22-26.

Fields, B. S. & Benson, R. F. (2002). Legionella and legionnaires' disease: 25 years of investigation. *Clinical Microbiology Reviews*, *15*, 506-526.

Finke, B., Stahl, B., Pfenninger, A., Karas, M., Daniel, H. & Sawatzki, G. (1999). Analysis of high-molecular-weight oligosaccharides from human milk by liquid chromatography and MALDI-MS. *Analytical Chemistry*, *71*, 3755-3762.

Gilboa-Gaber, N. (1982). *Pseudomonas aeruginosa* lectins. *Methods in Enzymology*, *83*, 378-385.

Gnoth, M. J., Rudloff, S., Kunz, C. & Kinne, R. K. H. (2001). Investigations of the *in vitro* transport of human milk oligosaccharides by a Caco-2 monolayer using a novel high performance liquid chromatography-mass spectrometry technique. *Journal of Biological Chemistry*, *276*, 34363-34370.

Gopal, P. K. & Gill, H. S. (2000). Oligosaccharides and glycoconjugates in bovine milk and colostrum. *British Journal of Nutrition*, *84(S1)*, S69-S74.

Gyorgy, P., Norris, R. F. & Rose, C. S. (1954). Bifidus factor. I. A variant of *Lactobacillus bifidus* requiring a special growth factor. *Archives of Biochemistry and Biophysics*, *48*, 193-201.

Hakkarainen, J., Toivanen, M., Leinonen, A., Frängsmyr, L., Strömberg, N., Lapinjoki, S., et al. (2005). Human and bovine milk oligosaccharides inhibit *Neisseria meningitidis* pili attachment *in vitro*. *Journal of Nutrition*, *135*, 2445–2448.

Hanson, L. A., Jalil, F., Ashraf, R., Bernini, S., Carlsson, B., Cruz, J.R., et al. (1991). Characteristics of human milk antibodies and their effect in relation to the epidemiology of breastfeeding and infections in a developing country. *Advances in Experimental Medicine and Biology*, *310*, 1–15.

Hardy, S. J., Christodoulides, M., Weller, R. O. & Heckels, J. E. (2000). Interactions of *Neisseria meningitidis* with cells of the human meninges. *Molecular Microbiology*, *36*, 817–29.

Idänpään-Heikkilä, I., Simon, P. M., Zopf, M., Vullo, T., Cahill, P., Sokol, K., et al. (1997). Oligosaccharides interfere with the establishment and progression of experimental pneumococcal pneumonia. *Journal of Infectious Diseases*, *176*, 704–12.

Kukkonen, K., Savilahti, E., Haahtela, T., Juntunen-Backman, K., Korpela, R., Poussa, T., et al. (2008). Long-term safety and impact on infection rates of postnatal probiotic and prebiotic (synbiotic) treatment: randomized, double-blind, placebo-controlled trial. *Pediatrics*, *122*, 8-12.

Kunz, C. & Rudloff, S. (2008). Potential anti-inflammatory and anti-infectious effects of human milk oligosaccharides. In Z. Bösze (Ed.), *Bioactive Components of Milk* (455-465). Springer.

Kunz, C., Rudloff, S., Baier, W., Klein, N. & Strobel, S. (2000). Oligosaccharides in human milk: structural, functional, and metabolic aspects. *Annual Review of Nutrition*, *20*, 600-722.

Kunz, C., Rudloff, S., Schad, W. & Braun, D. (1999). Lactose-derived oligosaccharides in the milk of elephants: comparison with human milk. *British Journal of Nutrition*, *82*, 391-399.

Lamsal, B. P. (2012). Production, health aspects and potential food uses of dairy prebiotic galactooligosaccharides. *Journal of the Science of Food and Agriculture*, *92*, 2020-2028.

Lane, J. A., Mehra, R. K., Carrington, S. D. & Hickey, R. M. (2010). The food glycome: a source of protection against pathogen colonization in the gastrointestinal tract. *International Journal of Food Microbiology*, *142*, 1-13.

Lesman-Movshovich, E. & Gilboa-Garber, N. (2003). *Pseudomonas aeruginosa* lectin PA-IIL as a powerful probe for human and bovine milk analysis. *Journal of Dairy Science*, *86*, 2276-2282.

Lesman-Movshovich, E., Lerrer, B. & Gilboa-Garber, N. (2003). Blocking of *Pseudomonas aeruginosa* lectins by human milk glycans. *Canadian Journal of Microbiology*, *49*, 230-235.

LoCascio, R. G., Ninonuevo, M. R., Freeman, S. L., Sela, D. A., Grimm, R., Lebrilla, C.B., et al. (2007). Glycoprofiling of bifidobacterial consumption of human milk oligosaccharides demonstrates strain specific, preferential consumption of small chain glycans secreted in early human lactation. *Journal of Agricultural and Food Chemistry*, *55*, 8914-8919.

Martín-Sosa, S., Martín, M. J., García-Pardo, L. A. & Hueso, P. (2003). Sialyloligosaccharides in human and bovine milk and in infant formulas: variations with the progression of lactation. *Journal of Dairy Science, 86*, 52-59.

Martín-Sosa, S., Martín, M. J. Hueso, P. (2002). The sialilated fraction of milk oligosaccharides is partially responsible for binding to enterotoxigenic and uropathogenic *Escherichia coli* human strains. *Journal of Nutrition, 132*, 3067-72.

McJarrow, P. & van Amelsfort-Schoonbbeek, J. (2004). Bovine sialyl oligosaccharides: seasonal variations in their concentrations in milk, and a comparison of the colostrums of Jersey and Friesian cows. *International Dairy Journal, 14*, 571-579.

McNiel, M. E., Labbok, M. H. & Abrahams, S. W. (2010). What are the risks associated with formula feeding? A re-analysis and review. *Birth, 37*, 50-58.

Nakamura, T., Kawase, H., Kimura, K., Watanabe, Y., Ohtani, M., Arai, I., et al. (2003). Concentrations of sialyloligosaccharides in bovine colostrum and milk during the prepartum and early lactation. *Journal of Dairy Science, 86*, 1315-1320.

Nassif, X. (2000). A furtive pathogen revealed. *Science, 287*, 1767–1768.

Nassif, X. (1999). Interactions between encapsulated *Neisseria meningitidis* and host cells. *International Microbiology, 2*, 133–6.

Newburg, D. S. (2009). Neonatal protection by an innate immune system of human milk consisting of oligosaccharides and glycans. *Journal of Animal Science, 87*, 26-34.

Newburg, D. S., Ruiz-Palacios, G. M., Altaye, M., Chaturvedi, P., Meinzen-Derr, J., Guerrero Mde, L., et al. (2004). Innate protection conferred by fucosylated oligosaccharides of human milk against diarrhea in breastfed infants. *Glycobiology, 14*, 253-263.

Newburg, D. S., Ruiz-Palacios, G. M. & Morrow, A. L. (2005). Human milk glycans protect infants against enteric pathogens. *Annual Review of Nutrition, 25*, 37–58.

Ninonuevo, M. R., Ward, R. E., LoCascio, R. G., German, J. B., Freeman, S. L., Barboza, M., et al. (2007). Methods for the quantitation of human milk oligosaccharides in bacterial fermentation by mass spectrometry. *Analytical Biochemistry, 361*, 15–23.

Obermeier, S., Rudloff, S., Pohlentz, G., Lentze, M. J. & Kunz, C. (1999). Secretion of 13C-labelled oligosaccharides into human milk and infant's urine after an oral (13C)galactose load. *Isotopes in Environmental and Health Studies*, *35*, 119-125.

Panesar, P. S., Kumari, S. & Panesar R. (2012). Biotechnological approaches for the production of prebiotics and their potential applications. *Critical Reviews in Biotechnology*, doi: 10.3109/07388551.2012.709482.

Perret, S., Sabin, C., Dumon, C., Pokorná, M., Gautier, C., Galanina, O., et al. (2005). Structural basis for the interaction between human milk oligosaccharides and the bacterial lectin PA-IIL of *Pseudomonas aeruginosa*. *Biochemistry Journal*, *389*, 325-332.

Picaud, J-C., Chapalain, V., Paineau, D., Zourabichvili, O., Borner, F. R. J. & Duhamel, J-F. (2010). Incidence of infectious diseases in infants fed follow-on formula containing synbiotics: an observational study. *Acta Paediatrica*, *99*, 1695–1700.

Prieto, P. A. (2005). *In vitro* and clinical experiences with a human milk oligosaccharide, lacto-N-neotetraose, and fructooligosaccharides. *FFI Journal*, *210*, 1018-1029.

Pujol, C., Eugène, E., de Saint Martin, L. & Nassif, X. (1997). Interaction of *Neisseria meningitidis* with a polarized monolayer of epithelial cells. *Infection and Immunity*, *65*, 4836–4842.

Purchiaroni, F., Tortora, A., Gabrielli, M., Bertucci, F., Gigante, G., Iarino, G., et al. (2013). The role of intestinal microbiota and the immune system. *European Review of Medical and Pharmacological Sciences*, *17*, 323-333.

Ramphal, R., Carnoy, C., Fievre, S. & Michalski, J-C. (1991). *Pseudomonas aeruginosa* recognizes carbohydrate chains containing type 1 (Gal□1-3GLcNAc) or type 2(Gal□1-4G1cNAc) disaccharide units. *Infection and Immunity*, *59*, 700-704.

Roth, D. E., Caulfield, L. E., Ezzati, M. & Black, R. E. (2008) Acute lower respiratory infections in childhood: opportunities for reducing the global burden through nutritional interventions. *Bulletin of the World Health Organisation*, *86*, 356-364.

Rudloff, S., Pohlentz, G., Borsch, C., Lentze, M. & Kunz, C. (2012). Urinary excretion of *in vivo* 13C-labelled milk oligosaccharides in breastfed infants. *British Journal of Nutrition*, *107*, 957-963.

Rudloff, S., Pohlentz, G., Diekmann, L., Egge, H. & Kunz, C. (1996). Urinary excretion of lactose and oligosaccharides in preterm infants fed human milk or infant formula. *Acta Paediatrica*, *85*, 598-603.

Scharfman, A., Degroote, S., Beau, J., Lamblin, G., Roussel, P. & Mazurier, J. (1999). *Pseudomonas aeruginosa* binds to neoglycoconjugates bearing mucin carbohydrate determinants and predominantly to sialyl-Lewis x conjugates. *Glycobiology*, *9*, 757-764.

Stahl, B., Thurl, S., Zeng, J. R., Karas, M., Hillenkamp, F., Steup, F., et al. (1994). Oligosaccharides from human-milk as revealed by matrix-assisted laser-desorption ionization mass-spectrometry. *Analitycal Biochemistry*, *223*, 218–226.

Stepans, M. B. F., Wilhelm, S. L., Hertzog, M. T., Rodehorst, T. K. C., Blaney, S., Clemens, B., et al. (2006). Early consumption of human milk oligosaccharides is inversely related to subsequent risk of respiratory and enteric disease in infants. *Breastfeed Medicine*, *1*, 207-215.

Strateva, T. & Yordanov, D. (2009). *Pseudomonas aeruginosa* – a phenomenon of bacterial resistance. *Journal of Medical Microbiology*, *58*, 1133-1148.

Teele, D. W., Klein, J. O. & Rosner, B. (1989). Epidemiology of otitis media during the first seven years of life in children in greater Boston: a prospective, cohort study. *Journal of Infectious Diseases*, *160*, 83-94.

Thomas, R. & Brooks, T. (2004a). Common oligosaccharides moieties inhibit the adherence of typical and atypical respiratory pathogens. *Journal of Medical Microbiology*, *53*, 833-840.

Thomas, R.J. & Brooks, T. J. (2004b). Oligosaccharide receptor mimics inhibit *Legionella pneumophila* attachment to human respiratory epithelial cells. *Microbial Pathogenesis*, *36*, 83-92.

Tlaskalová-Hogenová, H., Stepánková, R., Hudcovic, T., Tucková, L., Cukrowska, B., Lodinová-Zádníková, R., et al. (2004). Commensal bacteria (normal microflora), mucosal immunity and chronic inflammatory and autoimmune diseases. *Immunology Letters*, *93*, 97-108.

Tong, H. H., McIver, M. A., Fisher, L. M. & DeMaria, T. F. (1999). Effect of lacto-N-neotetraose, asialoganglioside-GM1 and neuraminidase on adherence of otitis media-associated serotypes of *Streptococcus pneumoniae* to chinchilla tracheal epithelium. *Microbial Pathogenesis*, *26*, 111–119.

Ukkonen, P., Varis, K., Jernfors, M., Herva, E., Joniken, J., Ruokokoski, E., et al. (2000). Treatment of acute otitis media with an antiadhesive oligosaccharide: a randomised, double-blind, placebo-controlled trial. *Lancet*, *356*, 1398–1402.

von Bismark, P., Schneppenheim, R. & Schumacher, U. (2001). Successful treatment of *Pseudomonas aeruginosa* respiratory tract infection with a sugar solution – a case report on a lectin based theurapetic principle. *Klinische Pädiatrie*, *213*, 285-287.

Winn, W. C. (1988). Legionnaires disease: historical perspective. *Clinical Microbiology Reviews*, *1*, 60-81.

World Health Organisation and Department of Nutrition for Health and Development. (2001). The optimal duration of exclusive breastfeeding: report of an expert consultation. WHO: Geneva.

In: Oligosaccharides
Editors: L. S. Schweizer, S. J. Krebs
ISBN: 978-1-62948-328-3

Chapter V

Alterations in the Glycosylation Pattern of Alpha-1-Glycoprotein May be Diagnostic for the Detection of Breast Cancer and/or Reducing the Effect of Chemotherapy *In Vivo*

Kate Doak[1,3]***, Jodi Flaws***[2,4]***, Yvonne Cruickshank***[1] ***and Kevin D. Smith***[1,3,*]

[1]School of Life, Sport and Social Sciences,
Edinburgh Napier University, Edinburgh, Scotland
[2]College of Veterinary Medicine,
Vet Med Basic Sciences Bldg., Urbana, IL, US
[3] Department of Bioscience, University of Strathclyde,
Glasgow, Scotland
[4] Department of Epidemiology and Preventive Medicine,
University of Maryland School of Medicine, Baltimore, MD, US

[*] Author to whom correspondence should be addressed; Email: k.smith@napier.ac.uk;
Tel.: +44-131-455-2509

Abstract

Alpha-1-acid glycoprotein AGP is an important binder of drugs in the plasma. Although the binding site is protein in nature, the glycosylation pattern present could still affect the overall conformation of the glycoprotein and could also have implications on access to the binding site. Drugs bound by plasma proteins have reduced efficacy as pharmacological effects are exerted by free unbound drug.

The present study aimed to determine if AGP could bind to tamoxifen, the gold standard of treatment for breast cancer, and if, through this binding, AGP could represent a mechanism of resistance to tamoxifen.

The glycosylation of AGP was also examined to determine if any disease-specific changes were present in breast cancer and if these disease-specific glycoforms displayed altered binding properties for tamoxifen.

The study used different cohorts of patients - healthy individuals, those with untreated breast cancer, those at risk of breast cancer and those with breast cancer being treated with tamoxifen therapy. AGP was isolated by a technique specifically chosen because it did not degrade the glycoprotein structure, thus reflecting the situation *in vivo*. Drug binding was examined by the use of an intrinsic fluorescence assay. Patients with untreated and tamoxifen treated breast cancer had significantly increased AGP levels compared to healthy individuals. The monosaccharide fucose, which is normally not present on AGP glycosylation in healthy plasma, was found in untreated breast cancer patients. Fucose was also found in some tamoxifen treated samples, which correlated with tamoxifen treatment, with those patients at the end of treatment displaying no fucose indicating no inflammation or disease, thus backing up the standard 5 year treatment period with tamoxifen. Patients with untreated breast cancer displayed a disease specific shift to bisiaylated glycan chains compared with healthy AGP. Tamoxifen treated breast cancer patients showed changes in the oligosaccharide structures over the time course of tamoxifen therapy, reverting to normal at a later time point – also indicated by the monosaccharide analysis. Untreated breast cancer patients were able to bind tamoxifen to a similar extent as healthy AGP. In contrast those at risk and those with tamoxifen treated breast cancer had significantly reduced binding. Although binding was minimal at therapeutic concentrations, the decrease in binding by breast cancer patients could lead to issues regarding toxicity as there would be more free drug in the plasma and therefore this concentration could exceed the therapeutic level.

1. Introduction

Although proteomic approaches to cancer biomarkers have yielded important results, recent research has also started to focus on post-translational modifications of proteins such as glycosylation. It has already been shown that glycosylated structures can be diagnostic in breast cancer, but this observation is limited to the overall degree of glycosylation and not alterations occurring in specific proteins (Chew et al., 2007; Leiserowitz et al., 2008).

Alpha-1-acid glycoprotein (AGP) is the most extensively studied of the acute phase proteins (APP). AGP has a molecular weight of 41-43kDa, of which 45% is carbohydrate and it is known to be a binder of drugs in the plasma (Fournier et al, 2000). AGP is a positive APP and under disease conditions the plasma concentration increases 2-5 fold. It has been found that there are no significant differences in plasma AGP concentration in healthy men and women (Yost and DeVane, 1985). In contrast to albumin concentration, AGP levels are not affected by age (Veering et al, 1990). Many studies have investigated changes to the levels of AGP under pathophysiological conditions (reviewed in Ceciliani and Pocacqua, 2007; Israili and Dayton 2001; Tesseromatis et al., 2011), and concluded that the concentration of AGP alone is not diagnostic for a particular pathological condition. The glycosylated structures of AGP have been shown to be specifically altered during diseased states and indeed subtle variations in glycosylation can be disease specific (Elliott et al., 1997; Anderson et al., 2002; Smith et al., 2002) in terms of the proportions and compositions of the oligosaccharide population (bi-, tri- and tetra-antennary chains) .

As previously, the majority of the research regarding AGP both generally, and specifically with respect to its relationship with cancer, has focused on the plasma levels of the glycoprotein in the disease state. Duche et al (2000) found that AGP levels are increased in cancer, with significant increases found in lung, breast and ovarian carcinomas, when compared to a healthy population. Conversely, cancer therapy can affect the levels detected; treatment, for breast cancer, using tamoxifen was been found to decrease the levels of AGP in the plasma (Kailajarva et al, 2000). However, not only does the plasma concentration of AGP alter during malignancy, but there are also changes in glycosylation. Investigations into the heterogeneity of cancer AGP have found an increase in bi-antennary glycan content with increased fucosylation and sialylation (Moule et al. 1987; Turner et al. 1985). More recently, Hashimoto et al. (2004) examined AGP fucosylation as a marker of cancer progression and prognosis. They found cancer patients had raised AGP levels compared to the control group, but that this level was not diagnostic. The study also found that there was a correlation between AGP glycoforms and disease progression and prognosis – patients with increased fucosylation and increased tri- and tetrantennary chains for a long period of time post surgery had a poorer prognosis. They found that branching and fucosylation were good markers for prognosis compared to tumour size and TNM status.

A physiological function of AGP is as a binder of particular drugs present in the plasma, thus reducing the free and active concentration of the drug. AGP mainly binds to basic drugs but is also able to bind acidic drugs (Kremer et al, 1988). In order for a drug to have a pharmacological effect, a minimum effective concentration must be obtained at the site of action. If a drug is bound to AGP then it is rendered inactive and not able to exert its pharmacological effects (Israili and Dayton, 2001). As a result the minimum effective concentration is not achieved and therefore drug therapy is ineffective. The drug binding site on AGP is peptide in origin, and is found between two oligosaccharide chains (Kute and Westphal, 1976). Although not directly involved in drug binding, it is thought that the degree of branching of the AGP oligosaccharide chains (and any disease specific changes) could affect the conformation of, or access to, the binding site. There have been a range of studies to date (reviewed in Kremer et al., 1988; Israili and Dayton, 2001) that have both supported and disputed this theory.

Tamoxifen has been the gold standard of breast cancer treatment for over 40 years. It is a competitive inhibitor of the binding of oestrogen to the oestrogen receptor and is also classified as a selective estrogen receptor modulator (SERM), (Jordan, 2007). Although given to both pre- and postmenopausal women tamoxifen is only effective in ER positive patients (Silva and Zurrida, 2005) regardless of menopausal status, age and lymph node status. The effect of a reduction in incidence of contralateral breast cancer is related to the duration of treatment. Benefits of tamoxifen therapy are maintained for around 5 years after treatment has finished (Rose, 2008).

Although tamoxifen has been highly successful, many patients either do not respond to treatment or respond initially but then go on to develop resistance to the therapy. Many theories have been postulated (MacGregor and Jordan, 1998) e.g. the overexpression of HER2/*neu* (Dorssers, 2001). An alternative mechanism for tamoxifen resistance could be the binding of the drug to plasma proteins such as AGP *in vivo*. It is hypothesised that AGP could bind tamoxifen in the plasma and therefore reduce the efficacy of or cause resistance to tamoxifen in breast cancer patients to AGP from healthy donors. Previous work by Paterson et al (2003) examined the binding of various metabolic products of tamoxifen to AGP. The study found that there was minimal binding at therapeutic concentrations of the metabolites.

The present study aimed to determine if AGP isolated from breast cancer patients and individuals identified as "at risk" of breast cancer, had altered binding properties compared to AGP from healthy individuals. The study also sought to determine if the oligosaccharide chains on AGP, isolated from individuals with breast cancer, are altered compared with AGP isolated from healthy individuals and if these alterations correlate with the drug binding ability of AGP. Altered glycosylation is likely to be of extreme importance as the binding site on AGP lies between glycosylation sites. Thus an increase in oligosaccharide branching could alter the conformation of the protein and therefore the binding site, or access to it.

1.1. Materials

1.1.1. Patient Plasma Samples

All plasma samples were obtained from female patients with full ethical consent and permission. One set of plasma samples, (which was a gift from Prof. David George, Glasgow Western Infirmary), were obtained from women with untreated (UT) breast cancer.

The patients who have not had any form of cancer therapy or surgery have an average age of 67.25 years. There was no correlation between age and the grade of cancer present.

Additionally, two populations of plasma samples were obtained from Dr. Jodi Flaws, University of Maryland. One population consisted of individuals who were deemed "at risk" (AR) of breast cancer and the other consisted of individuals with breast cancer (TT). All patients received tamoxifen therapy (details given in table 1). Both the populations receiving tamoxifen therapy have 50% of samples obtained from African-American and 50% obtained from Caucasian individuals.

The average age for the "at risk" population is 58.9 years. The average age for the "at risk" African-American and Caucasian patients is 62.2 and 55.6 years respectively.

The average time on tamoxifen therapy is 25.6 months for "at risk" individuals, with the African-American and Caucasian individuals having an average of 28.8 and 22.4 months therapy respectively. Thus the African-American cohort are older and have been on tamoxifen therapy longer than the Caucasian individuals.

Table 1. Patient demographic details for individuals who received tamoxifen therapy

Patient I.D.	Ethnicity	Age at Date of Visit (years)	Time on Tamoxifen (Months)
At Risk (AR)1	African-American	52	11
AR2	African-American	48	26
AR3	African-American	76	27
AR4	African-American	58	27
AR5	African-American	77	53
AR6	Caucasian	49	15
AR7	Caucasian	68	19
AR8	Caucasian	48	19
AR9	Caucasian	51	20
AR10	Caucasian	62	39
Average AR	-	58.9	25.6
Tamoxifen Treated (TT)1	African-American	45	4
TT2	Caucasian	42	4
TT3	African-American	53	8
TT4	Caucasian	69	8
TT5	African-American	43	14
TT6	Caucasian	48	14
TT7	African-American	50	26
TT8	Caucasian	66	26
TT9	African-American	69	51
TT10	Caucasian	57	54
Average TT	-	54.2	20.9

With respect to the tamoxifen treated breast cancer (TT) population, the average age is 54.2 years. Focusing on racial groupings, the average age is 52 and 56.4 years for African-American and Caucasian individuals respectively. The average time on tamoxifen is 20.9 months, with African-American and Caucasian individuals having an average therapy of 20.6 and 21.2 months respectively.

1.1.2. Determination of AGP Concentration in Plasma

The level of AGP in plasma samples were determined by the Department of Pathology at Leicester Royal Infirmary using an immunoturbidimetric method.

1.1.3. Alpha-1-Acid Glycoprotein Isolation

Polyethylene glycol 3350, Reactive Blue 2 Sepharose, potassium chloride, Trizma base, sodium azide, potassium thiocyanate, sodium chloride and sodium acetate were obtained from Sigma, Poole, UK. Q-sepharose Fast Flow and Red Sepharose CL-6B were obtained from Pharmacia, Little Chalfont, UK. Poly-prep disposable columns were obtained from Biorad, Hemel Hempstead UK. Centricon YM-10cetrifugal filter devices were purchased from Millipore (UK) Ltd, Watford, UK.

1.1.4. Drug Binding Assay

Normal human AGP, Dimethyl sulphoxide (DMSO), chlorpromazine hydrochloride, Dulbecco's Phosphate buffered saline were obtained from Sigma, Poole, UK. Tamoxifen Citrate was purchased from MP Biomedicals Ltd, London, UK. Microtitre plates were obtained from Nunc^{TM}, Denmark. SpectraMax GeminiXS fluorescence microplate reader was purchased from Molecular Devices, Sunnyvale, CA, USA. KaleidaGraph 3.51 was purchased from Synergy Software, Reading, PA, USA. For the drug binding assay carried out using cuvettes, a Shimadzu RF-5301PC spectrofluorophotometer, from Shimadzu Corporation, Kyoto, Japan. A 10x4mm quartz cuvette was obtained from Merck, UK.

1.1.5. High pH Anion Exchange Chromatography

1.1.5.1. Monosaccharide Analysis

HPLC grade trifluoroacetic acid (TFA) and constant boiling 4M hydrochloric acid (HCL) were obtained from Perbio Science UK Ltd, Cheshire, UK. Monosaccharide standards (fucose, glucosamine, galactose and mannose), Reacti-vials and Dowex-50H$^+$ were purchased from Sigma, Poole, UK. High pH Anion Exchange Chromatography (HPAEC) was performed using a DX-600^{TM} system supplied by Dionex, Camberley, UK. The DX-600^{TM} system consisted of a GP50 gradient pump, ED40 electrochemical detector and an anion self-regenerating suppressor controlled by a Dell OptiPlex GX110 using PeakNet^{TM} software. The pulsed electrochemical waveform used in the detection was Time = 0s, E = +0.05V; 0.29s, +0.05V;

0.49s, +0.05V, 0.5s, +0.05V; 0.51s, +0.6V; 0.6s, +0.6V; 0.61s, -0.6V; 0.65s, -0.6V; 0.66s, 0.05V.. Monosaccharide analysis was performed on a CarboPac™ PA-100 analytical (4x250mm) and a guard column (4x50mm), were supplied by Dionex. HPLC grade water was obtained from Rathburn Chemical Ltd, Walkerburn, Scotland, UK. Sodium Hydroxide 50%w/v was purchased from VWR International Ltd, Poole UK.

1.1.5.2. Oligosaccharide Analysis

PNGase F and buffers (NP40 and NE buffer G7) were supplied by New England BioLabs (New England, USA). HPAEC was carried out on a DX500 system supplied by Dionex (Camberley, UK). The DX-500™ system consisted of a GP50 gradient pump, ED40 electorchemical detector and an anion self-regenerating suppressor controlled by a Vtech 486SX25 (Viglen, UK) using PeakNet™ software. The pulsed electrochemical detection was the same as that described for the DX-600™ system (section 2.1.6.1). Oligosaccharide analysis was performed on a CarboPac™ PA-100 analytical (4x250mm) and a guard column (4x50mm), were supplied by Dionex. HPLC grade water was obtained from Rathburn Chemical Ltd, Walkerburn, Scotland, UK. Sodium Hydroxide 50%w/v was purchased from VWR International Ltd, Poole UK. Sodium acetate was purchased from Sigma (Poole, UK). Ethanol was obtained from Bamford Laboratories Limited, Norden Rochdale, UK.

1.1.6.3. Sialic Acid Analysis

Neuraminidase and NE buffer G1 were supplied by New England Biolabs. The HPAEC system was the same as for oligosaccharide analysis, (section 2.1.6.2).

1.2. Methods

1.2.1. Determination of AGP in Plasma

The AGP content of serum was measured by an immunoturbidimetric method using AGP antiserum. The AGP present in the sera formed an immunoprecipitate, which was measured spectrophotometrically at 340nm. The AGP concentration in the sample was determined from a standard curve.

1.2.2. Isolation of AGP

1.2.2.1. Polyethylene Glycol Precipitation

Polyethylene Glycol (PEG) 3350 was added to the plasma sample to give a final concentration of 40% w/v, mixed for 10 minutes at room temperature and left overnight at 4°C. It was then centrifuged at 20,000g $^{-1}$ for 30 minutes and the supernatant removed, placed in a clean eppendorf and then frozen. The pellet was discarded.

1.2.2.2. Low Pressure Chromatography

A Poly-prep disposable chromatography column was packed with the appropriated chromatographic resin (i.e. Reactive blue 2 sepharose or Q-sepharose Fast flow or Red sepharose) to a volume of approximately 5ml. Columns were stored in 10% ethanol and at 4°C to preserve them for future use. The appropriate elution buffer was run through the column to allow the column to equilibriate, using a peristaltic pump at a flow rate of 0.5ml per minute. A Pharmacia LKB optical unit measured the change in absorbance at 280nm and a Servogor 120 chart recorder recorded the peak. The chart recorder was run at 12cm per hour.

Initially PEG-precipitated plasma samples were applied to the Reactive blue 2 sepharose column. The elution buffer (50mM Trizma/0.1mM KCl/0.02% Azide pH 7) was applied with the sample. The peak was collected in 15ml centrifuge tubes. Any albumin in the sample was retained by the Reactive blue 2 sepharose column and was removed by the passage of desorption buffer in order to regenerate the column. The collected sample was dried in the rotary evaporator (under a vacuum) to a volume of approximately 2ml. The Q-sepharose Fast flow column was equilibriated using 20mM Trizma/75mM NaCl pH6.5 solution. The 2ml sample was run through the Q-sepharose column, with a salt gradient (75mM – 500mM NaCl/20mM Trizma pH6.5). The peak was collected. 75mM NaCl was used to regenerate the column. The sample was dried to approximately 2ml. The 2ml sample was run through the Red-sepharose resin, with the elution buffer (30nM NaOAc). The peak obtained represented the AGP and this was collected. The column was regenerated by desorption buffer. The sample was dried to approximately 2-3ml.

1.2.2.3. Desalting

The sample was applied to a Centricon® centrifugal filter device to remove the salt and concentrate the AGP.

This was placed in the centrifuge at 4000rpm. When the AGP was completely dry, the Centricon® device was inverted, water added and then spun at 4000rpm. The desalted glycoprotein was deposited in the bottom of the eppendorf.

1.2.3. Determination of AGP Concentration

A set of normal (Sigma) AGP standards over a concentration of 0-3mg/ml at 0.5mg/ml intervals was prepared. Dilutions of AGP, from a stock solution of 5mg/ml, were prepared using DPBS.

The absorbance of each solution was measured at 280nm, using a Shimadzu spectrophotometer and cuvette with pathlength 10x4mm. A standard curve was prepared of Sigma AGP concentration (mg/ml) versus absorbance at 280nm.

1.2.4. Drug Binding Assay

1.2.4.1. Cuvette Method

AGP was dissolved in D-PBS to a stock concentration of 5mg/ml. Tamoxifen citrate was prepared to 5mM in DMSO and stored at room temperature. A concentration range (1μM – 5mM) of this drug was prepared using DPBS. A standard curve was prepared. 100μl of each drug concentration was added to 100μl of 5mg/ml AGP solution and the final volume made up to 1ml with DPBS, thus diluting the drug and AGP concentrations 1 in 10. The fluorescence was read using a Shimadzu Spectrofluorimeter RF-5301PC and cuvette with pathlength 10x4mm. The fluorescence emission spectrum was recorded between 300 to 400nm, with an excitation wavelength of 280nm. The percentage reduction in peak fluorescence at 335nm (percentage quenching) was calculated for each drug concentration compared to the solution of AGP alone.

The percentage quenching value was plotted against drug concentration and the data fitted using the quadratic binding equation of Parikh et al (2000) using KaleidaGraph 3.51 software.

$$\text{Percentage quenching} = C_1 \frac{(S-\sqrt{S-4DT-PT})}{2} - C_2 D_T$$

TT 7 (26 months)

TT 9 (51 months)

Healthy volunteer

where D_T = total drug concentration, P_T = total protein concentration, $S = D_T + P_T + K_d$, and C_1, C_2 and K_d are unknown constants, determined by least square curve fitting.

1.2.4.3. Determination of Extent of Binding to Tamoxifen

Isolated AGP from patient plasma samples was used at a concentration of 0.5mg/ml. 10μl of stock tamoxifen citrate (concentration 100μM in DMSO, stored at room temperature) was added and the final volume made up to 1ml with DPBS. The fluorescence was read using a Shimadzu Spectrofluorimeter RF-5301PC and cuvette with pathlength 10x4mm. The fluorescence emission spectrum was recorded between 300 to 400nm, with an excitation wavelength of 280nm. A further two concentrations of tamoxifen were analysed. 10μl and 30μl of stock tamoxifen citrate was added and each time the fluorescence read.

1.2.5. High pH Anion Exchange Chromatography (HPAEC)

1.2.5.1. Acid Hydrolysis of AGP

125μg of AGP was added to a reacti-vial and sealed with a Teflon disc. To this 100μl of 2M Trifluroacetic acid (TFA) and 50μl of 4M Hydrochloric acid was added. The reacti-vial was placed in a heating block for 4 hours at 100°C. A Pasteur pipette was plugged with glass wool and packed with activated Dowex-50 H^+ cation exchange resin to a bed volume of 1ml. The column was washed with HPLC grade water. The AGP sample was added to the column and washed through with 2ml of HPLC grade water. The solution eluted was collected in eppendorf tubes and then dried to completion under a vacuum.

1.2.5.2. Monosaccharide Analysis of Hydrolysed AGP

The analysis of monosaccharides was by high pH anion exchange chromatography with pulsed ampertomic detection. Monosaccharides from AGP were reconstituted in 20μl of HPLC grade water plus 5μl of internal standard, 2-deoxygalactose. The sample was applied to the column for separation. Separation was run for 30 minutes with 3%1MNaOH /97%HPLC grade water as the eluents at a flow rate of 1mL/min. Following this, the column was regenerated with 50%1MNaOH/50%HPLC grade water for 10 minutes. Normal conditions were then restored with 3%1MNaOH/97%HPLC grade water. Monosaccharide standards were used to make calibration curves, with a linear relationship, – amount of monosaccharide vs peak area. The equation of the line was used to quantify monosaccharides present in AGP samples. The monosaccharides from samples were identified by comparison of the elution time of a particular peak with that of the internal standard and also by comparison to a set of standard monosaccharides.

1.2.5.3. Enzyme Digestion for Oligosaccharide Analysis

100µg of AGP was reconstituted in 100µl of HPLC grade water. The AGP solution was then denatured by heating at 100°C for approximately 3 hours. On completion, the solution was dried under a vacuum and then reconstituted in 20µl HPLC grade water. 10µl of NP-40 and NE Buffer G7 and 5U of PNGase F were added to the denatured glycoprotein. The reaction mixture was incubated for 24 hours at 37°C. A further 5U of PNGase F was added and the reaction was incubated for 24 hours at 37°C. Ice cold ethanol was added to the reaction mixture to give a final concentration of 40%v/v and this was incubated overnight at -20°C. The reaction mixture was then centrifuged at 14,000g for 30 minutes to pellet the peptide fraction and leaving the oligosaccharyl residues in the supernatant. The supernatant was then removed and dried to completion under a vacuum.

1.2.5.4. Oligosaccharide Analysis of Hydrolysed AGP

The analysis of oligosaccharides obtained from AGP was by HPAEC with PAD. Oligosaccharides were reconstituted in 25µl of HPLC grade water. The sample was then applied to the CarboPacTM 100 column for separation. The column was equilibrated with 10% NaOH/5% NaOAc/85% HPLC grade water, for 10 minutes. Separation was carried out for 55minutes with a linear gradient with a final composition of 10%NaOH/17%NaOAc/73%HPLC grade water, at a flow rate of 1mL/min. The column was regenerated for 10 minutes with 50% NaOH/50% HPLC grade water. Normal conditions were restored as 10 % NaOH/5% NaOAc/85% HPLC grade water. An oligosaccharide library was run as a reference with which to compare AGP samples to.

1.2.5.5. Enzyme Digestion of AGP for Sialic Acid Analysis

30µg of AGP was incubated, with 5µl of NE buffer G1 and 1µl of neuraminidase enzyme and the total reaction volume made up to 50 µl with HPLC grade water, for 24 hours at 37°C. Ice cold ethanol was added to the reaction mixture to give a final concentration of 40%v/v and this was incubated overnight at -20°C. The reaction mixture was then centrifuged at 14,000g for 30 minutes to pellet the peptide fraction and leaving the neuraminic acid residues in the supernatant. The supernatant was then removed and dried to completion under a vacuum.

1.2.6. Statistical Analysis

The Pearson correlation coefficient (R^2) was determined using Micosoft Excel 2000. Correlations range from –1 to 1, with a correlation range of 0.7-1 representing a strong linear relationship and a correlation of 1 representing a perfect linear fit. KaleidaGraph® 3.5 software was used to analyse the statistical significance between different groups of data.

2. Results

2.1. AGP Plasma Concentration

Table 2 details the AGP concentration present in each sample.

Plasma from healthy donors was found to have an AGP concentration of 0.53mg/ml. The untreated breast cancer population had an average AGP concentration of 1.03mg/ml. Within the "at risk" population, the mean AGP concentration was 0.66mg/ml. The African-American and Caucasian subgroups had an average concentration of 0.6mg/ml and 0.73mg/ml respectively. The tamoxifen treated breast cancer patient population had a mean AGP concentration of 1.07mg/ml. The African-American and Caucasian subgroups have an average concentration of 1.30mg/ml and 0.85mg/ml respectively.

Table 2. Plasma levels of AGP (mg/mL) in patient samples

Patient I.D.	AGP Level (mg/ml)
Healthy plasma	0.53
UT1	1.20
UT2	1.12
UT3	0.94
UT4	0.86
Average UT	1.03
AR1	0.5
AR2	0.68
AR3	0.62
AR4	0.67
AR5	0.54
AR6	0.55

Table 2. Plasma levels

Patient I.D.	AGP Level (mg/ml)
AR7	0.68
AR8	0.73
AR9	0.61
AR10	1.08
Average AR	0.66
TT1	0.91
TT2	0.95
TT3	1.63
TT4	0.64
TT5	1.53
TT6	0.84
TT7	1.16
TT8	0.72
TT9	1.22
TT10	1.12
Average TT	1.07

2.2. Monosaccharide Composition of AGP

The monosaccharide components of the oligosaccharide chains from the AGP present in each sample were analysed by HPAEC after acid hydrolysis. An example of a monosaccharide trace is shown in Figure 1.

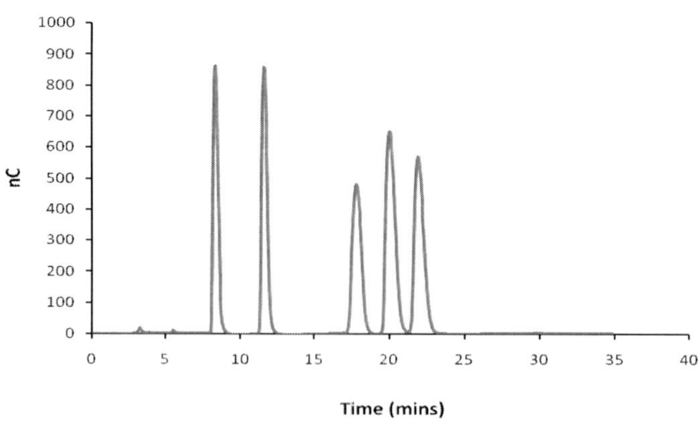

Figure 1. Monosaccharide composition analysis by HPAEC.

The peaks were identified by comparison to an internal standard and the ratios compared to a known mixture of monosaccharide standards, specifically chosen because they are the normal constituents of N-linked oligosaccharide chains. Calibration curves were prepared and were used to calculate how much of a particular monosaccharide was present in each sample. Results of the analysis of patient samples are shown in Table 3. The mean values for the untreated population were 0.61 mole/mole AGP, 5.78 mole/mole AGP, 4.59 mole/mole AGP and 2.81 mole/mole AGP for fucose, glucosamine, galactose and mannose respectively. Mean values for the at risk population were 0.19, 2.78, 1.89 and 1.39 mole/mole AGP and for the tamoxifen treated population 0.21, 3.80, 3.20 and 3.01 mole/mole AGP compared with 0, 2.37, 1.83 and 1.78 mole/mole AGP found in normal plasma.

2.3. Oligosaccharide Analysis of AGP Samples

The oligosaccharide profile of the AGP samples was analysed by HPAEC. This technique allows oligosaccharide structures to be separated by their charge (which equates to the number of negatively charged terminal sialic acids on each chain) and thereafter within each charge band, on the basis of size and isomeric differences. The retention times are as follows: monosiaylated oligosaccharides elute between 10-20 minutes, bisiaylated structures elute between 20-30 minutes, trisiaylated between 30-40 minutes and tetrasiaylated between 40-50 minutes. Within each charge bands there are isometric differences. A HPAEC profile for the oligosaccharides released from AGP from a healthy volunteer is shown in Figure 2.

Figure 3 shows profiles from untreated breast cancer. There are many distinct differences, both qualitative and quantitative, in the oligosaccharide profiles as cancer progresses in severity.

Within each charge band, the variations in retention times will be due to the sialic acid linkages present and the presence or absence of fucose residues. All grades of cancer show a strong presence of bisiaylated oligosaccharide chains. Additionally there is another peak present in this charge band. This could be a structure with 3 or 4 branches but with only 2 sialic acid residues present.

These structures are therefore breast cancer specific and form the basis of a glycan biomarker. All samples show tetrasiaylated structure eluting earlier compared to the oligosaccharide library, due to the presence of fucose on the structures.

Table 3. Monosaccharide composition of individual AGP samples (expressed as mol/mol AGP)

Sample	Fucose	Glucosamine	Galactose	Mannose
Commercial	0.52	8.03	7.17	6.51
Normal	0	2.37	1.83	1.78
UT1	0.62	6.70	5.46	4.14
UT2	1.22	5.33	2.73	1.52
UT3	0.29	4.43	4.31	2.64
UT4	0.3	6.68	5.87	2.93
Average UT	0.61	5.78	4.59	2.81
AR1	0	2.40	2.17	0.89
AR2	0	0.76	0.47	0.53
AR3	0.53	3.34	2.21	1.53
AR4	0.36	4.1	3.01	2.25
AR5	0.7	4.96	3.23	2.20
AR6	0	2.01	1.29	0.92
AR7	0.38	5.85	3.96	3.09
AR8	0	1.72	0.83	0.77
AR9	0	0.52	0.43	0.25
AR10	0	2.19	1.31	1.52
Average AR	0.19	2.78	1.89	1.39
TT1	0.32	4.94	3.63	3.08
TT2	0.73	3.75	3.15	3.17
TT3	0	4.18	2.12	1.60
TT4	0.19	5.00	4.11	4.09
TT5	0	1.85	1.55	1.34
TT6	0.57	4.62	5.33	4.59
TT7	0	2.17	2.14	1.88
TT8	0.33	4.74	4.07	3.76
TT9	0	1.49	0.98	0.65
TT10	0	5.27	4.90	5.93
Average TT	0.21	3.80	3.20	3.01

UT 1 and UT 4 cancer samples are comparable especially when focusing on the trisiaylated charge band, where there are two peaks observed. These peaks are present in similar proportions in these two samples. However both the UT 2 and UT 3 samples show different proportions of these peaks when compared to the UT 1 and 4 samples. The UT 2 and 3 samples show these peaks in approximately equal proportion, whereas the UT 1 and 4 samples show a ratio of approximately 1 to 4. Sample UT 3 shows trisiaylated structures eluting marginally earlier than the oligosaccharide library, perhaps indicating the presence of fucose on the structures.

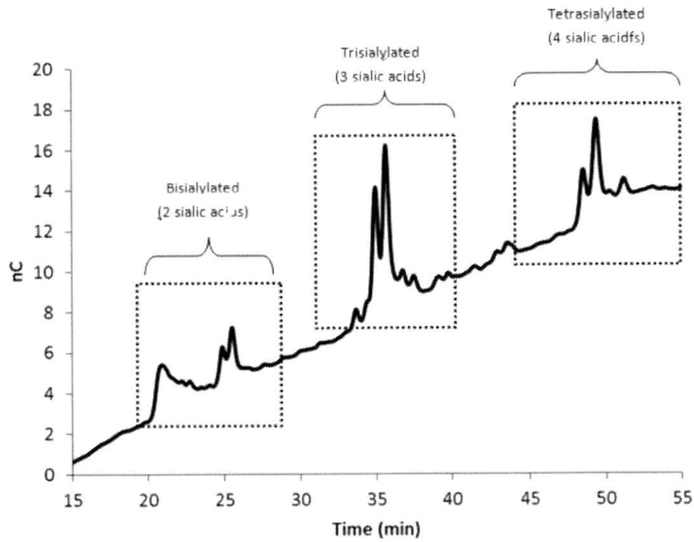

Figure 2. HPAEC profile of oligosaccharides from AGP released from a healthy volunteer.

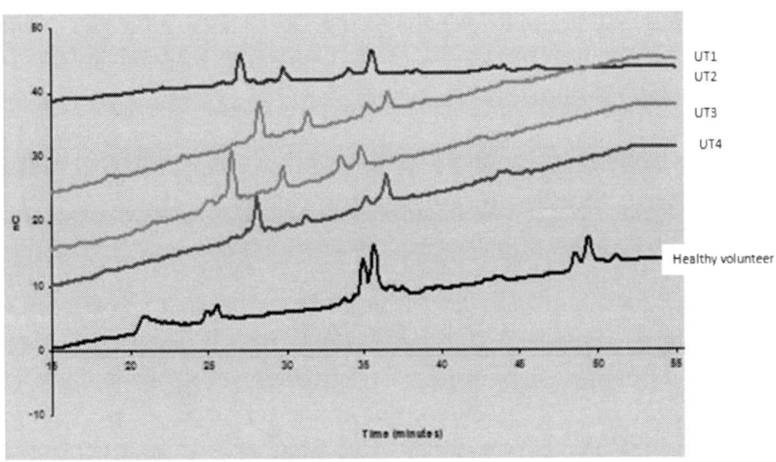

Figure 3. HPAEC profile of oligosaccharide structures present on AGP released from samples obtained from an untreated breast cancer population.

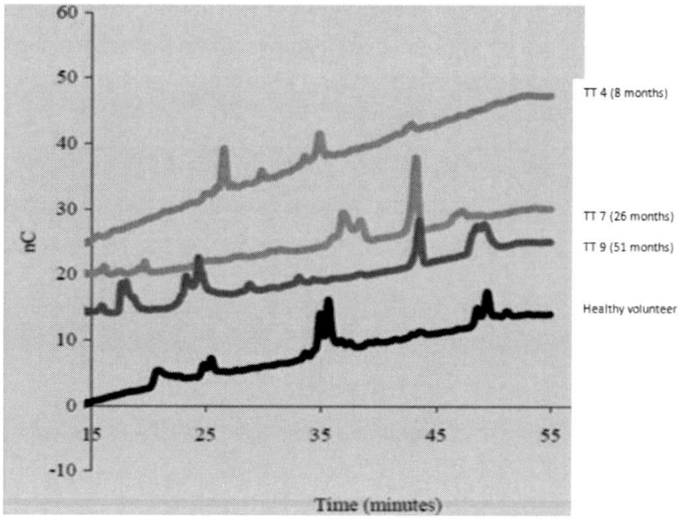

Figure 4. HPAEC profile of oligosaccharide structures present on AGP released from samples obtained from a tamoxifen treated breast cancer population.

Figure 4 shows tamoxifen treated breast cancer sample glycosylation. Analysis of the tamoxifen treated population shows glycan structures reverting to a 'normal' appearance over time. Of particular interest is the trisiaylated tetra-antennary structures that are present in approximately the first 2 years of tamoxifen therapy but are no longer present at a later time point in therapy. This reversion to 'normal' glycosylation was also indicated with the monosaccharide analysis for this sample population, backing up the 5 year tamoxifen therapy period.

2.4. AGP-Drug binding Fluorescence Assay

The extent that a drug is capable of being bound by a plasma protein was determined using measuring the intrinsic fluorescence of proteins, which occurs due to the presence of tryptophan residues. Binding of a drug to a plasma protein is indicated by a decrease in fluorescence known as quenching. The fluorescence of AGP solutions, due to the presence of three tryptophan residues, was measured before and after the addition of various concentrations of drug and the difference in fluorescence was calculated. The ability of AGP, isolated from healthy individuals, those with untreated breast cancer, at risk and breast cancer patients, to bind tamoxifen was evaluated.

The concentrations of drug used in these experiments were chosen with respect to the therapeutic range. It was found that a linear relationship existed for the % quenching with the various concentrations of tamoxifen citrate. Tables 4, 5 and 6 along with Figures 5, 6 and 7 show the results obtained.

Table 4. % quenching values for AGP isolated from normal individuals, and from the untreated breast cancer population with tamoxifen citrate

Sample	% Quenching for 1μM Tamoxifen Citrate	% Quenching for 2μM Tamoxifen Citrate	% Quenching for 5μM Tamoxifen Citrate	Correlation Co-efficient
Healthy	8.96	16.97	26.07	0.94
UT1	11.54	18.50	33.58	0.99
UT2	10.83	14.49	24.18	0.99
UT3	9.3	16.66	26.75	0.96
UT4	8.5	15.46	26.76	0.97
Average UT	10.04	16.28	27.82	-

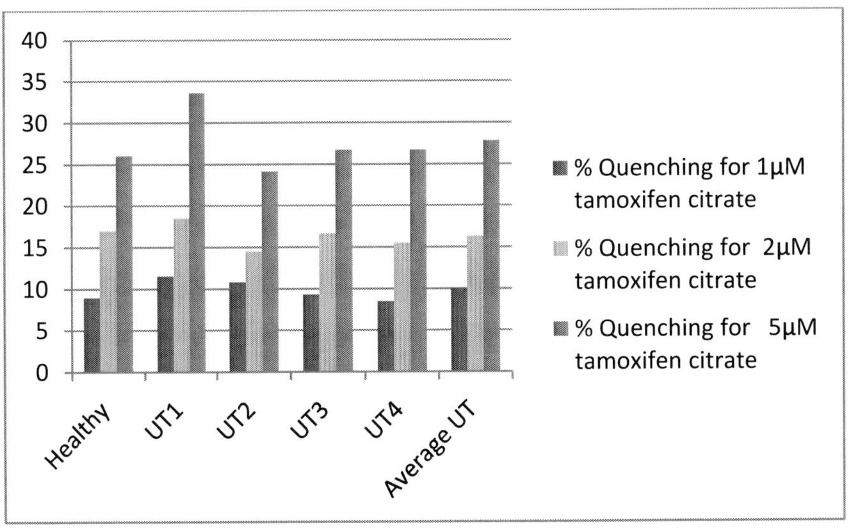

Figure 5. % quenching values for AGP isolated from normal individuals, and from the untreated breast cancer population with tamoxifen citrate.

Table 5. % quenching values for AGP isolated from the "at risk of" breast cancer population with tamoxifen citrate

Sample	% Quenching for 1µM tamoxifen citrate	% Quenching for 2 µM tamoxifen citrate	% Quenching for 5 µM tamoxifen citrate	Correlation coefficient
Healthy	8.96	16.97	26.07	0.94
AR1	10.38	13.90	20.47	0.99
AR2	7.89	16.17	17.66	0.62
AR3	2.55	4.44	15.44	0.99
AR4	8.36	10.30	14.38	0.99
AR6	9.67	12.53	20.56	0.99
AR8	11.70	11.93	17.00	0.96
AR9	6.26	6.88	16.96	0.96
AR10	0	11.29	21.19	0.70
Average AR	4.36	10.93	17.96	

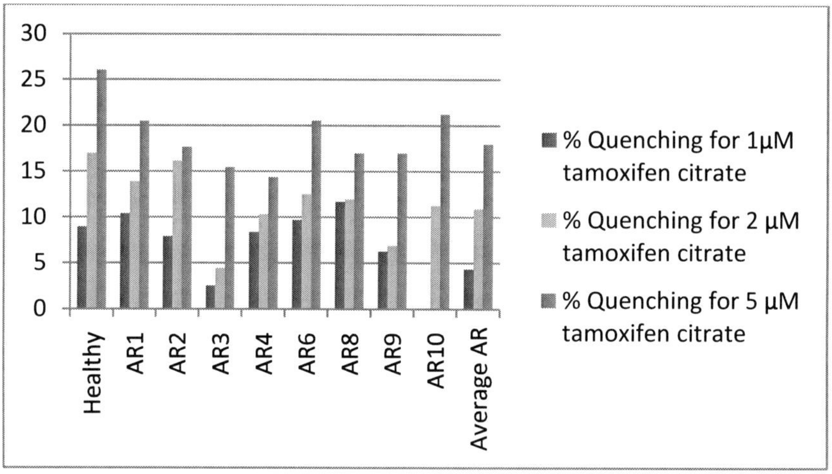

Figure 6. % quenching values for AGP isolated from the "at risk of" breast cancer population with tamoxifen citrate.

Table 6. % quenching values for AGP isolated from the breast cancer population treated with tamoxifen with tamoxifen citrate

Sample	% Quenching of 1µM tamoxifen citrate	% Quenching of 2 µM tamoxifen citrate	% Quenching of 5 µM tamoxifen citrate	Correlation coefficient
Healthy	8.96	16.97	26.07	0.94
TT4	9.83	15.35	21.11	0.93
TT5	10.62	13.56	18.16	0.98
TT6	6.29	8.82	18.21	0.99
TT7	7.23	10.41	13.01	0.89
TT8	7.33	16.80	20.91	0.76
TT9	7.05	9.44	14.75	0.99
Average TT	8.06	12.39	17.69	-

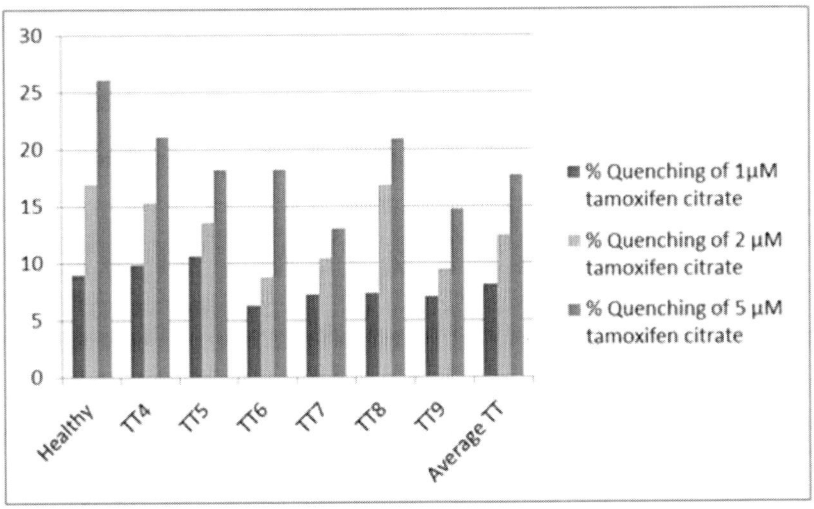

Figure 7. Graphical representation % quenching values for AGP isolated from the breast cancer population treated with tamoxifen with tamoxifen citrate.

Conclusion

Increased levels of AGP have been detected in the plasma of patients with breast cancer (Duche et al, 2000) and have been shown to increase with disease progression (Thompson et al, 1983). Alexander et al., (2004), analysed proteins found in nipple aspirate fluid (NAF). They found that patients with breast cancer had higher levels of AGP in NAF. Thus, the prognosis of breast cancer appears to be linked to the APR in general and AGP levels in particular.

The present study found that the plasma concentration of AGP in at risk individuals was significantly increased ($p<0.05$) compared to normal individuals. This indicates that although the "at risk" subjects appear healthy, they have a form of underlying inflammatory response present, leading to this rise in AGP concentration observed. On average the untreated and tamoxifen treated breast cancer population had significantly increased levels of AGP compared to healthy donors ($p<0.05$). The concentrations observed are similar to those reported by Duché et al. (2000) who found that individuals with breast cancer, who had undergone chemotherapy, had a 2.5 fold increase in AGP concentration compared with the control group. On sub-dividing the "at risk" and tamoxifen treated breast cancer populations, it was found that there were no significant differences ($p>0.05$), in the AGP concentration in African-American women compared with Caucasian women.

In terms of the effect of cancer therapy on AGP concentrations, Kailajarva et al. (2000) reported that treatment with tamoxifen returns AGP levels to normal with 5-6 weeks of therapy, with a further decrease at three months. Thus the plasma AGP concentrations measured in this study may not be representative of an acute phase response induced by breast cancer. However the untreated breast cancer population showed virtually the same concentrations of plasma AGP, thus differing with the findings of Kailajarva et al (2000). It further proves that protein levels alone are not specific enough to serve as a diagnostic marker for breast cancer. This study also found that there was no correlation between time on tamoxifen and AGP concentrations in the plasma.

In the tamoxifen treated breast cancer group, the African-American patients have higher (but not statistically significant) AGP concentrations and have been on tamoxifen for a shorter period of time, 20.6 months, compared to the Caucasian subgroup who had on average received tamoxifen therapy for 21.2 months. This very small difference in treatment period is probably not enough to explain the difference in AGP levels observed. Any potential differences between ethnic groups have been highlighted as it has been found that the risk of dying from breast cancer differs with racial origin. It has been indicated that tamoxifen is more effective in Caucasian women, compared to African-Americans (Flaws and Bush, 1998). This evidence would support findings in this study that the African-American breast cancer patients on tamoxifen have higher AGP concentrations and this could be due to the tamoxifen being less effective in this cohort.

Also it has been established that age is a determining factor in the response to tamoxifen therapy. Sheth et al., (2003), found a positive correlation between age and concentration of tamoxifen in the plasma, with older women at an increased risk of side effects. This study found that there was no correlation between age and AGP plasma concentration in any of the patient populations.

The concentration of AGP as a clinical parameter is not specific enough to distinguish breast cancer from any other inflammatory disease. Also, AGP concentrations are not correlated to the time on tamoxifen therapy, thus monitoring AGP concentration alone would not be sufficient to monitor patient response to tamoxifen therapy. Therefore more evidence is required to make a diagnosis of breast cancer and also to monitor a patients response to therapy with tamoxifen, thus this study focussed on alterations to the glycosylation of AGP and specifically if these changes are of functional significance.

It is known that AGP has 5 N-linked oligosaccharide chains (Fournier et al, 2000) which, in terms of monosaccharide composition, should result in varying proportions of the monosaccharides fucose, galactose, glucosamine (indicating the presence of N-acetylglucosamine) and mannose in each AGP sample. On average the monosaccharides present on the AGP isolated from "at risk" (AR) patient are very similar to that found in healthy plasma. However the AR population shows significantly higher ($p<0.05$) levels of fucose compared to normal plasma. This indicates that although the at risk group appear to be healthy individuals, there may be an underlying condition such as inflammation, that may even be asymptomatic.

The untreated breast cancer population shows significantly higher ($p<0.05$) levels of the monosaccharides glucosamine and galactose, compared with those found in healthy plasma which tends to indicate a higher degree of branching of the glycans chains. Of particular importance is the presence of fucose in the untreated breast cancer samples. Fucose is not present on healthy individuals AGP, although the difference between the two groups was found not to be statistically significant ($p>0.05$), and thus is a marker of disease and the APR. All samples in this category have fucose present, indicating the presence of an APR, which would be associated with breast cancer. The average fucose content is higher than that of the tamoxifen treated breast cancer population, although not statistically significant ($p>0.05$). This could simply be due to the fact that this group of patients has not received any treatment for their breast cancer. The untreated samples also show significantly higher ($p<0.05$) levels of all monosaccharides compared with the at risk group, indicating a difference in glycosylation structure. Untreated samples also show significantly increased ($p<0.05$) levels of glucosamine compared with tamoxifen treated breast cancer patients, indicating the presence of higher branching structures present on AGP in the untreated population.

Tamoxifen treated breast cancer patients show significantly increased ($p<0.05$) levels of all the monosaccharides compared with AGP isolated from healthy individuals, indicating disease specific glycosylation is present in this population. Fucose is only found in four samples in this group. Sample TT2 has a high fucose content, 0.73mole/mole AGP, compared to normal plasma. Sample TT2 had received tamoxifen for 4 months when the blood sample was taken. This sample suggests the idea that early on in tamoxifen treatment, patient AGP is going to be abnormal, thus reflecting the presence of the APR in response to the cancer and also the surgery to remove the tumour. In contrast, Samples TT9 and TT10 have no fucose present. These patients had received 51 and 54 months tamoxifen treatment respectively, out of a total treatment time of 5 years, so both patients were almost finished the recommended course of treatment. The lack of fucose in these samples suggests that the glycosylation of the AGP is back to "healthy" profile following the onset of cancer perhaps suggesting that the cancer is in remission. This finding would support the clinical recommendations of a 5 year treatment course of tamoxifen to prevent a recurrence of breast cancer.

Oligosaccharide profiles were determined using HPAEC, which allows oligosaccharides to be separated effectively and is able to take into account single linkage and single residue differences (Behan and Smith, 2011). A major advantage of HPAEC is that no prior derivitisation is required and that picomole amounts can be detected. HPAEC provides useful information about the oligosaccharide structures present on AGP and allows the development of an oligosaccharide 'fingerprint'. In particular the technique allows the changes in glycosylation of AGP to be observed that occur in breast cancer and specifically the differences between untreated patients and patient who have received tamoxifen therapy. This analysis of oligosaccharide chains allowed the effect of tamoxifen treatment to be observed over the course of standard 5 year treatment period. AGP from tamoxifen-treated breast cancer patients showed altered glycosylation, which changed over the course of tamoxifen therapy, with a return to the profile expected for 'healthy' individuals towards the end of treatment. AGP from untreated breast cancer patients displayed a shift to bi-siaylated oligosaccharide chains compared with AGP form healthy individuals.

At physiological pH, AGP is capable of high affinity, low capacity binding of basic and neutral drugs (Israili and Dayton, 2001; Fournier, 2000). The binding of a drug to AGP would reduce the free drug concentration present in the plasma to a level below that of the minimum effective concentration, thus making the drug incapable of exerting a therapeutic effect. This study made use of the intrinsic fluorescence of AGP imparted by tryptophan residues found in the protein backbone of AGP. When a drug binds to the protein the peak fluorescence is quenched (reduced) because the drug masks the fluorescent emission of tryptophan. If binding does not occur, quenching is not observed. There are several advantages of this technique over dialysis methods. These include the sensitivity, rapidity and accuracy of fluorescence spectroscopy compared to other methods (Epps et al, 1999). The bound and unbound fractions do not have to be separated and their concentrations determined separately as they do with dialysis methods. Additionally hydrophobic drugs may bind to dialysis membranes (Parikh et al, 2000).

Tamoxifen has been very successful in the treatment of breast cancer and thus has been the gold standard in terms of treatment for the last 30 years. However resistance to the treatment has increasingly been reported. Therefore it was important to determine if tamoxifen binding to AGP was responsible for resistance to therapy. The present study investigated any potential tamoxifen-AGP interactions. The AGP under investigation was isolated from healthy individuals, those at risk of breast cancer, those with untreated breast cancer and subjects with breast cancer receiving tamoxifen therapy.

Concentration dependent quenching was observed with all populations of AGP with increasing concentrations of tamoxifen. The increase in percentage quenching was statistically significant ($p<0.05$) in all cases except between 1μM and 2μM tamoxifen binding to AGP for at risk patients. No statistical significance ($p>0.05$) was found on comparing the untreated breast cancer population with AGP from normal individuals at any of the concentrations of tamoxifen examined. In contrast, the "at risk" and tamoxifen treated breast cancer populations showed statistically significantly ($p<0.05$) lower binding with the 2μM and 5μM concentrations of tamoxifen compared to normal individuals. This means that there will be more free drug present in the plasma and thus could lead to toxicity as the plasma levels of tamoxifen could potentially exceed the therapeutic range. The untreated breast cancer population showed significantly ($p<0.05$) higher binding compared to the at risk population for both the 2 and 5 μM tamoxifen concentrations. Comparing the untreated breast cancer population with the tamoxifen treated population, only binding between AGP and 5μM tamoxifen showed any statistical significance, with the untreated breast cancer population displaying significantly higher ($p<0.05$) percentage quenching. This implies that the ability of AGP to bind tamoxifen changes at a later stage in breast cancer progression. This is possibly due to the upregulation of the APR resulting in a change to the glycosylation present on AGP.

Studies with the "at risk" patient population found no correlation between the time on tamoxifen and drug binding at any of the three drug concentrations. This was to be expected because patients do not have breast cancer therefore there should not be disease induced changes in the glycosylation to affect drug binding. The glycosylation of this patient population was found to be similar to that of healthy plasma. In contrast, studies with the tamoxifen treated breast cancer population revealed a positive correlation between time on tamoxifen and drug binding at all three drug concentrations, with the strongest correlation found with 5μM of tamoxifen. It was found that there is a decrease in drug binding with increasing time on tamoxifen. This correlates with alterations in the glycosylation. Patients who had received only a few months of tamoxifen therapy had more highly altered AGP glycosylation than those who had received treatment for longer periods of time. This correlates with the findings of the drug binding experiments in that patients with more altered glycosylation are able to bind tamoxifen differently than patients with glycosylation that more closely resembled that of healthy plasma.

In conclusion, the present study aimed to determine if AGP could represent a mechanism of resistance in breast cancer chemotherapy through physically binding to tamoxifen and whether this was correlated with glycosylation pattern. Compared to healthy plasma all patients show increased plasma AGP concentrations, which decreases with increased severity of disease. In terms of glycosylation, the presence of the monosaccharide, fucose, varies with the prognosis of breast cancer and there are also alterations in the populations of oligosaccharide chains. In terms of AGP from the healthy population, AGP bound to tamoxifen citrate, as evidenced by the quenching of fluorescence, to a degree which increased with increasing concentration. AGP from breast cancer patients, both untreated and tamoxifen treated, showed less binding than healthy AGP. Although binding was minimal at therapeutic concentrations, the decrease in binding by breast cancer patients could lead to issues regarding toxicity as there would be more free drug in the plasma and therefore this concentration could exceed the therapeutic range. These findings could lead to the development of a breast cancer biomarker which would aid early diagnosis and thus improve prognosis and could also be used to determine drug effectiveness.

References

Alexander, H., Stegner, A.L., Wagner-Mann, C., Du Bois, G.C., Alexander, S. and Sauter, E.R., (2004) Proteomic analysis to identify breast cancer biomarkers in nipple aspirate fluid. *Clinical Cancer Research,* 10, 7500-7510.

Anderson, N., Pollacchi, A., Hayes, P., Therapondos, G., Newsome, P., Boyter, A. and Smith, K. (2002) A preliminary evaluation of the differences in the glycosylation of alpha-1-acid glycoprotein between individual liver diseases. *Biomedical Chromatography*, 16: 365-372.

Behan, J. and Smith, K.D. (2011) The analysis of glycosylation: A continued need for high pH anion exchange chromatography. *Biomed. Chrom.,* 25(1-2):39-46.

Ceciliani, F. and Pocacqua, V. (2007) The acute phase protein α_1-acid glycoprotein: a model for altered glycosylation during diseases. *Current Protein and Peptide Science*, 8: 91-108.

Chew H.K, Miyamoto, S., An, H., Rocke, D. Lebrilla, C. (2007). Serum glycan analysis in metastatic breast cancer. *J. Clin. Oncol.,* 25, 11504.

Dorssers, L.C.J., van der Flier, S., Brinkman, A., van Agthoven, T., Veldscholte, J., Berns, E.M.J.J., Klijn, J.G.M., Beex, L.V.A.M. and Foekens, J.A. (2001) Tamoxifen resistance in breast cancer: elucidating mechanisms. *Drugs*, 61, 1721-1733.

Duché, J., Urien, S., Simon, N., Malaurie, E., Monnet, I. and Barré, J. (2000) Expression of the genetic variants of human alpha-1-acid glycoprotein in cancer. *Clinical Biochemistry*. 33: 197-202.

Elliot, M.A., Elliot, H.G., Gallagher, K., McGuire, J., Field, M. and Smith, K.D. (1997) Investigation into the Concanavalin A reactivity, fucosylation and oligosaccharide microheterogeneity of α_1-acid glycoprotein expressed in the sera of patients with rheumatoid arthritis. *Journal of Chromatography B*, 688: 229-237.

Epps, D.E., Raub, T.J., Caiolfa, V., Chiari, A. and Zamai, M. (1999) Determination of the affinity of drugs toward serum albumin by measurement of the quenching of the intrinsic tryptophan fluorescence of the protein. *J. Pharm. Pharmacol.,* 51, 41-48.

Flaws, J.A. and Bush, T.L. (1998) Racial differences in drug metabolism: an explanation for higher breast cancer mortality in blacks? *Medical Hypotheses*, 50, 327-329.

Fournier, T., Medjoubi-N, N. And Porquet, D. (2000) Alpha-1-acid glycoprotein. *Biochimica et Biophysica Acta*, 1482: 157-171.

Hashimoto, S., Asao, T., Takahashi, J., Yagihashi, Y., Nishimura, T., Saniabadi, A.R., Poland, D.C.W., van Dijk, W., Kuwano, H., Kochibe, N. and Yazawa, S. (2004) α_1-acid glycoprotein fucosylation as a marker of carcinoma progression and prognosis. *Cancer*, 101: 2825-2836.

Israili, Z.H. and Dayton, P.G. (2001) Human alpha-1-glycoprotein and its interactions with drugs. *Drug Metabolism Reviews*, 33: 161-235.

Jordan V.C. (2007) SERMs: meeting the promise of multifunctional medicines. *Journal of the National Cancer Institute*, 99, 350-356.

Kailajärvi, M., Ahokoski, O., Virtanen, A., Salminen, E. and Irjala, K. (2000) Early effects of adjuvant tamoxifen therapy on serum hormones, proteins and lipids. *Anticancer Research*, 20, 1323-1328.

Kremer JM, Wilting J, and Janssen LH (1988) Drug binding to human alpha1-acid glycoprotein in health and disease. *Pharmacol. Rev.,* 40: 1-47.

Kute T and Westphal U. (1976) Steroid-protein interactions. XXXIV. Chemical modification of alpha1-acid glycoprotein for characterization of the progesterone binding site. *Biochim. Biophys. Acta.,* 420 195-213.

Leiserowitz GS, Lebrilla C, Miyamoto S, An HJ, Duong H, Kirmiz C, Li B, Liu H, Lam KS. (2008). Glycomics analysis of serum: a potential new biomarker for ovarian cancer? *Int. J. Gynecol. Cancer,* 18, 470-5.

Macgregor, J.I. and Jordan, V.C. (1998) Basic guide to the mechanisms of antiestrogen action.*Pharmacological Reviews*, 50, 151-196.

Moule, S.K., Peak, M., Thompson, S. and Turner, G.A. (1987) Studies of the sialylation and microheterogeneity of human serum α_1-acid glycoprotein in health and disease. *Clinica Chimica Acta*., 166: 177-185.

Parikh, H.H., McElwain, K., Balasubramanian, V., Leung, W., Wong, D., Morris, M.E. and Ramanathan, M. (2000) A rapid spectrofluorimetric technique for determining drug-serum protein binding suitable for high-throughput screening. *Pharmaceutical Research*, 17, 632-637.

Paterson, S.C., Lim, C.K. and Smith, K.D. (2002) Analysis of the interaction between alpha-1-acid glycoprotein and tamoxifen and its metabolites. *Biomedical Chromatography*, 17,143-148.

Rose C. (2008) Increasing protection after tamoxifen: insights from the extended adjuvant aromatase inhibitor trials *Journal of cancer research and clinical oncology*, 134, 7-17.

Sheth, H.R., Lord, G., Tkaczuk, K., Danton, M., Lewis, L.M., Langenberg, P., Lim, C.K. and Flaws, J.A. (2003) Aging may be associated with concentrations of tamoxifen and its metabolites in breast cancer patients. *Journal of Womans Health*, 12, 799-808.

Silva, O.E. and Zurrida, S. (2005) Breast Cancer A Practical Guide, 3rd edtion

Smith, K.D., Pollacchi, A., Field, M. and Watson, J. (2002) The heterogeneity of the glycosylation of alpha-1-acid glycoprotein between the sera and synovial fluid in rheumatoid arthritis. *Biomedical Chromatography*. 16: 261-266.

Tesseromatis C., Alevizou A., Tigka E. and Kotsiou A. (2011) Acute-Phase Proteins: Alpha -1- Acid Glycoprotein Acute Phase Proteins - Regulation and Functions of Acute Phase Proteins ISBN 978-953-307-252-4 Edited by: Francisco Veas Publisher: InTech.

Thompson, D.K., Haddow, J.E., Smith, D.E. and Ritchie, R.F. (1983) Elevated serum acute phase protein levels as predictors of disseminated breast cancer. *Cancer*, 51, 2100-2104.

Turner, G.A., Skillen, A.W., Buamah, P., Guthrie, D., Welsh, J., Harrison, J. and Kowalski, A. (1985) Relation between raised concentrations of fucose, sialic acid and the acute phase proteins in serum from patients with cancer: choosing suitable serum glycoprotein markers. *Journal of Clinical Pathology*. 38: 588-592.

Yost, R.L. and DeVane, C.L., (1985). Diurnal variation of alpha 1-acid glycoprotein concentration in normal volunteers.*Journal of Pharmaceutical Sciences*, 74, 777-779.

In: Oligosaccharides
Editors: L. S. Schweizer, S. J. Krebs
ISBN: 978-1-62948-328-3

Chapter VI

Production of Oligosaccharides with Low Molecular Weights, Secoiridoids and Phenolic Glycosides from Thermally-Treated Olive By-Products

Juan Fernández-Bolaños**, *Fátima Rubio-Senent*, *Antonio Lama-Muñoz*, *Aránzazu García* and *Guillermo Rodríguez-Gutiérrez
Instituto de la Grasa,
Consejo Superior de Investigaciones Científicas (CSIC),
Sevilla, Spain

Abstract

Alperujo was hydrothermally pre-treated to solubilize the oligosaccharides from the cell wall material of olive pulp. A protocol for the isolation of oligosaccharides of low molecular weight has been developed, which involves thermal treatment, mild acid hydrolysis, separation by ultra-filtration, adsorption chromatography and size exclusion chromatography using Superdex Peptide HR. Three groups of

* Corresponding author: Tel: 34-954691054; Fax 34-954691262; E-mail: jfbg@cica.es.

fractions with common characteristics were obtained and studied in two samples of alperujo from different seasons, origin and variety of olives. In the first group, neutral and acidic oligosaccharides with a low DP (4-10) bound or not to phenol-containing compounds and with antioxidant activity were identified. In the second group, several secoiridoid glycosides esterified to phenolic compounds and their aglycones were also detected. In the third group, a significant fraction of di- and trisaccharides of glucose or glucose-xylose bound to phenol residues (hydroxytyrosol and tyrosol) that could be considered antioxidant-soluble fiber with bioactive properties was also isolated. All of them were identified by HPLC, CG, and ESI-MS and in certain cases by ^{1}H and ^{13}C MNR. The similar behavior obtained for both samples confirms the protocol followed as a useful tool for bioactive compound isolation.

Keywords: Alperujo, steam treatment, secoiridoids, xylo-oligosaccharides, xyloglucan, arabinans, antioxidants, phenolic compounds

1. Introduction

Alperujo is a semisolid, residual by-product of virgin olive oil produced by a continuous, two-phase extraction system. In Spain, the annual production of this by-product is approximately 2.5-6 million tons, depending on the season. From an environmental point of view, alperujo represents a serious problem due to its highly polluting organic load which limits its biodegradation because of its high toxicity. The combustion of alperujo as a fuel is a commonly used procedure to eliminate its harmful effects on the environment. In recent years, many management options have been proposed for the treatment and valorization of alperujo. One of the most interesting is a hydrothermal treatment developed and patented by our research group [1] already implemented at industrial scale. This treatment is conducted in a continuous reactor using steam at high pressures and temperatures (150-170ºC, 5-8 kg/cm^2) and allows for easy separation of the solid and liquid phases and for recovery of the bioactive compounds and valuable fractions from each phase. In this steam treatment, an auto-hydrolytic process occurs and solubilizes part of the alperujo. It allows for the recovery of added-value compounds in the water-soluble fraction to create a fraction which is rich in valuable phenols (hydroxytyrosol, 3,4-dihydroxyphenylglycol and others) and sugars leaving a solid residue with oil which has been enriched with minor components [2, 3].

Recent studies have shown that a substantial part of the hemicelluloses present in alperujo could be also recovered as oligosaccharides by steam treatment or auto-hydrolysis of the water-soluble fraction [4]. Cell wall material for olive pulp comprises a number of molecular components, with considerable quantities of cellulose (30%), pectic polysaccharides (39%) and hemicellulosic polymers rich in xylans and glucuronoxylans (14%), xyloglucans (15%) and mannans (2%) [5, 6, 7], which could provide a wide range of functional components.

The phenolic compounds hydroxytyrosol, tyrosol or secoiridoid derivatives are abundant in olive pulp but were not identified in the olive cell wall [8]. However, it has been reported that the interaction between polysaccharides and the hydrophilic compounds present in the olive paste may be involved in the apparent loss in the phenolic compounds from the oil during crushing and malaxation [9]. Thermal treatment can provide a large number of bioactive compounds, including oligosaccharides and phenolics associated with polysaccharides, with potential applications in the food or nutraceutical industry. The optimization and integration of the process for revaluing alperujo, which in this case can also be used for the production of this type of oligosaccharides for intermediate pharmaceutical or food use, could be an important economical improvement.

There is increasing interest in generating non-digestible oligosaccharides that would promote the growth of beneficial bacteria in the large intestine, providing protection from infection and maintaining the normal function of the gastro-intestinal system [10, 11]. Small oligomers of galacturonic acid (with degrees of polymerization [DP] from 2 to 7) were also found responsible for inhibiting the adherence of bacteria to epithelial cells, the initial and crucial stage of infection [12]. Xylo-oligosaccharides act as pre-biotics, maintaining gastrointestinal health and reducing the risk of colon cancer [13]. These pre-biotics reduce cholesterol levels, improve the biological availability of calcium and have beneficial effects on type 2 diabetes mellitus; in addition, they cause the apoptosis of human leukemia cells [14]. The acidic xylo-oligosaccharides show antibacterial activity against Gram-positive bacteria, especially against *Helicobacter pylori* [15] (Christakopoulus et al., 2003) and direct mitogenic activity, indicating a potential inmunostimulatory [16] (Nabarlatz, Montané, Kardosova, Bekesová, Hribalová, & Ebringerová, 2007).

The acidic and neutral oligosaccharides, together as secoiridoid compounds and mono- and disaccharides linked to phenolic compounds, released from alperujo that was steam-treated in a bath pilot reactor (capacity of 100 L), following mild acid hydrolysis has been reported in a previous paper [17, 18]. Process definition and results obtained on different raw material and on a larger scale for the ultra-filtration steps are presented here and compared with previous results.

2. Materials and Methods

Raw Material

Two samples of olive pomace or "alperujo" (a semisolid residue composed of skin, pulp, ground stone and olive seeds) from different olive cultivars and consecutive seasons were directly collected from the two-phase centrifugal system decanter before recovery of the olive-pomace oil. One of them (Sample1) was supplied by the local pomace-processing mill (Oleícola el Tejar, Córdoba, Spain), in January, at the halfway mark of the olive oil production season (2009-2010), and the other (Sample2) was taken in an experimental olive oil mill plant located at the Instituto de la Grasa (CSIC) in Seville (Spain), in December, also in the half-way mark of the olive oil production season (2010-2011) in the province of Seville.

Thermal Treatments or Steam Processing of Alperujo

The hydrothermal treatment has been patented [1], and it was performed using a steam treatment reactor prototype designed by our research group at the Instituto de la Grasa (Seville, Spain). The reactor has 100 L capacity and a stainless steel reservoir that can operate at temperatures between 50 and 190°C and at a maximum pressure of 1.2 MPa.

Twenty kilogram samples of alperujo were treated with saturated steam for 15 min at 170°C (Sample 1) and 60 min at 170ºC (Sample 2). Heating of the alperujo was performed by direct injection of pressure and temperature steam enhancing the contact between the steam and the alperujo. After treatment the steam inlet valve was closed and the pressure reduced to atmospheric pressure at a controlled rate. The wet material was centrifuged at 4700 g (Comteifa, S.L., Barcelona, Spain) to separate the solid and liquid phases.

Phenol Extraction

At the end of the treatment, an aliquot of the liquid phase (10 L) was concentrated to 1 L by rotary evaporation a vacuum at 37ºC; five portions of 200 mL were extracted with 500 mL volumes of ethyl acetate (refluxed at 77ºC) for 5-6 h in a continuous extraction from a heavier liquid (water) to a lighter one (ethyl acetate) in order to separate the phenolic compounds from the carbohydrate components. The aqueous solution was concentrated and lyophilized.

Acid Hydrolysis and Ultra-Filtration through a Membrane of 3000 Da

In order to increase the proportion of oligosaccharide of low molecular weight in the initial lyophilized liquid, a mild chemical hydrolysis with 0.5 N HCl was carried out [17].

In Sample 1 a small volume (100 mL) of hydrolyzed fraction was ultra-filtered at room temperature using an Amicon 8400 stirred cell (Millipore Corp., Bedford, MA. USA) through 3000 Da molecular weight cut off membranes (cellulose regenerated). The solutions were washed with water at 40ºC until 500 mL of permeate were collected.

In Sample 2, hydrolyzed fractions (1000 mL) were ultra-filtered at room temperature using a Prep/Scale®-TFF Cartridge holder XX42PS001 with Prep/Scale®-TFF Cartridge 3000 Da regenerated cellulose (Millipore Corp., Bedford, MA, USA). The solutions were washed with 5000 mL of water at 40ºC and concentrated under vacuum at 40ºC until obtaining a final volume of 150 mL.

Adsorption XAD Chromatography

Fractionation and purification of the neutral and pectic oligosaccharides were performed by adsorption on Amberlite XAD-16 resin (Figure 1). 25 mL of hydrolyzed sample (Sample 1) were injected onto a column (35 x 3.0 cm) of Amberlite XAD-16 and eluted at 1.5 mL/min with water (200 mL) and 20% (v/v) methanol/water (200 mL), successively. In Sample 2 the fraction obtained after ultra-filtration with a membrane of 3000 Da and concentrated to 700 mL was injected onto a column (74.5 x 8.0 cm) of Amberlite XAD-16 and eluted with water (10 L) and methanol/water 20% (v/v) (7 L) and 50% (v/v) (5 L), successively.

The carbohydrates and uronic acids in each eluted fraction were assayed with the anthrone-sulphuric acid colorimetric assay [19] and the m-hydroxybiphenyl method [20], respectively.

Quantification of Individual Neutral Sugars by Gas Chromatography (GC) after Converting Them into Alditol Acetates

Individual neutral sugars were analyzed from duplicate samples with and without initial TFA hydrolysis (2N TFA at 121ºC for 1h) prior to reduction, acetylation and analysis by GC [21]. Inositol was used as an internal standard. A Hewlett-Packard 5890 series II chromatograph, fitted with a 30 m x 0.25 mm fused silica capillary column (SP-2330 from Supelco, Bellefonte, PA), was employed. The oven temperature program used was as follows: 180ºC for 7 min and then increased to 220ºC at 3ºC/min (15 min). The carrier gas was helium, which had a flow rate of 1 mL/min. The injector temperature was 250ºC, and the FID detector temperature was 300ºC. The split ratio was 1/100.

Because oligosaccharides were converted into monosaccharides during the post-hydrolysis treatment, the increase in the concentration of monosaccharides caused by post-hydrolysis provided a measurement of the concentration of oligosaccharides. All determinations were made by triplicate.

Ultra-Filtration through a Membrane of 1000 Da

The fractions eluted from Amberlite XAD-16 resin with water and methanol/water were subjected to ultra-filtration through a 1000 Da molecular weight cut-off membrane on a fiber unit similar to the one described above for Samples 1 and 2, respectively. In Sample 1 four fractions were obtained: Fraction A (<1000 Da), Fraction B (between 1000-3000 Da), eluted with water; Fraction C (<1000 Da) and Fraction D (between 1000-3000 Da), eluted with 20% methanol. In Sample 2 six fractions were obtained: Fraction A (<1000 Da), Fraction B (between 1000-3000 Da), eluted with water; Fraction C (<1000 Da) and Fraction D (between 1000-3000 Da), eluted with methanol 20%; Fraction E (<1000 Da) and Fraction F (between 1000-3000 Da), eluted with methanol 50%.

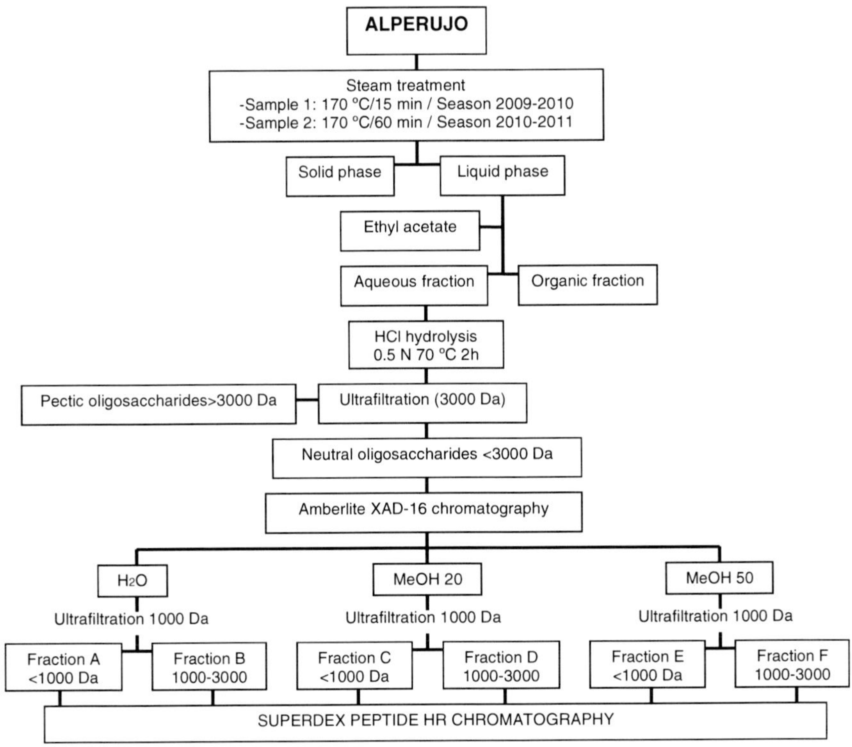

Figure 1. Scheme of the fractionation process performed to obtain oligosaccharides with a low degree of polymerization.

Determination of the Total Phenolic Content and Antioxidant Activity

Total phenols

The phenolic content was measured by the Folin-Ciocalteu method [22], and expressed as grams of gallic acid equivalents per kilogram of fresh alperujo.

Antiradical activity, DPPH (2,2-diphenyl-1-picrylhydrazyl)

Free-radical scavenging capacity was measured using the DPPH method described in a previous work [23]. The method is based on the measurement of the free-radical scavenging capacity of the antioxidant against the stable radical $DPPH^{\cdot}$. An iMark^{TM} microplate absorbance reader model 550 (Bio-Rad, Hercules, CA, USA) was used for the absorbance measurements. $DPPH^{\cdot}$

has an absorption band at 515 nm which disappears upon reduction by antiradical compounds. For each fraction and standard, the decrease in absorbance (expressed as a percentage of the initial absorbance) was plotted against the concentration of the antioxidant in the reaction mixture. The radical-scavenging capacity of each fraction was expressed as EC_{50} (effective concentration, mg/mL), which is the amount of antioxidant necessary to decrease the initial absorbance by 50%.

Reducing power

The reducing power assay was carried out according to the procedure described in a previous work [23]. The antioxidative potentials of the fractions were estimated for their ability to reduce $FeCl_3$. All of the analyses were made using a microplate reader and the absorbance was measured at 490 nm. The assay was calibrated using Trolox. To express the results, a calibration curve was established by plotting A_{490} against known concentrations of Trolox (0.059 to 0.56 mg/mL). The reducing power (P_R) was expressed as Trolox equivalents (TE) in mg/mL from the equation as determined from linear regression.

Size Exclusion Chromatography on Superdex Peptide HR

After the separation of the monosaccharides and fractionation of the oligosaccharides by adsorption chromatography, the fractions A, B, C and D of Sample 1 and the fractions A, B, C, D, E and F of Sample 2 were fractionated by size exclusion chromatography on two Superdex Peptide HR 10/30 (30 x 1 cm) columns (Pharmacia Biotech, Uppsala, Sweden) connected in line to a Jasco LC-Net II/ADC (Easton, MD, USA) HPLC system. Samples (100 µL) containing 10 mg of oligosaccharide/mL were applied several times to the column and eluted with 100 mM ammonium acetate buffer (pH 5) and water in Samples 1 and 2, respectively, at a flow rate of 0.5 mL/min in both cases. The peaks were monitored by a Jasco MD-1550 Diode Array detector (DAD) and a Jasco RI-1530 Refraction Index detector. Column calibration was performed with a variety of standards, including galacturonic acids (RT= 69.6 min) (purchased from Fluka) and maltose (RT= 66.5 min), maltotriose (RT= 64.5 min), maltotetraose (RT= 63.2 min), maltopentaose (RT= 60.5 min), maltohexaose (RT= 59.1 min) and maltoheptose (RT= 57.0 min) (purchased from Sigma); glucose (RT= 69.5 min), and compounds of a series of cellobiose (RT= 66.5 min), cellotriose (RT= 63.5 min), cellotetraose (RT=

61.1 min), cellopentaose (RT= 59.0 min) and cellohexaose (RT= 56.2 min) (purchased from Megazyme) for neutral oligosaccharides.

Electrospray Ionization Mass Spectrometry

Neutral oligosaccharides collected from the Superdex Peptide column fractions were identified by their electron impact mass data, which were collected on a quadrupole mass analyzer (ZMD4, Micromass, Waters Inc; Manchester, U.K.). Electrospray ionization (ESI) mass spectra were obtained at an ionization energy of 50 and 100 eV in negative mode and at 50 eV in positive mode; the capillary voltage was 3 kV, the desolvation temperature was 200°C, the source temperature, was 100°C and the extractor voltage was 12V. The flow was maintained at 0.3 mL/min in splitless mode (UV detector MS) for each analysis.

A Teknokroma Tracer Extrasil ODS-2 column (250 mm x 4.6 mm, i.d. of 5 μm) was used. The mobile phase was 0.01% trichloroacetic acid in water and acetonitrile using the following gradient over a total run time of 55 min: 95% A initially, 75% A at 30 min, 50% A at 45 min, 0% A at 47 min, 75% A at 50 min, and 95% at 52 min until the run was completed.

NMR Spectroscopy

^{1}H and ^{13}C NMR spectra were recorded on a Bruker AMX300 spectrometer operating at 300 and 75 MHz, respectively. Samples were collected from Superdex Peptide column fractions at 62-68 min, which coincided with the standard oligosaccharides with DP from 4-2 in the calibration sample (first group), and fraction eluted with the monosaccharide standard, after solvent (negative peak) and another eluted after solvent (second group) was examined in acetylated form in CD_3OD and $CDCl_3$ respectively. Chemical shifts (δ) are expressed in parts per million (ppm) values relative to the central solvent peak TMS, tetramethylsilane, $(CH_3)_4Si$ (δ = 0,00 ppm). The experiments were performed at 25°C. *Acetylation*: Fractions were evaporated to dryness and 20 mg were dissolved in pyridine (2 mL) and cooled in an ice bath. Acetic anhydride (2 mL) was added to this solution and the mixture was left over night at 4° C. The resulting solution was evaporated to dryness and co-evaporated four times with water (2 mL) and toluene (2 mL) to remove the pyridine. The solid was dissolved in 0.5 mL of water. *Sample preparation*:

The acetylated samples (0.5 mL) were lyophilized and dissolved in isotopically enriched $2D_2O$ and lyophilized to replace exchangeable protons with deuterons. The residue was dissolved in methanol-d and chloroform-d and transferred to a 5 mm of NMR tube.

3. Results and Discussion

Extraction and Isolation of Oligosaccharides and Sugars Linked to Phenolics from Alperujo and their Antioxidant Activity

A steam treatment was applied to two different samples of alperujo collected from two different seasons. One of them was treated with saturated steam for 15 min at 170ºC and other for 60 min at 170ºC. In both samples the hydrothermal treatment of moist alperujo samples was followed by ethyl acetate extraction of the liquid phase in order to separate the phenolic compounds from carbohydrate components. A subsequent mild acid hydrolysis with 0.5 N HCl at 70ºC for 2 h was carried out to increase the proportion of oligosaccharides of low molecular weight. The overall fractionation procedure described in a previous paper [17] is presented in Scheme 1 (Figure 1). The mild hydrolysis with HCl did not hydrolyze most of the pectic polysaccharides and neutral oligosaccharides were isolated after an ultra-filtration through a 3000 Da membrane. The production of oligosaccharides was evaluated by the measurement of oligomers in the fraction below 3000 Da, calculated for the difference between those sugars which are completely free and those that are present as oligosaccharides. The monosaccharide compositions of the permeate and retentate of the ultrafiltration were determined by the total hydrolysis of the oligosaccharides with 2 N of TFA at 120ºC for 2 hours [24].

Analysis of the carbohydrates (Table 1) in fraction > 3000 Da revealed the presence mainly of pectic polysaccharides composed of galacturonic acid (46%) with galactose (28%) and rhamnose (9%). The fraction between < 3000 Da showed that they contained monosaccharides and partial hydrolysis products from xylan, xyloglucan, arabinan, arabinogalactan, glucomannan or mannan from the cell wall of the olive pulp.

However, in this work we concentrated on the study of this last fraction < 3000 Da that resulted from mild acid hydrolysis, when the oligosaccharides were efficiently separated from an important proportion of the monosaccharides by a simple chromatography step on XAD-16 resin and eluted with water.

Table 1. Composition of oligosaccharides (neutral sugars and uronic acids) in the fractions obtained after ultrafiltration through a 3000 Da membrane. The results are expressed in percentage with respect to the total content

	Sample 1		Sample 2	
	>3000 Da	< 3000 Da	>3000 Da	< 3000 Da
Uronic acids	45,96[a]±5,49[b]	n.d	18,92±1,61	8,02±0,23
Rhamnose	8,91[c]±0,85	1,42±0,08	2,96±1,28	6,32±0,11
Fucose	0,90±0,08	n.d	0,11±0,04	0,25±0,07
Arabinose	10,45±0,04	4,77±0,40	29,59±1,53	27,60±0,11
Xylose	2,74±0,22	0,79±0,08	27,91±1,21	17,42±1,62
Mannose	n.d [d]	27,90±2,70	1,89±0,09	1,80±0,08
Galactose	28,07±2,47	1,83±0,16	11,03±0,83	5,93±0,70
Glucose	2,97±0,22	63,59±6,20	7,60±0,39	32,67±1,57

[a] Uronic acids were determined following the method described by Blumenkrantz and Asboe-Hansen (1973).

[b] Standard deviation (n=3).

[c] Neutral sugars were quantified as alditol acetates by GC.

[d] n.d., not detected.

In Sample 1 four fractions (A, B, C, and D) were obtained after chromatography on Amberlite XAD-16 resin and after ultrafiltration through a 1000 Da molecular weight cut off membrane (Table 2). The fraction obtained by elution with water fraction A (< 1000 Da) and B (between 1000 and 3000 Da) were rich in glucose and mannose, with substantial proportion of arabinose in fraction A and arabinose and galactose in fraction B; the fractions obtained by elution with 20% methanol/water (C and D) contained predominantly glucose but also mannose and xylose.

In Sample 2 six fractions (A-F) were obtained since it was necessary to increase the proportion of MeOH to 50% for the complete elution of carbohydrates from Amberlite XAD-16 (Table 2) probably due to certain interactions of phenolic compounds with the resin. However, each fraction presented a certain content in phenolic compounds (Figure 2a) and they were tested *in vitro* assays to evaluate their radical scavenging activity and reducing power capacity using hydroxytyrosol (HT), α-tocopherol and Trolox as a positive control (Figure 2b, c). The results show that fractions eluted with 50

% MeOH were the fraction with a higher antiradical activity with an EC50 of 2.2 mg/mL for fraction F (between 1000 and 3000 Da) and 4.5 mg/mL for fraction E (< 1000 Da) with an EC50 for HT and α-tocopherol of 0.05 and 0.15 mg/mL, respectively. The antiradical capacity of all fractions had a high correlation with the phenolic content (r = 0.9579), following a lineal model (data not shown). The reducing power was low for all fractions and the statistical analysis confirmed that fractions A, C, D and E were not significantly different with α-tocopherol, and the fraction F had a higher reducing capacity than α-tocopherol. A high correlation with the phenolic content was also obtained (r = 09646), not following a lineal model in this case but an "inverse of x" (data not shown).

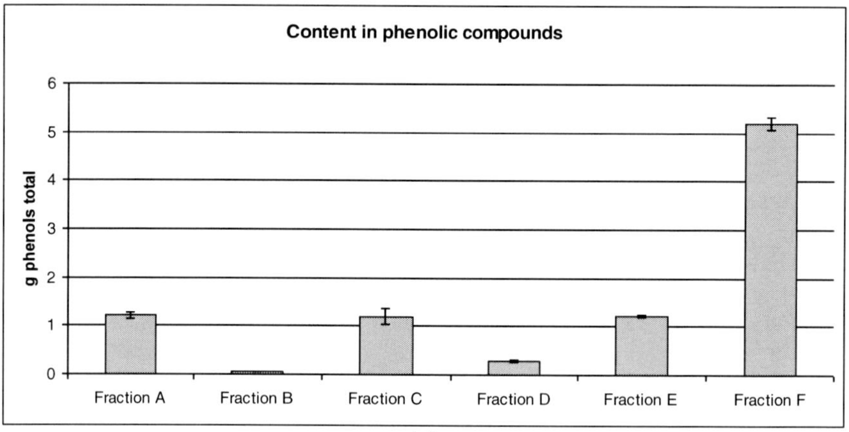

(a)

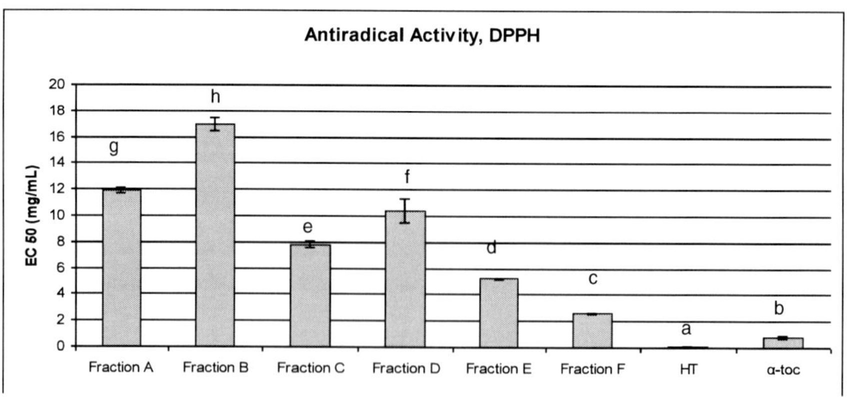

(b)

Figure 2. (Continued)

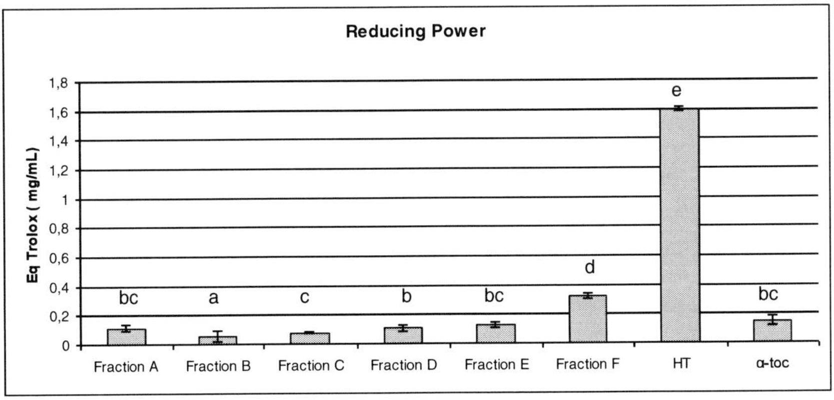

(c)

Figure 2. Content of total phenols (grams) presents in the different fractions obtained by elution with water and methanol 20 and 50% from XAD-16 chromatography and ultrafiltration for Sample 2. Radical scavenging capacity on DPPH (b) and reducing power (c) of each sample and two standards (hydroxytyrosol and α-tocopherol), data are presented as the means ± SDs. Different letters indicate significantly different results ($p < 0.05$).

Isolation and Separation of Three Groups of Compounds by Size Exclusion Chromatography

Four and six new fractions eluted from Amberlite XAD resin for Sample 1 and Sample 2, respectively were further separated and purified by gel filtration using a Superdex Peptide HR 10/30 resin column based on their molecular weight, although this resin also separates molecules on the basis of their interaction with phenolics. The refractive index elution profile obtained from the gel filtration column for Sample 2 is shown in Figure 3. In fact, we have confirmed that peaks eluted after the standard monosaccharide sample and solvent. Fractions corresponding to the peaks of refractive index were pooled and analyzed for their sugar content. In the case of faction F (between 1000 and 3000 Da), with higher antioxidant activity, the peaks were collected and the total phenolic content together with DPPH radical-scavenging ability and reducing power were determined (Table 3, Figure 4) for evaluating the contribution of each peak to antioxidant activity. These results imply a synergistic effect between fractions since any individual fraction has less activity than the former. Both antioxidant assays were correlated with phenolic content, with $r = 0.8781$ for antiradical activity, by lineal model, and a better correlation for reducing power (r = 0.8918), described by an "inverse of x" regression model. Therefore it appears that peak 4, with higher contents in phenolic compounds had greater activity while the peak of larger DP with lower phenolic contents and higher in oligosaccharides showed the lowest ability.

Table 2. Sugar composition of each fraction obtained after elution with water and methanol 20 and 50 % from XAD-16 chromatography and ultrafiltration through a 1000 Da membrane for each sample analyzed[a]

	Sample 1			
	A	B	C	D
Rhamnose	1,7±0,2[b]	3,8±0,1	0,9±0,1	3,6±0,3
Fucose	0,5±0,1	n.d[c]	n.d	n.d
Arabinose	13,4±0,3	8,3±0,5	2,0±0,2	6,1±0,5
Xylose	3,5±0,1	4,5±0,2	3,4±0,1	9,8±0,8
Mannose	19,9±1,4	17,2±1,5	8,5±0,8	9,9±0,9
Galactose	3,0±0,3	10,7±0,7	1,0±0,1	6,6±0,6
Glucose	58,1±2,7	55,5±1,7	84,3±8,3	64,0±6,3
% oligosaccharide	10,5	6,8	12	10,7

	Sample 2					
	A	B	C	D	E	F
Rhamnose	3,1±0,2	5,5±0,4	3,3±0,2	2,7±0,5	5,2±0,4	5,5±0,4
Fucose	--	1,4±0,1	0,2±,0,1	0,5±,1	0,4±,1	--
Arabinose	12,8±0,4	53,4±0,8	8,1±0,5	52,9±2,5	4,0±0,4	17,6±0,9
Xylose	3,5±0,1	15,1±0,2	0,9±0,1	25,3±1,4	8,0±0,8	9,7±0,5
Mannose	29,9±0,4	--	5,7±0,5	1,8±0,2	4,4±0,4	3,0±0,2
Galactose	1,4±0,1	5,5±0,3	0,5±0,1	2,7±0,1	0,4±0,1	18,2±1,8
Glucose	49,7±0,2	19,2±0,7	81,3±1,9	14,0±0,8	77,7±1,6	46,1±3,6
% oligosaccharide	26,6	42,3	33,1	60,5	19,1	40,9

[a] The composition is determined by gas chromatography as alditol acetate. The results are expressed in percentage with respect to the total content.
[b] Standard deviation (n=3).
[c] n.d., not detected.

In Sample 1 the results of its elution profile were similar to Sample 2 as were described in a previous paper [17]. In both cases, despite being of different samples and different treatment, we have found three groups of fractions with common characteristics.

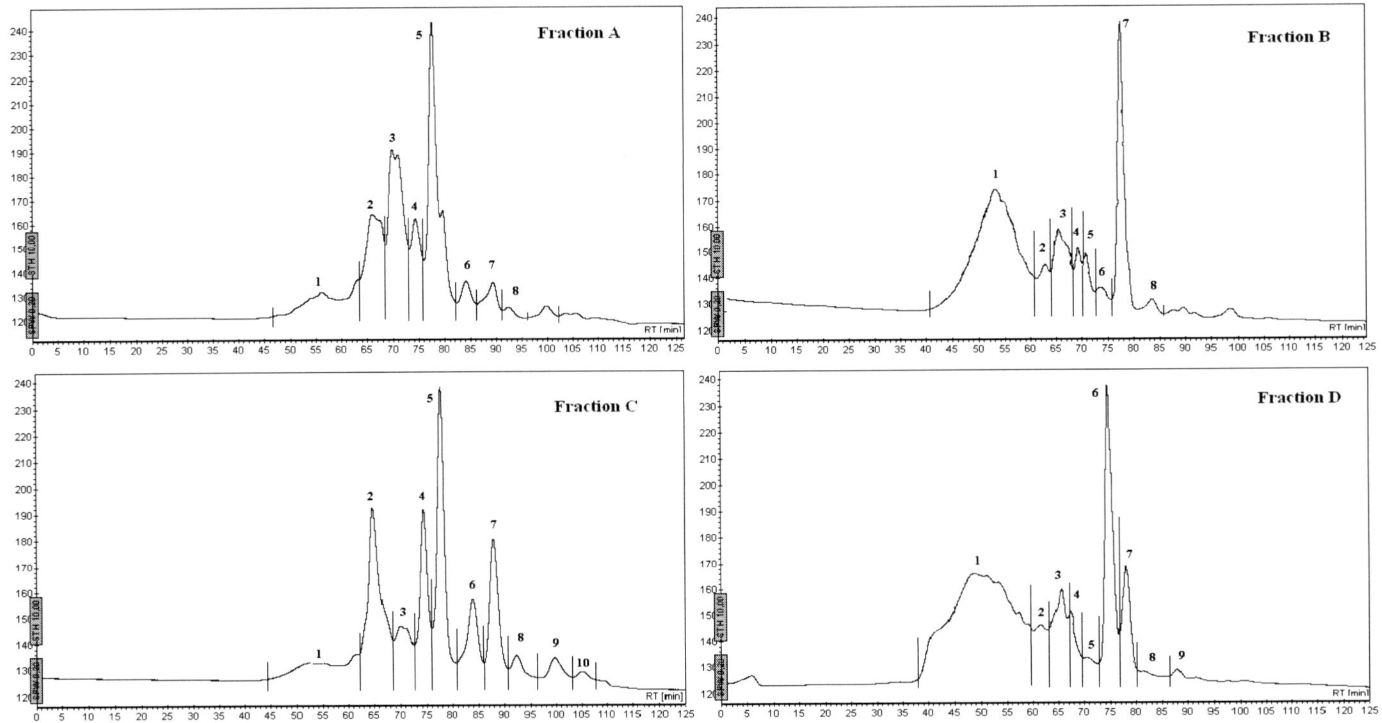

Figure 3. (Continued).

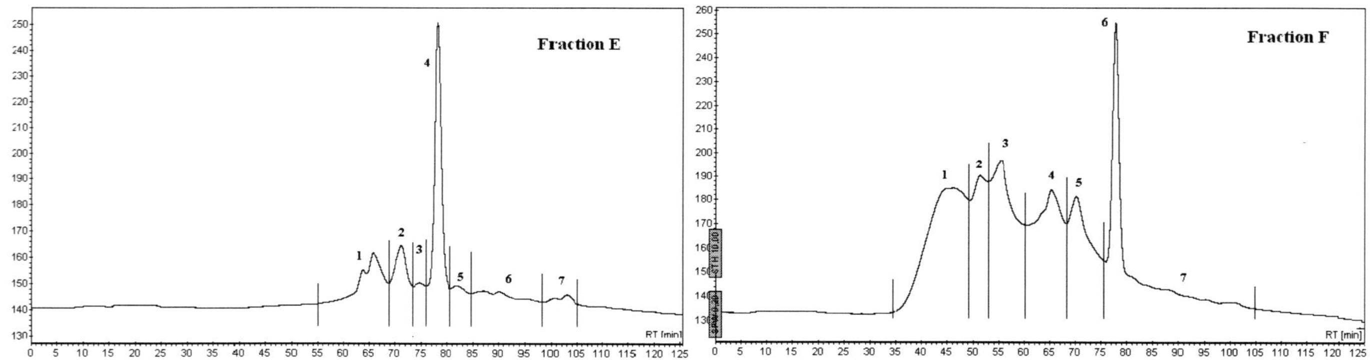

Figure 3. Refractive index profiles by size exclusion chromatography of the oligosaccharides obtained in fractions A, B, C, D, E and F from Sample 2.

Table 3. Composition of different subfractions separated by size exclusion chromatography from fraction F of Sample 2

	% (g/100 g extract dry) Sample 2-Fraction F			
	1+2	**3**	**4**	**5**
% phenols	12,1	12,26	29,68	18,66
% uronic acids	7,62	n.d.[a]	n.d.	n.d.
% neutral sugasr	45,24	12,08	24,68	21,05
	% molar neutral sugar			
Rhamnose	10,93[b]	13,27	9,18	10,46
Fucose	0,48	1,92	11,14	1,55
Arabinose	30,34	25,81	9,23	6,36
Xylose	43,13	16,43	6,01	2,95
Mannose	1,19	n.d.	5,62	10,73
Galactose	4,77	4,63	1,36	n.d.
Glucose	9,17	37,94	67,46	66,26

[a] n.d., not detected.

[b] The composition is determined by gas chromatography as alditol acetate. The results are expressed in percentage with respect to the total content.

One of them which included a fraction rich in neutral and acidic oligosaccharides and a low UV absorbance or no absorbance corresponds to molecules with a degree of polymerization (DP)from 10 to 3. The second group of fractions with theoretically smaller DP, next to standard oligosaccharides with DP from 4 to 2, showed similar absorption spectra with a maximum absorbance near 240 nm, characteristic of secoiridoid core [25], and a slight shoulder at 268-274 nm, characteristic of phenols (Figure 5a). This fact suggests that secoiridod glycosides esterified to phenolic compounds were detected. The third group of fractions included fractions with a retention time from Superdex Peptide that coincide with the monosaccharide standard in the calibration sample, or with the solvent (Rt = 78 min) and those fractions that eluted after the solvent because of interactions with the gel. Their absorption spectra differed from the previous group because they showed maximum absorbencies near 280 nm and 228 nm (Figure 5b), indicative of the presence of carbohydrates bound to phenolic compounds.

Characterization of Three Groups of Fractions

The structural analysis of neutral oligosaccharides of the first group using electrospray ionization mass spectrometry (ESI-MS) was conducted. The mass spectrum of these fractions showed a variety of oligosaccharides which correspond mainly to a series that coincides for Samples 1 and 2. One of them with $[M+Na]^+$ for a neutral pentose-oligosaccharide composed of 3, 4, 5, 7 or 10 pentose residues and resulting in ions for the larger series at *m/z* 569, 701, 833, 965, 1097, 1229 and 1361. The loss of 132 Da was attributed to the successive loss of a pentose residue. The ESI mass spectra of neutral oligosaccharides (Pentose $_n$) for n = 7 and 5 are shown in Figure 6.

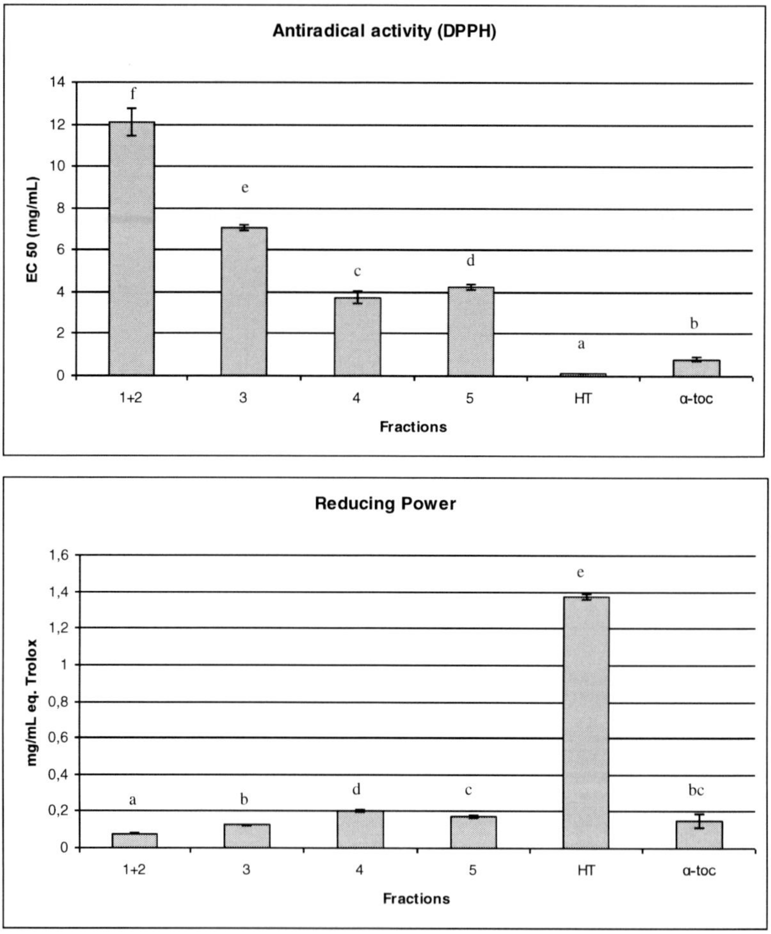

Figure 4. Radical scavenging capacity on DPPH and reducing power of each fraction isolated from Sample 2-Fraction F (F1+2, F3, F4 and F5) and two standards (hydroxytyrosol and α-tocopherol), date are presented as the means ± SDs. Different letters indicate significantly different result ($p < 0.05$).

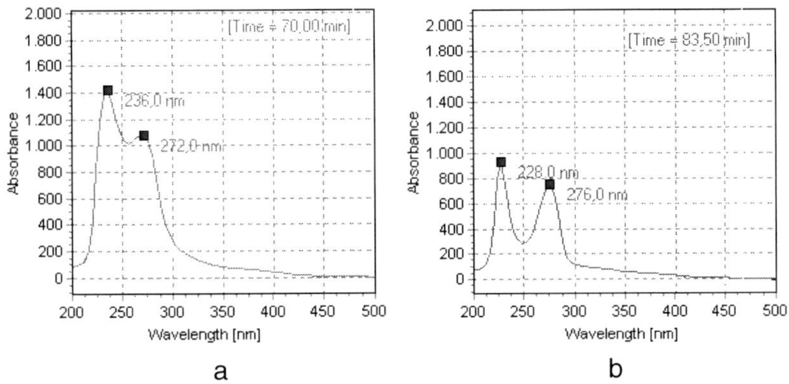

Figure 5. Ultraviolet absorption spectra for (a) fractions C5 and D4 from the Sample 1 (representative of the second group) have an absorption maximum near 240 nm and shoulder at 268-274 nm and (b) fractions C10 and D8 from the Sample 1 (representative of the third group) have two absorption maxima, one near 280 nm and a second at 228 nm.

A second series of abundant ions in both positive and negative modes was also observed in both samples. In the case of the Sample 1 was observed in mode positive ions at *m/z* 759 and 1023, which correspond to an increase of 58 Da on pentose$_5$ (*m/z 701)* and pentose$_7$ (*m/z* 965) and is attributed to the [M-H_2O+Na]$^+$ of the –methyl-D-glucuronic acid residue (-MeGlcA) of the Pent$_4$MeGlcA and Pent$_6$MeGlcA. In Sample 2 was observed in mode negative ions at *m/z* 983, 851, 719, 587, 455, 323 and 191 which correspond to Pent$_6$MeGlcA Ions that correspond to a series of Pent$_5$MeGlcA (Figure 7) and Pent$_4$MeGlcA were also found. In addition, only in the Sample 1 were ions at *m/z* 657 and 789 found, which revealed the presence of pentose-oligosaccharides that contain two GlcA residues (Pent$_2$ GlcA$_2$ and Pent$_3$GlcA$_2$) and the most abundant ions at *m/z* 957, which corresponded to the [M-H+2Na]$^+$ ion with the presence of both GlcA and MeGlcA as substitute residues in the same pentose-oligosaccharide (Pent$_4$GlcAMeGlcA) [17].

The analysis of the carbohydrates in these fractions (Table 4) revealed that the fractions with a neutral pentose-oligosaccharides series contained mainly arabinose, although some of them contained mainly xylose, while the fraction rich in acidic pentose-oligosaccharide contained mainly xylose. This fact suggests the presence of neutral arabinose- and xylose- oligosaccharide and acidic xylo-oligosaccharides, respectively, which is in accordance with the results previously reported in the alkali extract of olive pulp [26] (Reis et al. 2003).

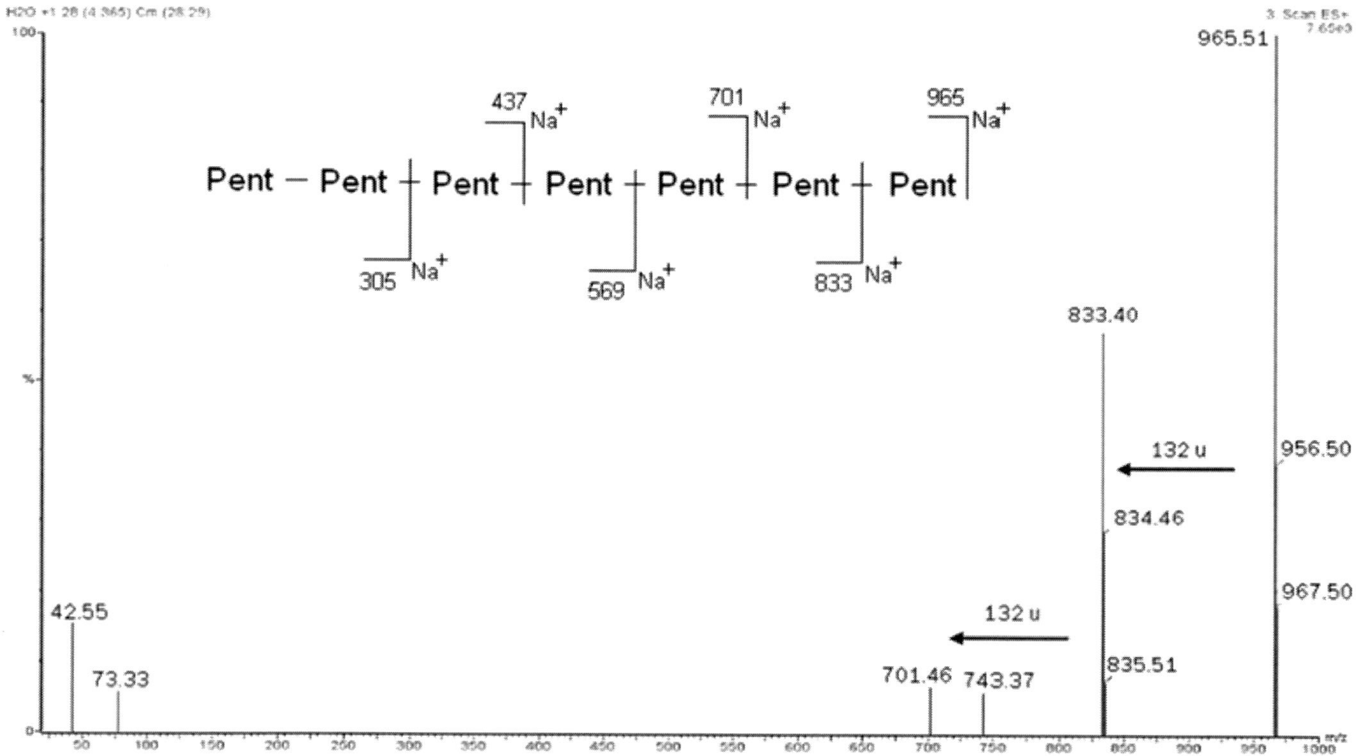
H2O +1 28 (4.365) Cm (28:29)
3: Scan ES+
7.65e3
Pent — Pent — Pent — Pent — Pent — Pent — Pent
305 Na+
437 Na+
569 Na+
701 Na+
833 Na+
965 Na+
965.51
956.50
967.50
833.40
834.46
835.51
132 u
132 u
701.46
743.37
42.55
73.33
100
%
0
m/z

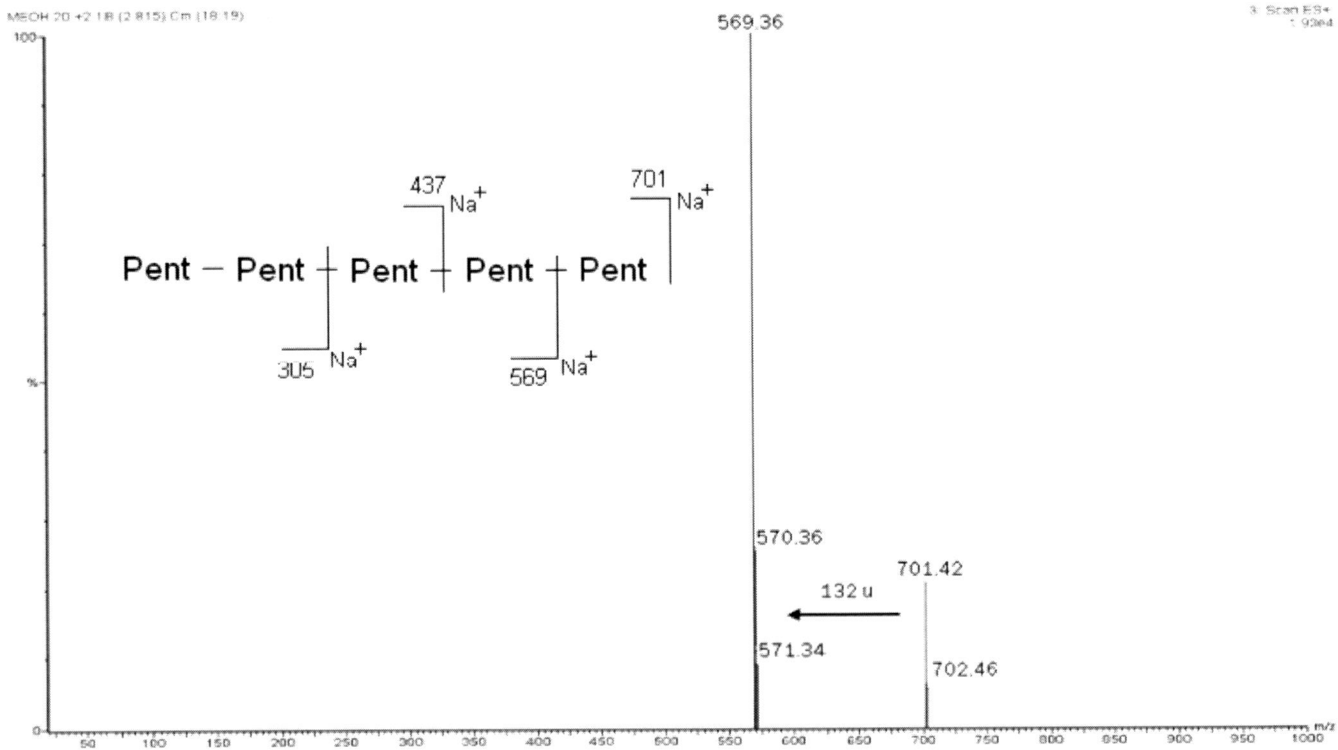

Figure 6. ESI mass spectra obtained for fractions A1, B1 and D1 of Sample 2 in positive mode, identified as neutral pento-oligosaccharides ($Pent_{7}$) (upper) and mass spectra obtained for fractions A2, B2 and D2 of Sample 2 in positive mode, identified as neutral pento-oligosaccharides ($Pent_{5}$) (lower).

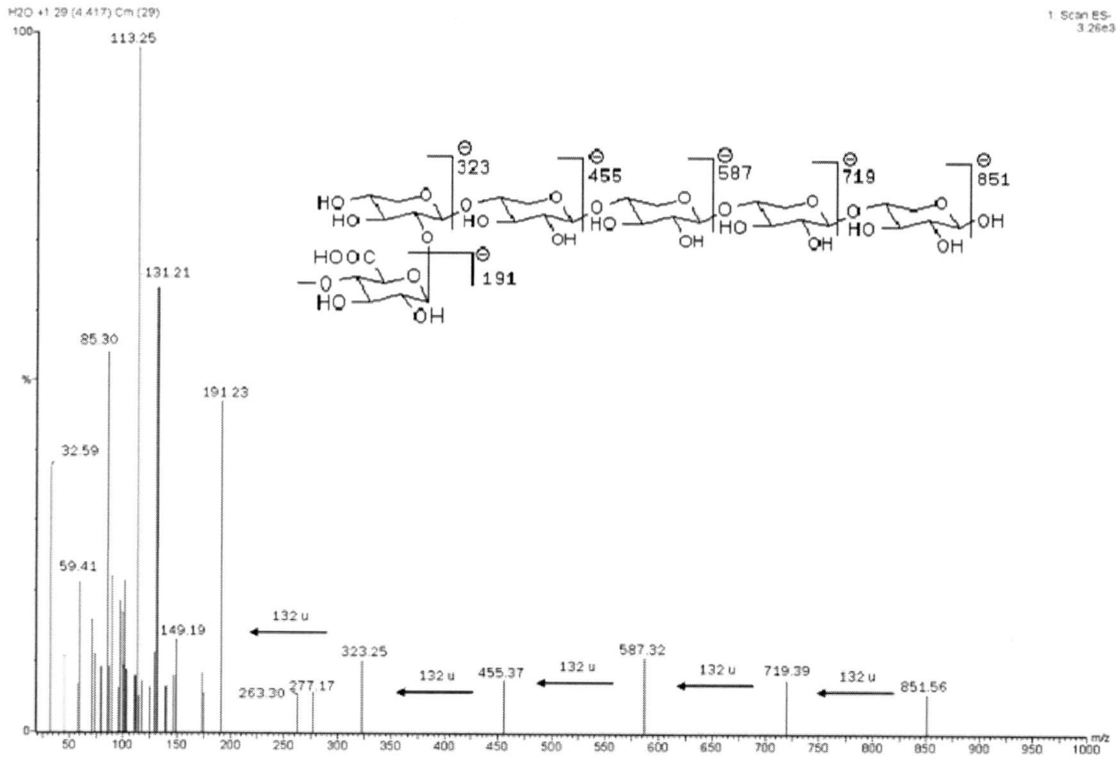

Figure 7. ESI mass spectra obtained in fraction D1 of Sample 1 and fraction C1 of Sample 2 in negative mode. Structure proposed for possible acidic pentose-oligosaccharide ($Pent_5MeGlcA$).

Table 4. Total neutral sugars in ultrafiltration fractions A, B, C and D, and sugar composition expressed as mol % of the oligomers of subfractions separated by size-exclusion or gel filtration chromatography for Sample 1 (upper) and Sample 2 (lower). Retention time (RT) and oligosaccharides identified in pools of each of subfractions

	Sample 1					
Fraction	**A**	**B**	**C**	**D**		
	Total sugars (mg/L)[a]					
1	4	29	33	119		
2	--	--	--	50		
3	--	--	--	22		
Fraction	**A1**	**B1**	**C1**	**D1**	**D2**	**D3**
RT (min)	49-61	49-59	50-60	46-55	55-59	59-62
Sugars	Oligosaccharides (% molar)[b]					
Rhamnose	12,8	11,6	12	3,9	8	5,7
Fucose	3,5	--	6,6	--	1,5	--
Arabinose	15,8	16,8	17,5	13,9	12,2	9,5
Xylose	10,4	11,3	18,4	20,9	21,2	17,9
Mannose	10,8	10	8	3,6	2,4	6,3
Galactose	19	24,1	5,9	22,4	13	6,6
Glucose	27,8	26,2	31,5	35,2	41,6	54
Oligosaccharides identified	Hex_7	Hex_6	Hex_6Pent_3	Hex_6Pent_3 $Pent_{4-10(c)}$	Hex_6Pent_3 Hex_5Pent_4	Hex_6Pent_3

Table 4. (Continued)

	Sample 2									
Fraction	**A**		**B**			**C**	**D**			
	Total sugars (mg/L)[a]									
1	48		36			47	104			
2	105		15			--	43			
3	--		52			--	66			
4	--		--			--	105			
Fraction	**A1**	**A2**	**B1**	**B2**	**B3**	**C1**	**D1**	**D2**	**D3**	**D4**
RT (min)	**45-62**	**62-68**	**41-61**	**61-64**	**64-68**	**44-62**	**38-60**	**60-63**	**63-66**	**66-69**
Sugars	Oligosaccharides (% molar)[b]									
Rhamnose	4,2	5,0	4,3	3,3	2,6	8,2	4,3	2,4	1,7	0,3
Fucose	0,1	0,2	0,2	--	--	0,7	0,2	0,3	--	0,2
Arabinose	60,9	15	58,4	36,5	19,9	72,0	67,2	20,9	6,0	1,8
Xylose	19	0,9	17,5	13,2	5,1	1,6	21,1	50,2	57,1	20,3
Mannose	2,9	7,9	0,4	11,5	9,2	14,3	0,4	3,6	0,6	0,3
Galactose	4,0	0,3	7,7	2,7	1,8	3,2	3,6	2,8	0,6	0,4
Glucose	8,8	63,5	11,5	32,9	67,4	--	3,2	19,9	33,1	46,7
Oligosaccharides identified	$Pent_7$ $Pent_5$l MeGlcA	$Pent_5$	$Pent_7$ $Pent_5$ MeGlcA	$Pent_5$	Hex_3	$Pent_6$ MeGlcA	$Pent_7$ $Pent_5$ MeGlcA	$Pent_5$ $Pent_4$ MeGlcA	$Pent_4$	Hex_3 $Pent_3$

[a] Colorimetric determination by the Anthrone method.

[b] Quantification by GC as alditol acetates prior to hydrolysis with 2 N TFA at 120 °C for 2 hours.

[c] Also in this fraction D1 was detected: $Pent_6$MeGlcA, $Pent_4$MeGlcA, $Pent_2MeGlcA_2$, $Pent_3MeGlcA_2$ and $Pent_4$GlcAMeGlcA

In addition, the examination of the mass spectra shows an ion at *m/z* 1409, which is attributed to the $[M+Na]^+$; the predominant ions at *m/z* 1247, 1115, 1085, 953 and 629 (Figure 8) were together the sugar composition in accordance with the results obtained in olive pulp xyloglucan with gross formation of $(Hexose)_6$ $(Pentose)_3$ [27]. In another fraction, the mass spectra revealed the presence of an ion pseudomolecular $[M+Na]^+$ at *m/z* 1380 that could correspond to a oligosaccharide with a structure like $(Hexose)_5(Pentose)_4$, which, based on the sugar composition (data not shown), could be a xyloglucan oligosaccharide found in the cell walls of olives [27]. Also, the spectra of fractions A and B of Sample 1 showed the ions at *m/z* 1175, 1013, 851 and 689 and 1013, 851 and 689, respectively, that corresponded at $(Hex)_7$ and $(Hex)_6$ with the loss of hexose residue of 162 Da in each series in positive mode (data not shown). Regarding the sugar composition, the oligosaccharides could be attributed to glucan, mannan or galactan chains. In Sample 2, an ion at *m/z* 527 in positive mode was observed, which corresponds to $(Hexose)_3$. The presence of 67.4% of glucose in this fraction confirmed the presence of $(Glucose)_3$, which was also proposed by Vierhuis et al. [27] .The MS spectrum of the de-protonated molecule $[M-H]^-$ with ion *m/z* 369 confirmed the presence of the dimer galacturonic acid-methyl-D-galacturonic acid in fraction A4, A5 and A6 of Sample 2.

The analysis of the fractions in the second group of Samples 1 and 2, gave a similar pattern of fragmentation of mass in both positive and negative modes, which suggest they all have a common structure. The mass spectra (Figure 9) show a predominant ion at *m/z* 407 which likely corresponds to a deprotonated molecule $[M-H]^-$, and the corresponding sodium $[M+Na]^+$ adduct at *m/z* 431 was observed in the positive mode. The appearance of the peak at *m/z* 569 was found only in some fractions of Sample 1 and Sample 2 and is consistent with the existence of one hexose unit, as suggested by the loss of 162 u from *m/z* 569, and an aglycone unit, as proven by the ions observed at *m/z* 407. These data together with the absorbance maximum at 240 nm are characteristic of a secoiridoid core, which suggests that the molecule under investigation might correspond to a glucosyl secoiridoid bearing a phenol unit. Its structure was recently proposed by our own research group and coincides with oleuropeinic acid [18]. The fragments in negative and positive mode have been proposed in Figure 10.

In addition, in Sample 2 several other derivatives from phenolic secoiridoids, which are closely correlated to oleuropein were isolated and identified by UV absorbance and spectral characterization. The mass spectra revealed the presence of the pseudo-molecular ion $[M-H]^-$ at *m/z* 555 and the ions at *m/z* 537 that appear as a result of the loss of water $[M-H_2O-H]^-$. The loss of glucose yield in the ions at *m/z* 393 which are characteristic of 10-hydroxyoleuropein has been previously described by Cardoso et al. [28]. This molecule could be the precursor of the oleuropeinic acid since it could be formed by the oxidation of hydroxyoleuropein.

Another component isolated showed an pseudomolecular ion $[M-H]^-$ at *m/z* 393 and the corresponding sodium $[M+Na]^+$ adduct at *m/z* 417 in positive mode (data not shown) which were consistent with a listroside derivative, identified in olive pulp by de la Torres-Carbot et al. [29]. An isomer of the oleuropein aglycon with ions at *m/z* 377, 197 and 153 and secologanoside with ions at *m/z* 389, 209 and 121 were also identified.

The analysis of the fractions in the third group showed the presence of glucosides and di- and trisaccharides linked to a hydroxytyrosol or tyrosol moiety. In Sample 1 a fraction containing an $[M+Na]^+$ ion at *m/z* 663 was isolated by Superdex Peptide chromatography . The molecule's proposed structure and fragmentation are shown in Figure 11. The resonance signals from the ^{1}H and ^{13}C NMR of the acetylated molecule are in accordance with the proposed structure (Figure 12) of this compound. After the consecutive loss of glucose (fraction contains 99.5% glucose), it was possible to observe the formation of the sodium adduct species at *m/z* 501 and 339. The analysis of the other fractions, by mass spectrometry, gave an $[M+Na]^+$ ion at *m/z* 501 and the corresponding deprotonated molecule, $[M-H]^-$ at *m/z* 477, in negative mode, and the groups of ions at *m/z* 315, 153 and 123, also present in the above molecule, which was consistent with the presence of a diglucosyl structure linked to hydroxytyrosol. The characteristic fragments of ions corresponding to a glucoside attached to hydroxytyrosol and tyrosol were also observed (data not shown).

Only in Sample 2 only was a molecule identified with a pseudo-molecular ion $[M-H]^-$ at *m/z* 477 and the species *m/z* 153 and 123 that could correspond to the molecule [Hydroxytyrosol-Hex-Pent]. The presence of 70% glucose and 15% xylose in this fraction led to the identification of [Hydroxytyrosol-Gluc-Xyl].

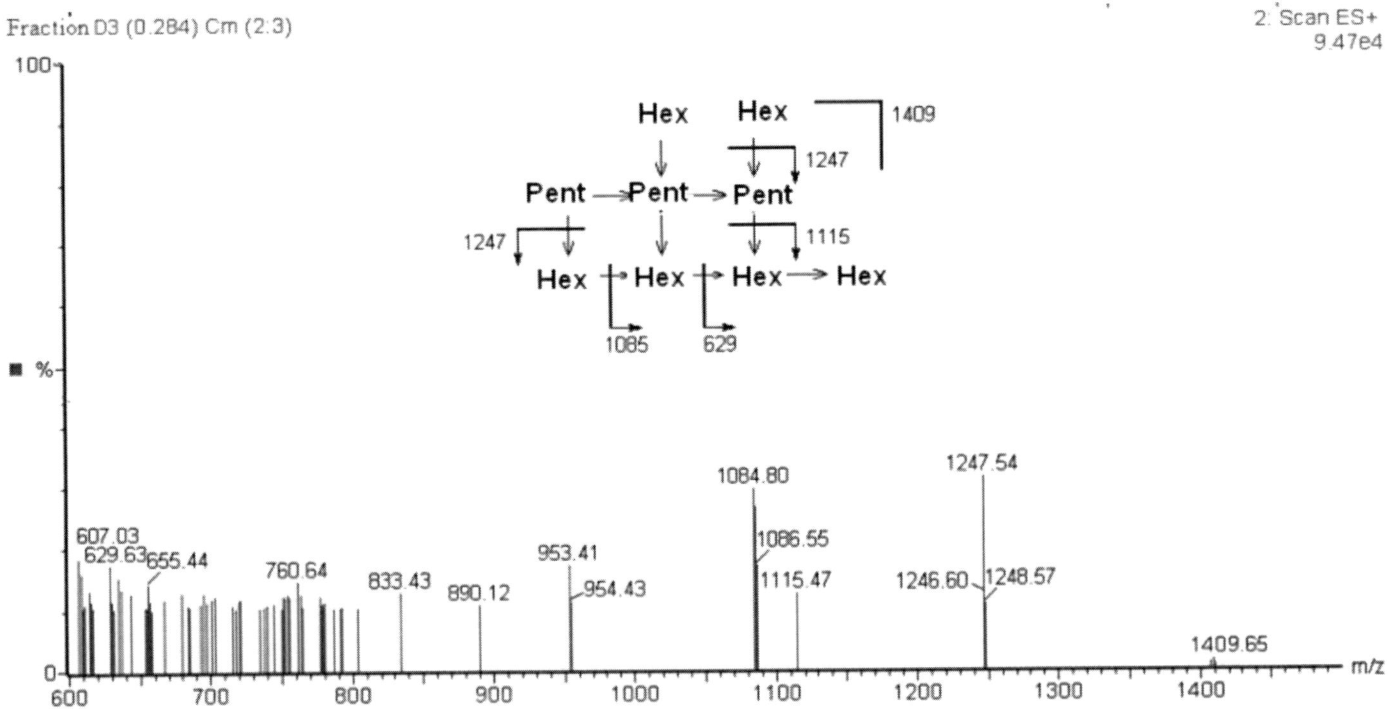

Figure 8. ESI mass spectra obtained for fractions C1, D1, D2 and D3 of the Sample 1 in positive mode and fragmentation scheme for the structure proposed of Hex_6Pent_3.

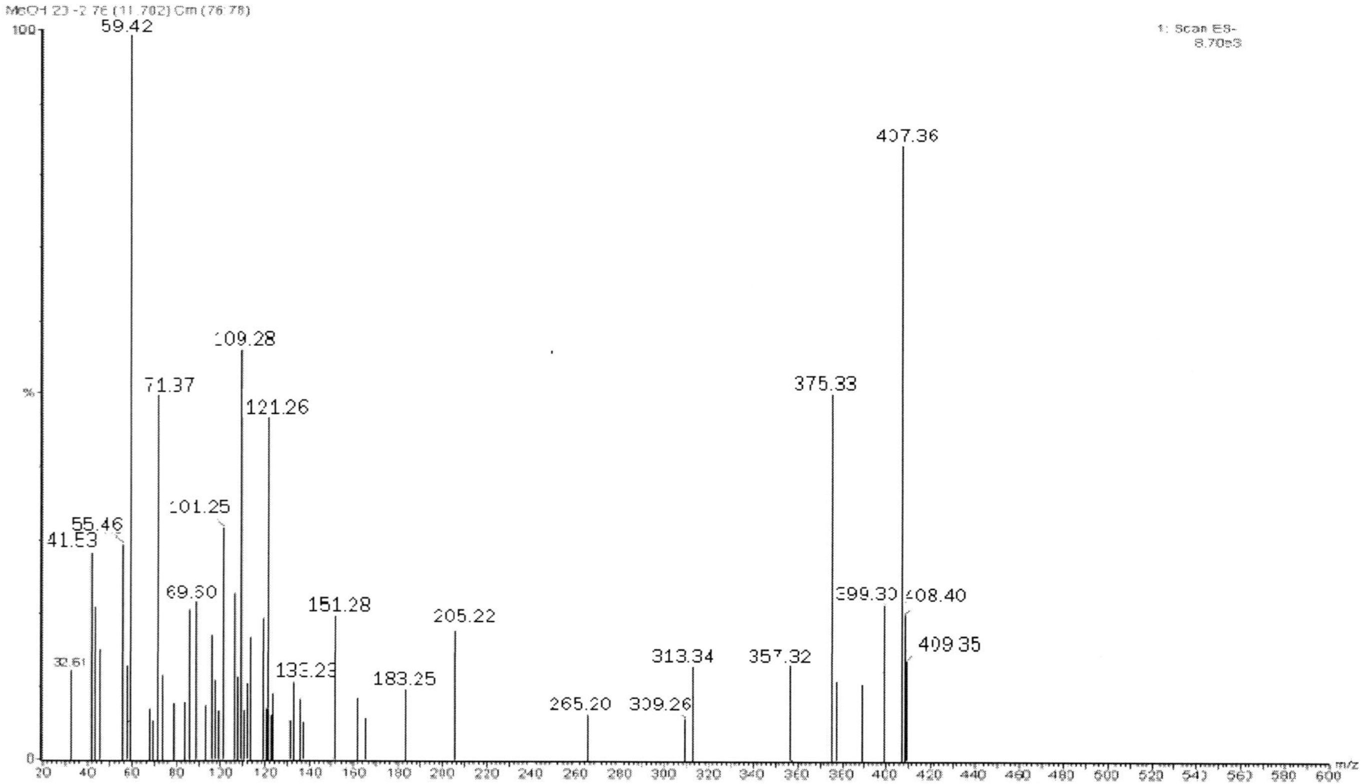

1: Scan ES-
8.70e3
59.42
407.36
109.28
71.37
375.33
121.26
101.25
55.46
41.53
69.30
399.30
408.40
151.28
205.22
409.35
32.61
133.23
183.25
313.34
357.32
265.20
309.26
100
%
0
20
40
60
80
100
120
140
160
180
200
220
240
260
280
300
320
340
360
380
400
420
440
460
480
500
520
540
560
580
600
m/z

(a)

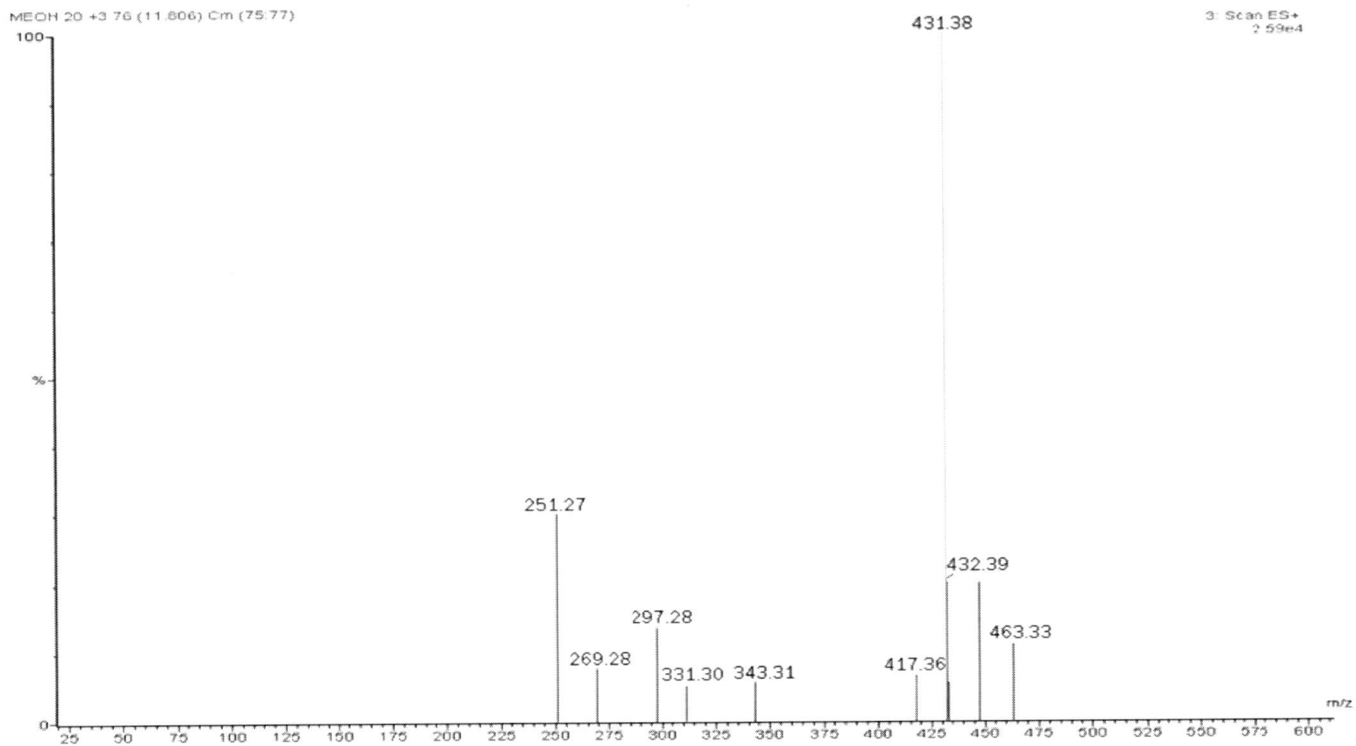

(b)

Figure 9. Mass spectra in full scan negative (a) and positive (b) identified as oleuropeinic acid in some of the fractions for the second group of Sample 1 and Sample 2.

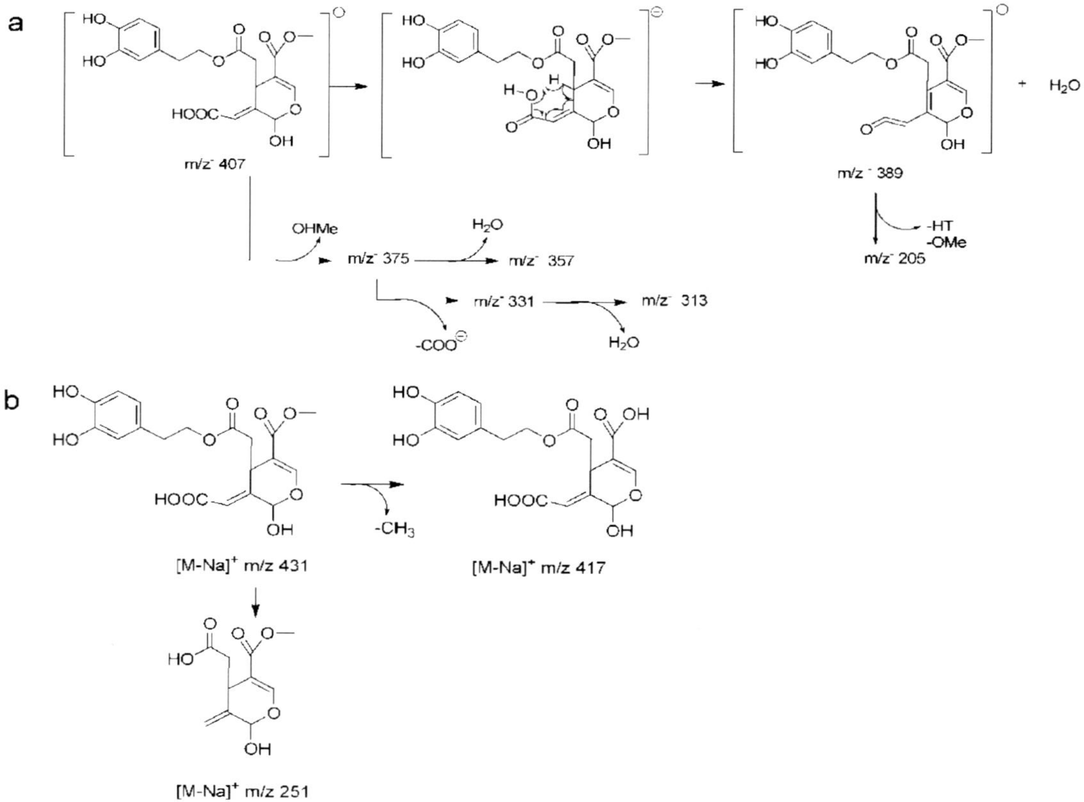

Figure 10. Scheme of fragmentation proposed for the oleuropeinic acid, in negative mode (a) and in positive mode (b).

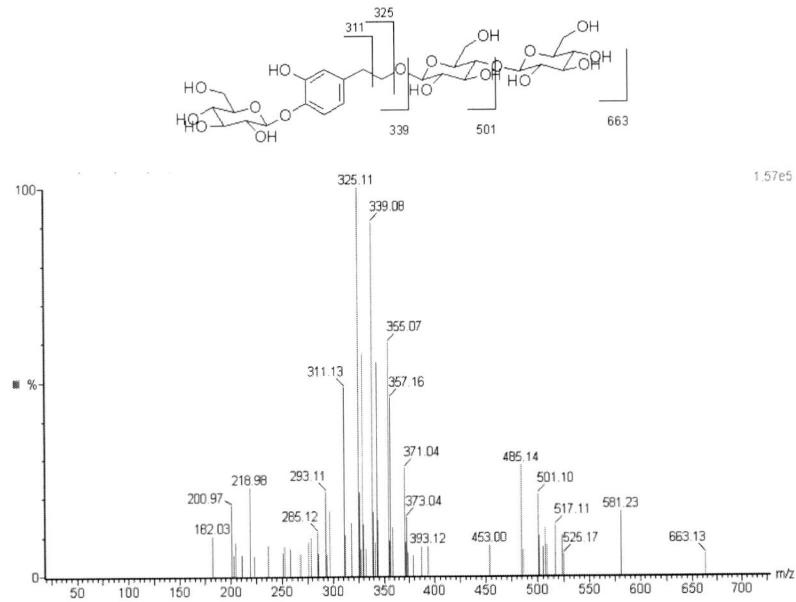

Figure 11. Mass spectra in positive mode obtained for the fraction D8 in Sample 1 and structure proposed according to the profile of fragmentation observed (all ions represented are sodium adducts).

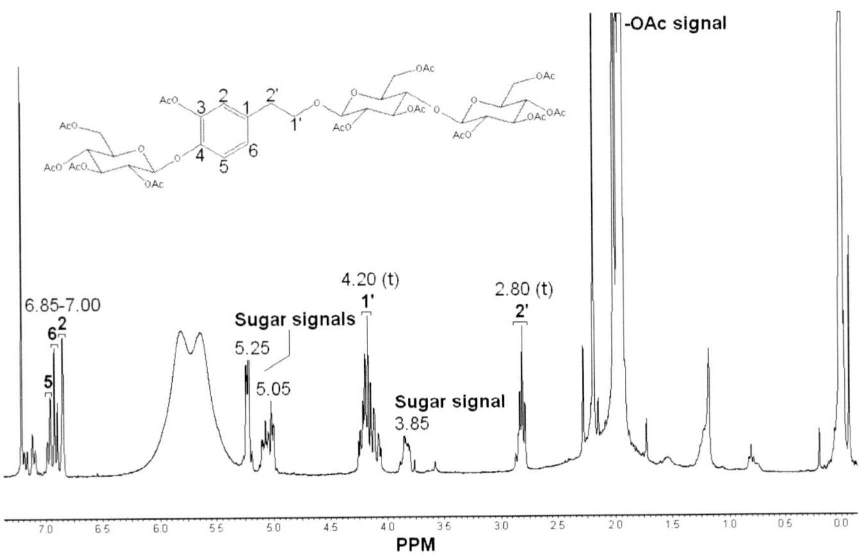

(a)

Figure 12. (Continued)

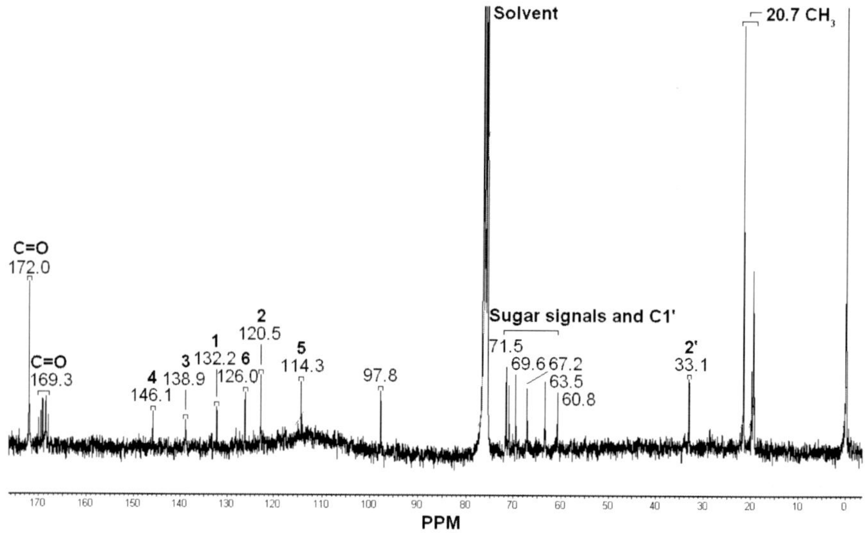

(b)

Figure 12. ^{1}H (a) and ^{13}C (b) NMR spectra (registered at 300 and 75 MHz, respectively) obtained for the acetylated simple. The chemical shifts are relative to TMS in $CDCl_3$.

Conclusion

The protocol for the isolation of oligosaccharides of low molecular weight developed for two samples of olive by-products or alperujo from different seasons, origin and variety of olives results in a reproducible and useful tool for the recovery of bioactive compounds. Hydrothermally treated alperujo represents a good source of not only bioactive phenols but also poly- and oligosaccharides with potential biological properties. Chemical hydrolysis has been performed to increase the neutral and acidic oligosaccharides with low molecular weight obtaining di-galacturonic acids, and different structures of neutral and acidic xylo-oligosaccharides, together with xyloglucan, glucans and arabinans.

The results obtained from the antioxidant test demonstrate that some fractions may be considered as new potentially useful bioactive compounds, especially those with the highest content in phenolic compounds, which are mainly responsible for the antioxidant capacity of the soluble fiber.

In addition, a group of fractions with oleosidic secoiridoid structures conjugated to phenolic compounds were identified as oleuropeinic acid, 10-hydroxyoleuropein, a ligstroside derivative, an isomer of oleuropein aglycon and secologonaside.

The presence of a significant fraction of di- and trisaccharides bound to phenolic compounds (hydroxytyrosol and tyrosol) could be considered soluble fiber and antioxidant soluble fibers with bioactive properties assigned to a phenolic moiety, respectively.

The fraction of neutral oligosaccharides or those bound to a phenol-containing compound can be fermented by colonic bacteria and may act as prebiotic compounds. The phenolic compound would be released and might exert the effect of scavenging free radicals and binding metal or to be absorbed, thus increasing their nutritional value.

Acknowledgments

We thank the Ministerio de Ciencia e Innovación of Spain for financial support (grant no. AGL2009-12352) and the Consejería de Innovación, Ciencia y Empresa de la Junta de Andalucía (Proyecto de Excelencia P08-AGR-03751).

References

[1] Fernández-Bolaños, J. Rodríguez-Gutiérrez, G. Lama-Muñoz, A. & Sánchez, P. (2010). *Dispositivo y procedimiento para el tratamiento de los subproductos de la obtención del aceite de oliva*. (Spanish Patent request No. P201031236).

[2] Rodríguez, G., Lama, A., Trujillo, M., Espartero, J. L. & Fernández-Bolaños, J. (2009). Isolation of a powerful antioxidant from Olea europaea fruit-mill waste: 3, 4-Dihydroxyphenylglycol. LWT- *Food Science and Technology*, *42*, 483- 490.

[3] Lama-Muñoz, A. Rodríguez-Gutiérrez, G. Rubio-Senent, F. Gómez-Carretero, A. & Fernández-Bolaños, J. (2011). A New Hydrothermal Treatment of Alperujo Enhances the Content on Bioactive Minor Components in Crude Pomace Olive Oil. *Journal of Agricultural and Food Chemistry*, *59*, 1115-1123.

[4] Rodríguez, G., Rodríguez, R., Guillén, R., Jiménez, A. & Fernández-Bolaños, J. (2007). Effect of steam treatment of alperujo on the enzymatic saccharification and in vitro digestibility. *Journal of Agricultural and Food Chemistry*, *55*, 136-142.
[5] Jiménez, A., Guillén, R., Fernández-Bolaños, J. & Heredia, (1994). A. Cell wall composition of olives. *Journal of Food Science*, *59*, 1192-1201.
[6] Jiménez, A., Rodríguez, R., Fernández-Caro, I., Guillén, R., Fernández-Bolaños, J. & Heredia, A. (2001). Olive fruit cell wall degradation of cellulosic and hemicellulosic polysaccharides during ripening. *Journal of Agricultural and Food Chemistry*, *49*, 2008-2013.
[7] Vierhuis, E. Schols, H. A. Beldman, G. & Voragen, A. G. J. (2001). Structural characterization of xyloglucan and xylans present in olive fruit (Olea europeae, cv Koroneiki). *Carbohydrates Polymers*, *44*, 51-62.
[8] Mafra, I. Barros, A. S. & Coimbra, M. A. (2006). Effect of black oxidising table olive process on the cell wall polysaccharides of olive pulp (Olea oleuopaea L. var. Negrinha do Douro). *Carbohydrates Polymers*, *65*, 1-8.
[9] Chiacchierini, E., Mele, G., Restuccia, D. & Vinci, G. (2007). Impact evaluation of innovative and sustainable extraction technologies on olive oil quality. *Trends in Food Science and Technology*, *18*, 299–305.
[10] Mussatto, S. I. & Mancilha, I. M. (2006). Non-digestible oligosaccharides: A review. *Carbohydrate Polymers*, *68*, 587-597.
[11] Gibson, G. R. & Roberfroid, M. B. (1995). Dietary modulation of the human colonic microbiota-introducing the concept of prebiotics. *Journal of Nutrition*, *125*, 1401-1412.
[12] Guggenbichler, J. P., Jurenitsch, J. & de Bettignies-Dutz, A. (1997). *Blocking the accumulation of germs on human cells.* (PCT/EP94/03006).
[13] Hsu, C. K., Liao, J. W., Chung, Y. C. & Chan, Y. C. (2004). Xylooligosaccharides and fructoligosaccharides affect the intestinal microbiota and precancerous colonic lesion development in rats. *Journal of Nutrition*, *134*, 1523-1528.
[14] Broekaert, W. F., Courtin, C. M., Verbeke, K., van de Wiedele, T., Verstraete, W. & Delcour, J. A. (2011). Prebiotic and other health-related effects of cereal-derived arabinoxylans, arabinoxylan-oligosaccharides, and xylooligosaccharides. *Critical Reviews in Food Science and Nutrition*, *51*, 178-194.

[15] Christakopoulos, P., Katapodis, P., Kalogeris, D., Kekos, D. Macris, B. J., Stamatis, H. & Skaltsa, H. (2003). Antimicrobial activity of acidic xylo-oligosaccharides produced by family 10 and 11 endoxylanases. *International Journal of Biological Macromolecules*, *31*, 171-175.

[16] Nabarlatz, D., Montané, D., Kardosova, A., Bekesová, S., Hribalová, V. & Ebringerová, A. (2007). Almond shell xylo-oligosaccharides exhibiting immunostimulatory activity. *Carbohydrate Research*, *342*, 1122-1128.

[17] Lama-Muñoz, A., Rodríguez-Gutiérrez, G., Rubio-Senent, F. & Fernández-Bolaños, J. (2012). Production, characterization and isolation of neutral and pectic oligosaccharides with low molecular weights from olive by-products thermally treated. *Food Hydrocolloids*, *28*, 1-13.

[18] Rubio-Senent, F., Lama-Muñoz, A., Rodríguez-Gutiérrez, G. & Fernández-Bolaños, J. (2013). Isolation and identification of phenoic glucosides from thermally treated olive oil byproducts *Journal of Agricultural and Food Chemistry*, *61*, 1235-1248.

[19] Dische, Z. (1962). Color reactions of carbohydrates. In Whistler, R. L., Wolfram, M. L., (Eds.), *Methods in Carbohydrates Chemistry* (477-512). New York: Academic Press.

[20] Blumenkrantz, N. & Asboe-Hansen, G. (1973). New method for quantitative determination of uronic acids. *Analytical Biochemistry*, *54*, 484-489.

[21] Englyst, H. N. & Cummings, J. H. (1984). Simplied method for the measurement of total nonstarch polysaccharides by gas-chromatography of constituent sugars as alditol acetates. *Analyst*, *109*, 937-942.

[22] Singleton, V. & Rossi, J. A. (1965). Colorimetric of total phenolics with phosphomolybdic-Phosphotungstic acid reagents. *American Journal of Enolology and Viticulture*, *16*, 144-158.

Rubio-Senent, F., Rodríguez-Gutiérrez, G., Lama-Muñoz, A. & Fernández-Bolaños, J. (2012). New phenolic compounds hydrothermally extracted from the olive oil byproduct alperujo and their antioxidative activities. *Journal of Agricultural and Food Chemistry*, *60*, 1175-1186.

[23] Fernández-Bolaños, J., Rodríguez, G., Gómez, E., Guillen, R., Jiménez, A., Heredia, A. & Rodríguez, R. (2004). Total recovery of the waste of two-phase olive oil processing: Isolation of added-value compounds. *Journal of Agricultural and Food Chemistry*, *52*, 5849-5855.

[24] Obied, H. K., Allen, M. S., Bedgood, D. R., Prenzler, P. D. & Robards, K. (2005). Investigation of Australian olive mill waste for recovery of

biophenols. *Journal of Agricultural and Food Chemistry*, *53*, 9911-9920.

[25] Reis, A., Domingues, R. M., Domingues, P., Ferrer-Correia, A. J., & Coimbra, M. A. (2003). Positive and negative electrospray ionisation tandem mass spectrometry as a tool for structural characterization of acid released oligosaccharides from olive pulp glucuronoxylans. *Carbohydrate Research*, *338*, 1497-1505.

[26] Vierhuis, E., Servili, M., Baldioli, M., Schols, H. A., Voragen, A. G. J. & Montedoro, G. (2001). Effect of enzyme treatment during mechanical extraction of olive oil on phenolic compounds and polysaccharides. *Journal of Agricultural and Food Chemistry*, *49*, 1218-1223.

[27] Cardoso, S. M., Guyot, S., Marnet, N., Lopes-da-Silva, J. A., Renard, C. & Coimbra, M. A. (2005). Characterisation of phenolic extracts from olive pulp and olive pomace by electrospray mass spectrometry. *Journal of Science. Food Agriculture*, *85*, 21-22.

[28] De la Torre-Carbot, K., Jauregui, O., Gimeno, E., Castellote, A. I. Lamuela-Raventós, R. M. & López-Sabater, M. C. (2005). Characterization and quantification of phenolic compounds in olive oils by solid-phase extraction, HPLC-DAD, and HPLC-MS/MS. *Journal of Agricultural and Food Chemistry*, *53*, 4331-4340.

In: Oligosaccharides
Editors: L. S. Schweizer, S. J. Krebs
ISBN: 978-1-62948-328-3

Chapter VII

Supramolecular Analytical Applications of Cyclodextrins and their Derivatives Using Fluorescence Spectroscopy

Abdalla A. Elbashir*
University of Khartoum, Faculty of Science, Chemistry Department, Khartoum, Sudan

Abstract

Cyclodextrins (CDs) are a family of structurally related natural products formed during bacterial digestion of cellulose. Theses cyclic oligosaccharides composed of α-(1, 4) linked glucopyranose subunits. Due to the chair conformation of the glucopyranose units, the cyclodextrins are shaped like a truncated cone rather than perfect cylinders. The most important feature of cyclodextrins is their ability to form inclusion complexes (host–guest complexes) with a very wide range of solid, liquid and gaseous compounds by a molecular complexation, that lead to enhancement of fluorescence intensity. During the last decade a considerable number of research works have been focused on the use of CDs to enhance fluorescence intensity of different analytes and to develop CD-induced spectrofluorimetric method. This chapter provides

* Correspondence to: Abdalla A.Elbashir. E-mail: hajaae@yahoo.com , aaelbashir@uofk.edu.

an overview of spectrofluorimetric methods based on host-inclusion complex for determination of various compounds.

Keywords: Cyclodextrins, inclusion complex, host-guest, spectrofluorimetric, supramolecular, analytical application

Abbreviations

Cyclodextrins, CDs; Methylated-β-CD, M-β-CD; Sulfobutylether β-cyclodextrin, SBE-β-CD; DE-β-CD, 2,6-di-O-ethyl-β-CD; hydroxy propyl β-cyclodextrin, H-β-CD; tri-O-methyl)-β-cyclodextrin, TM-β-CD; modified β-cyclodextrin, MAH-β-CD; thiabendazole, TBZ; sulphamethoxazole, SMTX; sulphaguanidine, SGN; mangiferin, MA; Mefenamic acid, MF; 5-(p-hydroxyphenyl)-10,15,20-tris-(4-chlorophenyl)porphyrin, *p*-HTClPP; MC, metacycline

1. Introduction

Cyclodextrins (CDs) are neutral and natural oligosaccharides that are prepared by the enzymatic degradation of starch. They have the property of forming inclusion complex with various guest molecules with suitable polarity and dimension because of their special molecular structure/hydrophobic internal cavity and hydrophilic external surface [1-5].

CDs are composed of several glucopyranose units. The native, α-, β-, and γ-CD contain six, seven, and eight glucose units, respectively. Their main physical properties are summarized in Table 1.

Table 1. The main properties of native cyclodextrins [Fanali, 2000]

	CD type		
	α	β	γ
Numbers of glucose units	6	7	8
Molecular mass	972	1135	1297
Inner diameter (nm)	0.57	0.78	0.95
Depth (nm)	0.78	0.78	0.78
Solubility (g/100 mL water)	14.5	1.85	23.2
pK_a	12.23	12.20	12.08

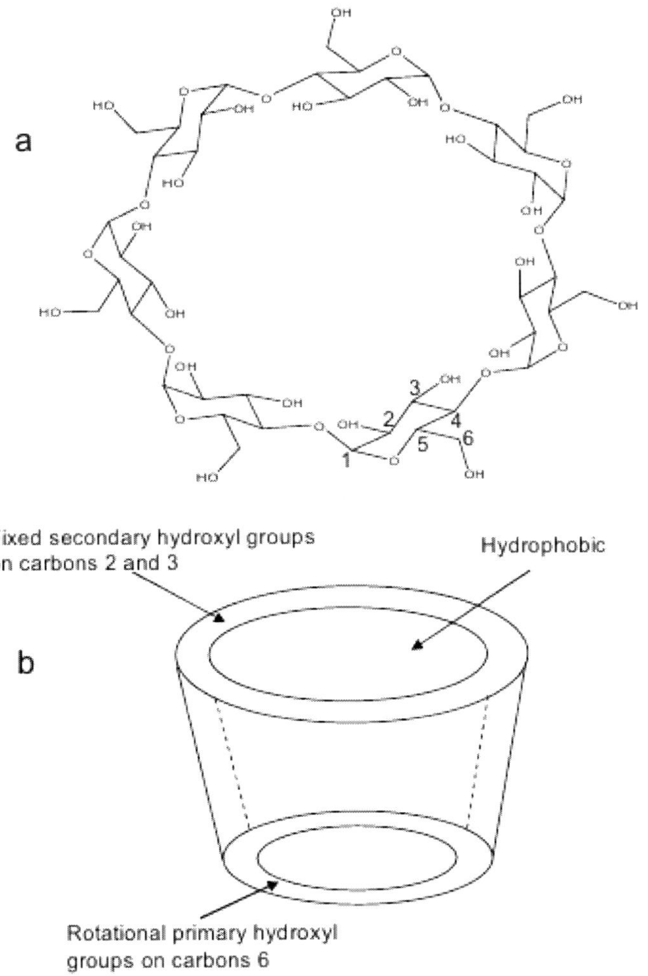

Figure 1. (a) Chemical structure of β-cyclodextrin with numbered carbons in one of the glucopyranose units and (b) its approximate shape.

As can be seen from the table, the physical properties of the three native CDs are quite different, e.g., cavity size, solubility, molecular mass etc. However they possess the same depth [6]. They have a shape of a truncated cone with an open cavity that is relatively hydrophobic that is able to host the molecule either completely, or at least its hydrophobic part (Figure 1b). The outside of the CD is relatively hydrophilic due to the presence of primary and secondary hydroxyl groups (positions 2, 3 and 6 of glucopyranose) (Figure 1a). The cyclodextrin structure provides a molecule shaped like a segment of a hollow cone which is capable of forming stable, supramolecular structures with various molecules fitting partially or completely in the host molecular cavity [7-10]. Such structure enables the CDs to capture various inorganic and organic molecules into their cavity to form host–guest inclusion complexes. This ability has been widely used in pharmaceutical industry to enhance solubility, stability and bioavailability of the drug molecules [11], foodstuff [12, 13], separation technique [14, 15], environmental protection [16, 17], and so on.

The inclusion complexes are stabilized by secondary bonds between the rim of the CD and substituent groups of the analyte. The hydroxyl groups present on the rim of CD can easily be modified by chemical reactions with various functional groups.

The most important property of inclusion compounds is that a host component can admit guest component into its cavity without any covalent bonds being formed. As inclusion is largely independent of the chemical properties of the guest molecules, the minimum requirement for inclusion complex formation is size compatibility between host and guest molecules, i.e., guest must fit, entirely or at least partially, into the cyclodextrin cavity Figure 2.

The presence of CD and correctly-sized organic molecule in aqueous solutions results generally in the formation of inclusion complexes [18-21]. It is well documented that even a partial binding to the CD cavity is accompanied by noticeable changes in the photophysical properties of the guest compound. Among other interesting features, fluorescence intensity enhancement [22] and chiral activity induction [23] are obtained upon addition of CD. Accordingly, better enantiometric separations by High Performance Liquid Chromatography (HPLC), [24, 25] Thin-Layer Chromatography (TLC) [26] or capillary zone electrophoresis (CZE), [27-30] have been reported for various compounds.

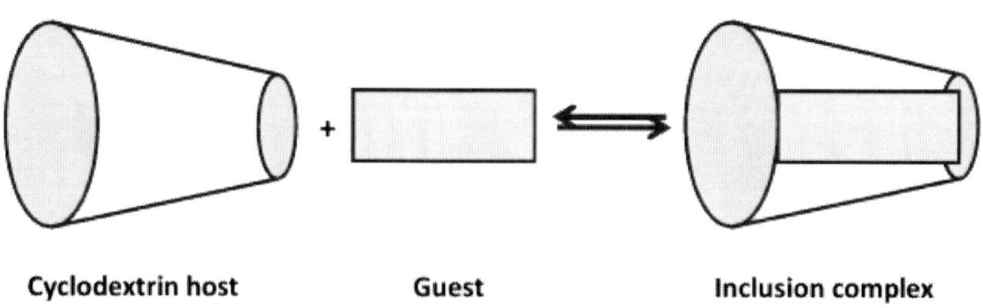

Figure 2. Schematic diagram of inclusion complex formation.

Considerable attention has been focused on the use of CDs in fluorescence applications. The first application of CDs has been reported by Kinoshita and co-workers for dansylated amino acids in the mid-1970s [21-33]. Fluorescence enhancements have been observed for a wide variety of other analyte molecules. On the other hand, the inclusion into the CDs depends on both the size and polarity of the guest; hence the specificity of the method can be extended in addition to sensitivity. Analysts have used this property of CDs , and a lot of methods based on the fluorescence of inclusion complexes with CDs have been proposed for the determination of several pharmaceutical drugs, pesticides, and metal ions [34-41].

Fluorescence spectroscopy is a powerful tool to study the host–guest molecular systems with both high sensitivity and high selectivity [42]. The remarkable advantages of fluorescence analysis are that it is capable of measuring much lower concentrations than spectrophotometric analysis, and it is potentially more selective because both the excitation and emission wavelength can be varied. The sensitivity of the fluorescence detection of host–guest molecular interaction depends upon the electronic or steric changes of the chromophore being responsible for the fluorescence behavior

In this review the analytical application of various CDs to enhance the fluorescence intensity of various analytes and the use of fluorescence signal enhancement to develop cyclodextrin-induced spectrofluorimetric methods for the determination of analytes was reviewed.

2. Analytical Application

2.1. Application Using Native Cyclodetrins

Spectrofluorimetric methods based on the inclusion complex formation between α-cyclodextrin or β-cyclodextrin and diclofenac was reported [43, 44]. Diclofenac forms a 1:1 inclusion complex with cyclodextrin with an association constant of 1.20x 10^3 M^{-1}. The quantification of diclofenac in the presence of α-CD or β-CD is obtained in the range 0-5 μg mL^{-1}. An application of the method to the determination of the diclofenac in pharmaceutical preparations was presented.

The inclusion complex formation between β-cyclodextrin and Vitamin K3 was studied spectrofluorimetrically by Nevado et al. [45]. The calibration graph was linear over the range 0.1-2.0 mg L^{-1} with limit of detection (LOD)

and limit of quantification (LOQ) were found to be 0.022 mg L^{-1} and 0.073 mg L^{-1}, respectively. The method was applied to pharmaceutical formulations.

A β-CD-enhanced spectrofluorimetric method for the determination of rubidate in the aqueous medium was developed and applied for detection of rubidate in the tablets with satisfactory results [46]. The improved sensitivity of the proposed method has been demonstrated comparing with that without β-CD. The linear relationship between the fluorescence intensity and rubidate concentration was obtained in the range from $2.7\text{-}10^{-2}$ to 3.0 mg mL^{-1}.

Determination of naproxen in pharmaceutical preparations by host-guest complexation with β- and γ-cyclodextrins was described [47]. The best signals were obtained when naproxen was in the presence of β-CD-cyclohexane-Tl(I), γ-CD-1,3-dibromopropane, Triton X-100-Tl(I) and SDS-Tl(I), respectively.

Spectrofluorimetric study on the interaction between ß-CD and dapsone in the absence and presence of linear alcohols was reported [48]. The fluorescence intensity was measured at 428 nm with excitation at 296 nm against a reagent blank. The linear relationship between the fluorescence intensity and dapsone concentration was obtained in the range of 3.39 to 1.50×103 ng mL^{-1} .

The supramolecular interaction between napropamide and β-cyclodextrin was investigated and applied to develop spectrofluorometric method [49]. Under the optimum conditions, the complex had excitation and emission maxima at 285 and 339nm, respectively. The linear range of the method was 3.7-1500 $ngml^{-1}$ with a detection limit of 1.1 $ngmL^{-1}$ was obtained. The proposed method was successfully used to determine napropamide in river water.

The complex formation between ibuprofen and ß-CD was studied by fluorescence measurements [50]. Based on host-guest complexation, a CD-enhanced spectrofluorimetric method for determination of ibuprofen was developed that is more sensitive than when β-CD is absent. A linear range of 0.1-2 μg mL^{-1} and a detection limit of 0.03 μg mL^{-1} were obtained. The method was found to superior over reported instrumental methods [51-54] in term of simplicity, sensitivity and cost. The method was successfully applied to the determination of ibuprofen in pharmaceutical preparations. Another spectrofluorimetric determination of ibuprofen in the presence of β-CD was a developed for serum samples at concentration levels between 5 and 70 μg mL^{-1}[55].

Highly sensitivity and selectivity spectrofluorimetric method was developed for the determination of tetramethrin in aqueous solution [56]. The method was based on the formation of inclusion complex between tetramethrin and β-cyclodextrin. The fluorescent product was measured at 295 nm after excitation at 356 nm. The method was found to be applicable in the linear range of 11-1500 ng mL^{-1} with a detection limit of 3.4 ng mL^{-1}. The proposed method was successfully used to determine trace amount of tetramethrin in river water.

Highly sensitive spectrofluorimetric method was reported for the determination of carbaryl in aqueous solution [57]. Under the optimum conditions, the complex had excitation and emission maxima at 278 nm and 332 nm, respectively. The linearity ranges were found to be 7.0-1500 ng mL^{-1} with a detection limit of 1.2 ng mL^{-1}. The proposed method was successfully applied to determine quantitatively of carbaryl in cottonseeds.

The host-guest complexation of celecoxib with β-CD was investigated by using fluorescence spectroscopy [58]. Based on the inclusion reaction, a sensitive spectrofluorimetric method for the determination of celecoxib was was developed. The measurement of relative fluorescence intensity was carried out at 390 nm with excitation at 270 nm. A linear relationship between the fluorescence intensity and celecoxib concentration was obtained in the range of 0.1-4.0 μg mL^{-1}. The detection limit was found to be 7.29 ng mL^{-1}. The method was successfully applied to the determination of celecoxib in capsules.

A new phosphorescence method for thiabendazole (TBZ) determination was established [59]. The method is based on supramolecular interaction of thiabendazole/β-cyclodextrin/triton X-100. The linear concentration range was determined to be 20-820 ng mL^{-1} with LOD 2.1 ng mL^{-1}. The proposed method has been successfully applied to the determination of TBZ in tap water, lake water and pineapples.

Highly sensitive and selective spectrophotometric method was developed for determination of dequalinium chloride (DQC) in capsules and serum by Tang et al [60]. The method is based on the inclusion reaction of dequalinium chloride (DQC) with disulfides linked β -CD dimer. The linear range was found to be 6.38–1.60×10^3 ng mL^{-1} with the LOD 1.9 ng mL^{-1}. The proposed method was superior to previously reported spectrofluorometric [61], HPLC [62] and CE [63] methods for determination of DQC in terms of its sensitivity and simplicity. Tang and co-worker also have developed spectrofluorimetric method for the determination of tranilast in bulk aqueous solution in the presence of ethylenediamine β-CD dimer, based on the enhancement of the fluorescence intensity of tranilast [64]. The linear range was 0-1.40×10^4 ng mL^{-1} with the detection limit 3.2 ng mL^{-1}. The proposed method was successfully applied to the determination of tranilast in serum.

Diez et al., have studied the complex formation between of sulphamethoxazole (SMTX) and sulphaguanidine (SGN) with β-cyclodextrin [65]. Based on the result obtained spectrofluorimetric method for determination SMTX and SGN in human urine samples was presented. The method was found to be applicable in the linear range of 0.09–0.5 μg mL^{-1} and 0.07–1.0 μg mL^{-1}, for SMTX and SGN respectively with a limit of detection (LOD) 0.05 and 0.04 μg mL^{-1}, respectively, were established. The results obtained by proposed method were compared with those obtained with a HPLC method [66] and no significant difference between the results was found.

A method for the detection of Aflatoxin B, in wheat using second order standard addition on the cyclodextrin-assisted fluorescence spectra with minimal sample preparation was presented [67]. The limit of detection of method was 0.9 μg kg^{-1} for the determination of AFB1 in wheat samples. The values obtained by the proposed method are statistically comparable to that obtained with the HPLC method [68].

Jalali and Afshoon have investigated the formation of inclusion complex between ketoconazole (KC) and β-cyclodextrin spectrophotometrically and enhancement in fluorescence intensity of KC was observed [69]. Based upon the enhancement effect observed, a sensitive spectrofluorimetric method for the detection of KC was proposed. The linear range was found to be 0.01–10 μg mL^{-1}. The method was applied to the detection of ketoconazole in pharmaceutical products and the results were satisfactory in comparison to the official method [70].

The synergetic effect of β-cyclodextrin with anion surfactant sodium dodecylbenzene sulfonate (DBS) on fluorescence intensity of metacycline was investigated [71]. Based on fluorescence enhancement a sensitive method for the determination of metacycline in pharmaceutical preparations was proposed. The calibration graph was linear over the range 1.7×10^{-7} to 8.8×10^{-5} $molL^{-1}$ and the detection limit was 7.4×10^{-9} $molL^{-1}$.

Valderrama, and Poppi, have presented direct quantification of enantiomers of propranolol and ibuprofen by formation of inclusion complex with β-cyclodextrin using molecular fluorescence spectroscopy and Parallel Factor Analysis (PARAFAC), followed by the standard addition method [72, 73]. The method was successfully applied for determination of propranolol and ibuprofen enantiomers in complex matrices such as human urine and plasma.

Development and validation of spectrofluorimetric method for determination of salbutamol (SAL) in different pharmaceutical formulation was described [74]. The method was based on formation of inclusion complex with β-cyclodextrin. The method was linear over concentration range 4-20 μg mL^{-1}. Based on the enhancement of the fluorescence intensity of 2-hydroxy-1-naphthoic acid (2H1NA), a spectrofluorimetric method for the determination of 2H1NA in bulk aqueous solution in the presence of β-CD was developed [75]. The linear relationship between fluorescence intensity and 2H1NA was obtained in the range of 9.00×10^{-7}-2.50×10^{-5} mol L^{-1} and the detection limit was 8.0×10^{-7} mol L^{-1}. The proposed method was successfully applied to determine 2H1NA in waste water.

Ng, and Narayanaswamy have investigated the host-guest interaction of between β-CD and N-phenyl-1-naphthylamine [76]. Based on enhancement in fluorescence intensity observed at 445 nm upon excitation of 334 nm spectrofluorimetric method for determination of N-phenyl-1-naphthylamine was proposed. The standard curves were linear over the concentration range up to 4.67×10^{-7} M with a limit of detection of 0.58 nM.

Recently Dani and Elbashir [77] have investigated the host–guest inclusion complex between β-CD and cephalexin spectrofluorometrically. The factors affecting the inclusion complex formation were carefully studied. A spectrofluorimetric method for the determination of cephalexin in solution in the presence of β-CD was developed based on the remarkable enhancement of the fluorescence intensity of cephalexin.

The supramolecular interaction of nizatidine (NIZ) and β-cyclodextrin (β-CD) has been studied by spectrofluorimetry. The results showed that β-CD reacted with NIZ to form an inclusion complex. Based on the significant enhancement of the fluorescence intensity of NIZ produced through complex formation, a simple, accurate, rapid and highly sensitive spectrofluorometric method for the determination of NIZ in aqueous solution in the presence of β-CD was developed [78]. The spectrofluorimetric analytical applications of native CDs are summarized in Table 2.

2.2. Application Using Derivatize Cyclodextrins

Native beta-cyclodextrin is very poorly soluble in water (1.85 g/100 mL, 25 °C) [79]. Cyclodextrins modified with hydrophilic substituents have enhanced solubilities in water compared to their native form [80]. While chemical modifications are often aimed at increasing the solubility of the cyclodextrin or of the inclusion complex, the presence of substituents may also contribute to the complexation of the host.

Table 2. Analytical application using native Cyclodextrins

Host	Guest	Linearity	Detection Ex/Em	Application	Ref.
, β-CD	Diclofenac	0-5 μg mL^{-1}	289 nm/362 nm	Pharmaceutical	[43,
				preparations	44]
α-CD	Vitamin K3	0.1-2.0 mg L^{-1}	313	Commercial	
			nm/407,417nm	preparation	[45]
β-CD	rubidate	$2.7\text{-}10^{-2}$-3.0 mg mL^{-1}	244 nm/430 nm	Tablets	[46]
β-CD	Naproxen	0.37–2.60 μg mL^{-1}	330 nm/353 nm	Pharmaceutical	[47]
β-CD	Dapsone	3.39-1.50x10^3ng mL^{-1}	296 nm/ 428 nm	preparations	
		3.7-1500 ngmL^{-1}	285 nm/ 393 nm	tablets and human	[48]
β-CD	napropamide			plasma	
		0.1-2 μg mL^{-1}	296 nm/428 nm	tea, human hairs	[49]
β-CD	ibuprofen	4.7–58 μg mL^{-1}	262 nm/ 288 nm	pharmaceutical	[50]
				preparations	
β-CD	ibuprofen	11-1500 ng mL^{-1}	295 nm/356 nm	serum	[51]
		7.0-1500 ng mL^{-1}	278 nm/332 nm	water	[56]
β-CD	Tetramethrin	0.1-4.0 μg mL^{-1}	270 nm/390 nm	cottonseeds	[57]
β-CD	Carbaryl	20-820 ng mL^{-1}	298 nm/ 481 nm	capsules	[58]
β-CD	Celecoxib	6.38–1.60×10^3 ng	240 nm /365 nm	tap water, lake water	[59]
β-CD	thiabendazole	mL^{-1}		and pineapples	
β-CD	(TBZ)	0–1.40 x10^4 ng mL^{-1}	345 nm/460 nm	capsules and serum	[60]
β-CD	Dequalinium	0.09–0.5 μg mL^{-1} and	264nm/274 nm	serum	[64]
dimer	chloride	0.07–1.0 μg mL^{-1}	340 nm	Urine	[65]
β-CD	Tranilast	0–18 μg kg^{-1}	-	Wheat	[67]
β-CD	SMTX and SGN	0.01–10 μg mL^{-1}.	288 nm/370 nm	pharmaceutical and	[69]
β-CD	aflátoxin B 1			serum	
β-CD	ketoconazole	1.7×10^{-7}-8.8×10^{-5}	360 nm/430 nm	pharmaceutical	[71]
β-CD	metacycline	molL^{-1}		preparations	
β-CD	propranolol		204 -364nm / 230-	human plasma and	[72]
β-CD	ibuprofen	4-20 μg mL^{-1}	500nm	urine	
β-CD	Salbutamol		254 -304 nm/312-	Pharmaceutical	[73]
β-CD	2H1NA	9.0×10^{-7}-2.50×10^{-5}	372 nm	formulation	
β-CD	N-phenyl-1-	mol L^{-1}	344nm/445nm	Water	[74]
	naphthylamine		279.6 nm/608.9nm	Rubber	[75]
β-CD	Cephalexin	0-4.67x10^{-7} M			[76]
		1–10 μg/mL	272 nm/418 nm	Pharmaceutical	[77]
β-CD	nizatidine	0.04-0.1μg mL^{-1}		formulation	
			314 nm/454 nm	Pharmaceutical	[78]
				formulation	

Derivative CDs were found to exhibits stronger inclusive ability than its native CDs. This strong inclusive ability can be understood that the substitution such as methyl, tri-o-methyl or hydroxyl propyl groups leads to the enlargement of the bigger opening and makes it easier for guest molecules to gain access to M-CD [81], TM-CD [71] or HP-CD [82] cavity and to have bigger inclusion constant. In this section the application of derivatize cyclodextrins for determination of various analytes will be shown.

The interaction between various cyclodextrins and melatonin was investigated spectrofluorimetrically [81]. A great increase of fluorescence signal was observed with methyl-β-cyclodextrin (M-β-CD). A spectrofluorimetric method for determination of melatonin was developed having a dynamic range 50-3000 ng ml^{-1} and a limit of detection of 10 ng mL^{-1} was obtained. The method was successfully used for the determination of melatonin in pharmaceutical preparations and urine.

The supramolecules of vitamin K3 (VK3) with β-CD and its derivatives were also investigated by Zhenming et al [82] using spectrofluorimetery. The inclusion constants of VK3 with CDs were determined, and the factors affecting the inclusion process such as CDs concentration and the pH were studied. A method of determining VK3 was developed with the linear range was 2.5×10^{-6} - 5.0×10^{-4} M.

Simple and sensitive spectrofluorimetric method for determination of fluoxetine in commercial formulations and biological fluid was proposed [83]. The method is based on the formation of inclusion complex between methylated- β-CD and fluoxetine in aqueous solution. The method was linear over the concentration 40-1000 μg L^{-1}. The method was simpler and cheaper than other previously reported methods [84, 85].

Jiang, and Hou have described a simple sensitive spectrofluorimetric method for the determination of metacycline (MC) [86]. The method was based on enhancement of the fluorescence intensity of the metacycline (MC)-Eu3+ complex at $\lambda = 612$ nm by Ethyl substituted β-cyclodextrin (DE-β-CD) and formation of inclusion complex MC-Eu3+-DE-β-CD. The linear concentration range was determined to be 2.20×10^{-7}-1.47×10^{-5} mol L^{-1} with limit of detection 3.60×10^{-8} mol L^{-1}. The method was successfully applied to determination of MC in s serum and urine samples.

Spectrofluorimetric method for determination of Vitamin B 12(VB 12) was described [87]. The method was based on enhancement of the fluorescence intensity of erythrosine sodium (ES) due to the formation of inclusion complex with Hydroxypropyl-β-cyclodextrin (HP-β-CD). The standard curves were linear over the concentration range of 0.0-2.1 x 10^{-5} mol L^{-1}. The detection limits were found to be 1.8×10^{-7} mol L^{-1}. The method was satisfactorily applied to the determination of VB 12 in injections.

Sensitive cyclodextrin-induced fluorescence method for the direct determination 5-(p-hydroxyphenyl)-10,15,20-tris-(4-chlorophenyl)porphyrin (*p*-HTClPP) was developed based on the enhancement of the fluorescence intensity of *p*-HTClPP in heptakis(2,3,6-tri-O-methyl)-β-CD (TM-β-CD) medium [88]. Linear relationship between the fluorescence intensity and the concentration of *p*-HTClPP was found to be in the range of 1.0×10^{-9} to 7.0×10^{-6} mol L^{-1}. The limit of detection 8.3×10^{-10} mol L^{-1} was obtained.

Analytical method for determination of α-cypermethrin based on CD-enhanced photochemically induced fluorescence (PIF) was developed [89]. The limits of detection (LOD) and limits of quantification (LOQ) were found to be 6 and 98 ng mL^{-1} respectively. The method was successfully applied to the analysis of tap water and Senegal natural water samples.

Huang et al., have established two spectrofluorimetric methods for determination mangiferin (MA) and meferamic acid (MF) [90,91]. The methods were based on the host-guest inclusion system of ethyl substituted β-cyclodextrin (DE-β-CD), and a modified β-cyclodextrin (MAH-β-CD), with MA and MF, respectively.

The linear ranges were 2.00×10^{-8}-7.00×10^{-6} mol L^{-1} and 2.00×10^{-8}-9.00×10^{-5} mol L^{-1} for MA and ME, respectively. The detection limit was 4.05×10^{-9} mol L^{-1} and 3.36×10^{-9} mol L^{-1}. The proposed methods were successfully applied to determine, MA and MF in serum, tablets, and urine. The spectrofluorimetric analytical applications of derivatize CDs are summarized in Table 3.

Table 3. Analytical application using derivative cyclodextrins

Host	Guest	Linearity	Detection Ex/Em	Application	Ref.
M-β-CD	Melatonin	50-3000 ng mL^{-1}	280 nm/355 nm	Pharmaceutical	[81]
			330 nm/353 nm	preparations and	
SBE-β-CD	Vitamin K3	2.5×10^{-6}-5.0×10^{-4} M	224 nm/294 nm	urine	[82]
			360 nm/430 nm	Tablets	
M-β-CD	Fluoxetine	40-1000 μg L^{-1}	529nm/549 nm		[83]
			334 nm/656	Pharmaceutical and	
DE-β-CD	Metacycline	2.20×10^{-7}-1.47×10^{-5}	282 nm/318 nm,	biological fluid	[86]
		mol L^{-1}	282nm/316nm	water	
HP-β-CD	Vitamin B 12	0.0~2.1×10^{-5} mol	406 nm /505 nm		[87]
	[*p*-HTClPP]	L^{-1}	334 nm / 462 nm	serum and urine	
TM-β-CD	α-cypermethrin	1.0×10^{-9} -7.0×10^{-6}			[88]
		mol L^{-1}		water	[89]
HP-β-CD	Mangiferin	21–8.333 ng mL^{-1},		pharmaceutical	
and β-CD	(MA)	166–8,330 ng mL^{-1}		preparations	[90]
DE-β-CD				serum, tablets, and	
	Mefenamic	2.0×10^{-8}-7.00×10^{-6}		urine	[91]
MAH-β-CD	acid (MF)	mol L^{-1}		serum, tablets, and	
		2.0×10^{-8}-9.00×10^{-5}		urine.	
		mol L^{-1}			

2.3. Application Using Cyclodextrins and Binding Agent

Tang et al [92, 93] have presented novel spectrofluorimetric methods for determination of zinc and cadmium with β-cyclodextrin-o-vanillin benzoylhydrazone and β-cyclodextrin-o-vanillin furfuralhydrazone, respectively. The fluorescent product was measured spectrofluorimetrically at emission 486 nm and excitation at 396 nm for zinc, and at emission 494 nm and excitation at 393 nm for cadmium. The method were found to be linear in range of 2.50 - 500 μgL^{-1}, with determination limit of 0.608 μgL^{-1} for zinc and from 3.0-500 μgL^{-1} with a detection limit of 0.80μgL^{-1} for cadmium. The method was applied to the determination of trace amounts of Zn^{2+} and Cd^{2+} in tea, human hairs, and mussel with satisfactory results.

A new sensitive and selective spectrofluorimetric method for the determination of copper (II) was proposed by Feng et al [94]. The method is based on the enhancement of fluorescent intensity by binding copper (II) with 1,3-bis(2-benzothiazolyl-diazoamino)benzene (BBTAB) and β-cyclodextrin (β-CD). Under the optimum condition the linear concentration range was determined to be 3.0×10^{-7} to 1.0×10^{-5} mol L^{-1}, and the detection limit is determined to be 1.2×10^{-8} mol L^{-1}. The method was applied determination of trace copper (II) in water samples. The results were compared with those obtained by the reported flame atomic absorption spectrometry (FAAS) [95] method.

The results have no significant differences in both cases. The spectrofluorimetric analytical applications of CDs and binding agent are summarized in Table 4.

Table 4. Analytical Application using cyclodextrins and binding agent

Host	Guest	Linearity	Detection Ex/Em	Application	Ref.
β-CD-o-vanillin benzoylhydrazone	Zinc	2.50 - 500 $\mu g\ L^{-1}$	396 nm/486 nm	tea, human hairs, and mussel	92,93
β-CD-o-vanillin furfuralhydrazone	Cadmium	3.0-500 μgL^{-1}	393 nm/ 494 nm	tea, human hairs, and mussel	[94]
H-β-CD	Copper (II)	3.0×10^{-7}-1.0×10^{-5} mol L^{-1}	389 nm/ 480 nm	water	[95]

Conclusion

We have presented a review on analytical applications of cyclodextrins and their derivatives using fluorescence spectroscopy. CDs or their derivatives have shown to be very effective in enhancement of fluorescence intensity. Many methods based on the enhancement of fluorescence intensity of the inclusion complexes with CDs have been proposed for the determination of several pharmaceutical drugs, pesticides, and metal ions.

Most published studies on CDs have been limited to native β-CD, and their derivatives this due to their suitable size more over increasing it solubility by derivatization. In recent years it have been shown that the molecular fluorescence and CD allied to chemo-metric methods such as (PARAFAC) can be used for the quantification of enantiomers in biological fluids.

References

[1] Szejtli J. *Cyclodextrins and their Inclusion Complexes*, Akademiai Kiado, Budapest, 1982.

[2] Scypinski S, Cline Love LJ. Cyclodextrin-induced room-temperature phosphorescence of nitrogen heterocycles and bridged biphenyls. *Anal. Chem.*, 1984; 46:331-6.

[3] Muñoz De La Peña A, Ndou TT, Zung JB, Greene KL, Live DH, Warner IM. Alcohol size as a factor in the ternary complexes formed with pyrene and β-cyclodextrin . *J. Am. Chem. Soc., 1991;* 113: 1572-7.

[4] Sano A, Takezawa M, Takitani S. Spectrofluorimetric determination of cyanide in blood and urine with naphthalene-2,3-dialdehyde and taurine. *Anal. Chim. Acta*, 1989; 225:351-8.

[5] Escandar GM, Muñoz De La Peña A. Room-temperature phosphorescence of acenaphthene in aerated solutions in the presence of bromoalcohols and γ-cyclodextrin. *Anal. Chim. Acta*, 1998; 370: 199-205.

[6] Fanali S. Enantioselective determination by capillary electrophoresis with cyclodextrins as chiral selectors. *Journal of Chromatography A,* 2000; 875:89-122.

[7] Hazekamp, A., & Verpoorte, R. Structure elucidation of the tetrahydrocannabinol complex with randomly methylated b-

cyclodextrin. *European Journal of Pharmaceutical Sciences*, 2006, 29, 340–347.

[8] Jullian, C., Miranda, S., Zapata-Torres, G., Mendizabal, F., & Olea-Azar, C.. Studies of inclusion complexes of natural and modified cyclodextrin with (+) catechin by NMR and molecular modeling. *Bioorganic & Medicinal Chemistry*, 2007, 15, 3217–3224.

[9] Liu, L. X., & Zhu, S. Y. Preparation and characterization of inclusion complexes of prazosin hydrochloride with b-cyclodextrin and hydroxypropyl-b-cyclodextrin. *Journal of Pharmaceutical and Biomedical Analysis*, 2006, 40, 122–127.

[10] Rajabi, O., Tayyari, F., Salari, R., & Tayyari, S. F. Study of interaction of spironolactone with hydroxypropyl-b-cyclodextrin in aqueous solution and in solid state. *Journal of Molecular Structure*, 2008, 878, 78–83.

[11] Tavornvipas, S., Hirayama, F., Arima, H., Uekama, K., Ishiguro, T., Oka, M., Hamayasu, K., Hashimoto, H.: 6-O-a-(4-O-a-Dglucuronyl)- D-glucosyl-b-cyclodextrin:solubilizing ability and some cellular effects. *Int. J. Pharm.* 249, 199–209 (2002).

[12] Padukka, I., Bhandari, B.D., Arcy, B.: Evaluation of various extraction methods of encapsulated oil from b-cyclodextrin-lemon oil complex powder. *J. Food Compos. Anal.* 13, 59–70 (2000).

[13] Szente, L., Szejtli, J.: Cyclodextrins as food ingredients. *Trends Food Sci. Technol.* 15, 137–142 (2004).

[14] Schneiderman, E., Stalcup, A.M.: Cyclodextrins: a versatile tool in separation science. *J. Chromatogr.* B 745, 83–102 (2000).

[15] Perrin, C., Fabre, H., Maftouh, M., Massart, D.L., Heyden, Y.V.: Robustness testing of chiral separations by capillary electrophoresis using highly-sulfated cyclodextrins. *J. Chromatogr. A* 1007, 165–177 (2003).

[16] Morillo, E., Pe´rez-Martı´nez, J.I., Gine´s, J.M.: Leaching of 2,4-D from a soil in the presence of b-cyclodextrin: laboratory columns experiments. *Chemosphere* 44, 1065–1069 (2001).

[17] Jozefaciuk, G., Muranyi, A., Fenyvesi, E.: Effect of randomly methylated b-cyclodextrin on physical properties of soils. *Environ. Sci. Technol.* 37, 3012–3017 (2003).

[18] Fro¨mming KH, Szejtli J, in: J.E. Davis (Ed.), *Topics in Inclusion Science: Cyclodextrin in Pharmacy*, Kluwer, Dordrecht, 1994.

[19] Wenz G. Cyclodextrins as building blocks for supramolecular structures and functional units. *Angew. Chem. Int. Ed. Engl.*, 1994; 33: 803-22.

[20] Schneider HJ. Mechanisms of molecular recognition: investigations of organic host–guest complexes. *Angew. Chem. Int. Ed. Engl.*, 1991; 30: 1417-1436.

[21] Li S, Purdy WC. Cyclodextrins and their applications in analytical chemistry. *Chem. Rev.*, 1992; 92: 1457-70.

[22] Maafi M, Mahedero MC, Aaron J.-J. Fluorimetric properties of a 2-hydroxypropyl-β-cyclodextrin: 9-methyl-benzo[a]phenothiazine inclusion complex in aqueous media. Analytical usefulness. *Talanta*, 1997; 44: 2193-99.

[23] Schuette Jodi M, Will A, Yvette Agbaria Rezik A, Warner Isiah M. Fluorescence characterization of the cyclodextrin/pyrene complex interaction with chiral alcohols and diols , *Appl. Spectrosc.*, 1995; 48: 581-6.

[24] Stalcup AM, Gahm KH. Chiral separation of chloroquine using heparin as a chiral selector in high-performance liquid chromatography. *Anal. Chem.*, 1996; 68: 2248–50.

[25] Armstrong DW, Yang X, Han SM, Menges RA. Direct liquid chromatographic separation of racemates with an. alpha.-cyclodextrin bonded phase. *Analytical,* 1987; 59: 2594-96.

[26] Armstrong DW, Faulkner Jr, JR, Han SM. Use of hydroxypropyl-and hydroxyethyl-derivatized [beta]-cyclodextrins for the thin-layer chromatographic separation of enantiomers and diastereomers *J. Chromatogr.*, 1988; 452: 323-30.

[27] Ward TJ. For capillary electrophoresis. *Anal. Chem.*, 1994; 66: 632A-40.

[28] Elbashir AA, Suliman FO. Computational modeling of capillary electrophoretic behavior of primary amines using dual system of 18-crown-6 and β-cyclodextrin. *Journal of Chromatography A*, 2011; 1218: 5344-51.

[29] Elbashir AA, Saad B, Ali ASM, Saleh MI, Aboul-Enein HY. Determination of quinocide as impurity in primaquine tablets by capillary zone electrophoresis. *Biomedical Chromatography*, 2009; 23: 464-71

[30] Elbashir AA, Suliman FEO, Saad B, Aboul-Enein HY. Capillary electrophoretic separation and computational modeling of inclusion complexes of β-cyclodextrin and 18-crown-6 ether with primaquine and quinocide . *Biomedical Chromatography*, 2010; 24: 393-8.

[31] Kinoshita T, Iinuma F, Tsuji A. Microanalysis of proteins and peptides. I. Enhancement of the fluorescence intensity of dansyl amino acids and

dansyl proteins in aqueous media and its application to assay of amino acids and proteins. *Chem. Pharm. Bull.*, 1974; 22: 2413–20.

[32] Kinoshita T, Iinuma F, Tsuji A Fluorescent labeling of proteins and a plasma membrane using cycloheptaamylose-dansyl chloride complex. *Anal. Biochem.,* 1974; 6: 632–7.

[33] Kinoshita T, Iinuma F, Atsumi K, Kanada Y, Tsuji A Microanalysis of proteins and peptides. V. Enhancement of fluorescence intensity of dansyl aminoacids on silica gel plates using cycloheptaamylose. *Chem. Pharm. Bull.*, 1975; 23: 1166–68.

[34] Garcia Sanchez F, Ramos Rubio AL, Cruces Blanco C, Hernandez Lopez M, Marquez Gomez JC, Carnero C. Enhanced fluorimetric determination of procaine in pharmaceutical preparations by aqueous β-cyclodextrin inclusion and non-aqueous competitive action. *Anal. Chim. Acta*, 1988; 205:139-47.

[35] Duran-Meras I, De la Pena AM, Salinas F, Caceres IR. Spectrofluorimetric determination of nalidixic acid based on host–guest complexation with γ-cyclodextrin. *Analyst*, 1994; 119: 1215-19.

[36] Marquez JC, Hernandez M, Garcia Sanchez F. Enhanced Spectrofluorimetric Determination of the Pesticide Warfarin by Means of the Inclusion Complex With β-Cyclodextrin. *Analyst*, 1990; 115: 1003-5.

[37] Raquel E, Galian Alicia V, Veglia and Rita H. de Rossi Cyclodextrin enhanced fluorimetric method for the determination of tryptamine. *Analyst*, 1998; 123: 1587–91.

[38] Frankewich RP, Thimmaiah KN, Hinze WL. Evaluation of the relative effectiveness of different water-soluble β-cyclodextrin media to function as fluorescence enhancement agents *Anal. Chem.*, 1991; 123: 2924-33.

[39] Garcia Sanchez F, Hernandez Mpez M, and Heredia A. Spectrofluorimetric determination of beryllium based on the inclusion complex of 1-amino-4-hydroxyanthraquinone with β-cyclodextrin as ligand. *Anal. Chim. Acta,* 1986; 187:147-55.

[40] Garcia Sanchez F, Hernandez Mpez M, and MBrquez Gdmez JC. Fluorimetric determination of scandium using the cyclodextrin - 1,4-dihydroxyanthraquinone inclusion complex. *Analyst*, 1987; 112:1037-404.

[41] Garcia Sanchez F, Hernandez Mpez M, and Marquez G6mez JC. Fluorimetric determination of gallium by means of a cyclodextrin inclusion compound as ligand. *Fresenius'* Z. *Anal. Chem.,* 1987; 328: 499-500.

[42] Valeur B, Leray I. Design principles of fluorescent molecular sensors for cation recognition Coord. *Chem. Rev.*, 2000; 205:3–40.

[43] Arancibia JA, Boldrini MA, Escandar GM. Spectrofluorimetric determination of diclofenac in the presence of α-cyclodextrin. *Talanta,* 2000; 52:261-68.

[44] Arancibia JA, Escandar GM. Complexation study of diclofenac with β-cyclodextrin and spectrofluorimetric determination. *Analyst*, 1999; 124: 1833-38.

[45] Nevado JJB, Pulgarín JAM, Laguna MAG. Spectrofluorimetric study of the β-cyclodextrin: Vitamin K3 complex and determination of vitamin K3. *Talanta*, 2001; 53: 951-959.

[46] Tang B, Ma L, Ma C. Spectrofluorimetric study of the b-cyclodextrin_rubidate complex and determination of rubidate by β-CD-enhanced fluorimetry. *Talanta*, 2002; 58: 841-8.

[47] Arancibia JA, Escandar GM. Determination of naproxen in pharmaceutical preparations by room-temperature phosphorescence. A comparative study of several organized media. *Analyst*, 2001; 126: 917-22.

[48] Li Ma, Bo Tang Chun Chu Spectrofluorimetric study of the ß cyclodextrin–dapsone–linear alcohol supramolecular system and determination of dapsone. *Analytica Chimica Acta*, 2002; 469: 273–83.

[49] Tang B, Liu W.-L, Wang Y, Chen Z.-Z. Studies on the supramolecular interaction between napropamide and β-cyclodextrin by spectrofluorimetry and its analytical application. *Analytica Chimica Acta*, 2004; 509:145-50.

[50] Manzoori JL, Amjadi M. Spectrofluorimetric study of host-guest complexation of ibuprofen with β-cyclodextrin and its analytical application. *Spectrochimica Acta - Part A: Molecular and Biomolecular Spectroscopy*, 2003; 59 :909-916.

[51] Ravisankar S, Vasudevan M, Gandhimathi M, Suresh B. Reversed-phase HPLC method for the estimation of acetaminophen, ibuprofen and chlorzoxazone in formulations. *Talanta,* 1998; 46: 1577-81.

[52] Garrigues S, Gallignani M, de la Guardia M. FIA-FT-IR determination of ibuprofen in pharmaceuticals. *Talanta,* 1993; 40:89-93.

[53] Sádecká J, Čakrt M, Hercegová A, Polonský J, Skačáni I. Determination of ibuprofen and naproxen in tablets. *J. Pharm. Biomed. Anal.*, 2001; 25: 881-91.

[54] Abdel-Hay MH, Korany MA, Bedair MM, Gazy AA. Colorimetric determination of seven nonsteroidal anti-inflammatory drugs using 2-

nitrophenylhydrazine hydrochloride. *Analytical Letters*, 1990; 23: 281-94.

[55] Hergert LA, Escandar GM. Spectrofluorimetric study of the β-cyclodextrin-ibuprofen complex and determination of ibuprofen in pharmaceutical preparations and serum. *Talanta*, 2003; 60: 235-46.

[56] Liu W, Tang B, Wang Y, Chen Z. Studies on the Supramolecular Interaction between Tetramethrin and β-Cyclodextrin by Spectrofluorimetry and Its Analytical Application. *Analytical Letters*, 2004; 37: 755-66.

[57] Liu W.-L , Tang B. Sensitized effect of β-cyclodextrin on the fluorescence in the determination of carbaryl. *Journal of Environmental Sciences*, 2004; 16: 829-31.

[58] Manzoori J.L, Abdolmohammad-Zadeh H, Amjadi M. Study on the inclusion complex between β-cyclodextrin and celecoxib by spectrofluorimetry and its analytical application. *Farmaco*, 2005; 60: 575-81.

[59] Tang B, Wang X, Wang G, Wang Y, Chen Z. Highly sensitive and selective room-temperature phosphorescence determination of thiabendazole by the supramolecular interaction of thiabendazole/β-cyclodextrin/tritonX-100. *Analyst*, 2005; 130:1038-45.

[60] Tang B, Liang H.-L, Xu K.-H, Mao Z, Shi X.-F, Chen Z.-Z. An improved synthesis of disulfides linked β-cyclodextrin dimer and its analytical application for dequalinium chloride determination by spectrofluorimetry. *Analytica Chimica Acta*, 2005; 554:31-6.

[61] Gagliardi L, Cavazzutti G, Tonelli D Determination of dequalinium chloride and related impurities in cosmetics and pharmaceuticals by reversed-phase HPLC. *Analytical Letters*, 1998; 31:829-839.

[62] Tang B. Jia BX, Shi XF. Flow injection spectrofluorimetric study of the supramolecular interaction between β-cyclodextrin and dequalinium chloride and its analytical application. *Spectrochim. Acta Part A*, 2004; 61:2203-09.

[63] Taylor RB, Toasaksiri S, Reid RG. Determination of antibacterial quaternary ammonium compounds in lozenges by capillary electrophoresis. *J. Chromatogr. A*, 1998; 798:335-43.

[64] Tang B, Liang H.-l, Tong L.-l, Li P. Synthesis of ethylenediamine linked β-cyclodextrin dimer and its analytical application for tranilast determination by spectrofluorimetry. *Bioorganic and Medicinal Chemistry*, 2006; 14: 3947-52.

[65] Diez NM, De La Peña, AM, García MCM, Gil DB, Cañada-Cañada F. Fluorimetric determination of sulphaguanidine and sulphamethoxazole by host-guest complexation in β-cyclodextrin and partial least squares calibration. *Journal of Fluorescence*, 2007; 17:309-318.
[66] Caballero RD, Carda-Broch S, Garc΄ıa-A΄ lvarez-Coque MC. Hydro-organic and micellar-organic reversed-phase liquid chromatographic procedures for the evaluation of sulphonamides in pharmaceuticals. *Analytical Letters,* 2001; 34:1189–203.
[67] Hashemi J, Kram GA, Alizadeh N. Enhanced spectrofluorometric determination of aflátoxin B 1 in wheat by second-order standard addition method. *Talánta*, 2008; 75:1075-81.
[68] Official Methods of Analysis of AOAC International, 17th ed., vol. 2, *Gaithersburg*, 2000, p. 35.
[69] Jalali F, Afshoon A. Spectrofluorimetric study and detection of ketoconazole in the presence of β-cyclodextrin. *Journal of Fluorescence*, 2008; 18:219-25.
[70] The United States Pharmacopeia (USP) *US pharmacopeia convention,* Inc., Rockville, MD, vol. 1, 2004;p 1056.
[71] Ao D.-G, Si, Q. Studies and application of fluorescence of metacycline enhanced by synergetic effect of β-cyclodextrin and DBS *Guang Pu Xue Yu Guang Pu Fen Xi/Spectroscopy and Spectral Analysis*, 2008;28: 2640-2643..
[72] Valderrama P, Poppi RJ. Determination of propranolol enantiomers in plasma and urine by spectrofluorimetry and second-order standard addition method. *Analytica Chimica Acta*, 2009; 651:31-5.
[73] Valderrama P, Poppi RJ. Second order standard addition method and fluorescence spectroscopy in the quantification of ibuprofen enantiomers in biological fluids. *Chemometrics and Intelligent Laboratory Systems*, 2011; 106:160-5.
[74] Pandya H N, Berawala H H, Khatri D M, Mehta PJ. Spectrofluorimetric estimation of salbutamol sulphate in different dosage forms by formation of inclusion complex with β-cyclodextrin. *Pharmaceutical Method* 2010; 1:49-53.
[75] Guo Y.-J, Zhang P, Chao J.-B, Shuang S.-M, Dong C. Study on the supramolecular system of 5-(p-hydroxyphenyl)-10,15,20-tris-(4-chlorophenyl)porphyrin with cyclodextrins and its analytical characteristics *Spectrochimica Acta - Part A: Molecular and Biomolecular* Spectroscopy, 2008; 71: 946-50.

[76] Ng SM, Narayanaswamy R. Spectrofluorometry study of β-cyclodextrin and N-phenyl-1-naphthylamine inclusion complex and its analytical application via artificial neural network. *Journal of Inclusion Phenomena and Macrocyclic Chemistry* 2012; 72:357-65.

[77] Randa Mohamed Ahmed Mohamed Dani, Abdalla Ahmed Elbashir, Host-guest Inclusion complex of β-cyclodextrin and Cephalexin and its Analytical Application. *International Journal of Pharmaceutical Chemistry Research.* 2, 2013.

[78] Elbashir, A.A., Ibrahim, M.A., Aboul-Enein, H.Y. Fluorimetric investigation of host-guest complexation of nizatidine with β-cyclodextrin and its analytical application. *Gazi University Journal of Science* 2013, 26:151-156.

[79] Qian L, Guan Y, Xiao H. Preparation and characterization of inclusion complexes of a cationic β-cyclodextrin polymer with butylparaben or triclosan. *Int. J. Pharm.* 2008; 357:244–51.

[80] Fakayode SQ, Swamidoss IM, Busch MA, Busch KW. Determination of the enantiomeric composition of some molecules of pharmaceutical interest by chemometric analysis of the UV spectra of guest-host complexes formed with modified cyclodextrins. *Talanta* 2005;65:838–45.

[81] Pola ML, Algarra M, Becerra A. Hernandez M. Cyclodextrin enhanced spectrofluorimetric determination of melatonin in pharmaceuticals and urine. *Analytical Letters*, 2000; 33:891-903.

[82] Zhenming D, Xiuping L, Guomei Z, Shaomin S, Jinghao P. Study on vitamin K3-cyclodextrin inclusion complex and analytical application *Spectrochimica Acta - Part A: Molecular and Biomolecular Spectroscopy*, 2003; 59 :2073-79.

[83] Berzas JJ, Alañón A, Lázaro JA. Cyclodextrin enhanced spectrofluorimetric determination of fluoxetine in pharmaceuticals and biological fluids. *Talanta*, 2002; 58:301–9.

[84] Starczewska B, Puzanowska-Tarasiewicz H, Baranowska K. Investigation and analytical application of the reactions of eriochrome cyanine R with fluvoxamine and fluoxetine. *J. Pharm. Biomed. Anal.*, 2000; 23:477-81.

[85] Yilmaz N, Özkan Y, Özkan SA, Biryol I, Aboul-Enein HY. High performance liquid chromatographic assay and drug dissolution studies of fluoxetine hydrochloride in capsule formulations. *J. Liq. Chromatogr. Relat. Technol.*, 2000; 23:1699-710.

[86] Jiang C, Hou F. Fluorescence study of the interaction between a modified β-cyclodextrin and metacycline-Eu3+ and its analytical application. *Analytical Letters*, 2004; 37:1885-95.

[87] Sun J, Zhu X, Wu M. Hydroxypropyl-β-cyclodextrin enhanced determination for the vitamin B 12 by fluorescence quenching method. *Journal of Fluorescence*, 2007; 17:265-270.

[88] Guo Y.-J, Zhang P, Chao J.-B, Shuang S.-M, Dong C. Study on the supramolecular system of 5-(p-hydroxyphenyl)-10,15,20-tris-(4-chlorophenyl)porphyrin with cyclodextrins and its analytical characteristics *Spectrochimica Acta - Part A: Molecular and Biomolecular Spectroscopy*, 2008; 71: 946-50.

[89] Mbaye M, Gaye Seye MD, Coly, A, Tine A, Aaron J.-J. Usefulness of cyclodextrin media for the determination of α-cypermethrin by photochemically induced fluorescence: Analytical applications to natural waters. *Analytical and Bioanalytical Chemistry*, 2009; 394:1089-98.

[90] Huang L, He J, Lu R, Ge X, GuoJ. Investigation on a host-guest inclusion system by β-cyclodextrin derivative and its analytical application. *Bioorganic and Medicinal Chemistry Letters*, 2011; 21:1113-9.

[91] Huang L, He J, Ge X, Lu R, Guo J. Fluorimetric investigation of supramolecular system by modified β-cyclodextrin and its analytical application. *Spectrochimica Acta - Part A: Molecular and Biomolecular Spectroscopy*, 2011; 78:1553-1559.

[92] Tang B, Zhang J, Chen Z. Synthesis and characterization of a novel cross-linking complex of β-cyclodextrin-o-vanillin benzoylhydrazone and its selective spectrofluorimetric determination of trace amounts of zinc. *Spectrochimica Acta - Part A: Molecular and Biomolecular Spectroscopy*, 2003;59 : 2519-26.

[93] Tang B, Zhang L, Zhang J, Chen Z.-Z, Wang Y. Synthesis of a novel host molecule of cross-linking-polymeric-β- cyclodextrin-o-vanillin furfuralhydrazone and spectrofluorimetric analysis of its identifying cadmium. *Spectrochimica Acta - Part A: Molecular and Biomolecular Spectroscopy*, 2004; 60:2425-31.

[94] Feng F, Liao X, Chen Z, Lin S, Meng S, Lu Z. Spectrofluorimetric study of the interaction of copper(II) and bis-heterocyclictriazene reagent in the presence of β-cyclodextrin. *Analytica Chimica Acta*, 2006; 575: 68-75.

[95] Ferreira SLC, Lemos VA, Moreira BC, Costa ACS, Santelli RE. An on-line continuous flow system for copper enrichment and determination by flame atomic absorption spectroscopy. *Anal. Chim. Acta* 2000; 403: 259-64.

In: Oligosaccharides
Editors: L. S. Schweizer, S. J. Krebs
ISBN: 978-1-62948-328-3

Chapter VIII

Utilization of Soy Oligosaccharides during Fermentation by Lactobacilli

Subrota Hati,[1,*] Shilpa Vij,[2] Surajit Mandal[2], Kaushik Khamrui[2] and J.B. Prajapati[1]
[1]Dairy Microbiology Department, S. M. C. College of Dairy Science, Anand Agricultural University, Anand, Inidia
[2]National Dairy Research Institute, Deemed University, Karnal, India

Abstract

Soymilk is the water extract of soybean. The major carbohydrates present in soy milk are sucrose, raffinose and stachyose and is free of lactose and cholesterol. The main problem of limiting the widespread consumption of soy products is their objectionable beany flavour and high levels of oligosaccharides, raffinose and stachyose, which can lead to flatulence when consumed in large amounts. An alternative approach would be the utilization of microorganisms possessing α-galactosidase activity which can hydrolyze the oligosaccharides during fermentation. This would provide simple sugars as substrate for the growth of lactic acid bacteria as well as reduce the level of oligosaccharides present in soy

* Correspondence author: Email id: subrota_dt@yahoo.com.

milk. Growth of lactic cultures in soymilk has been reported as a good growth medium. They produce small amounts of lactic acid, acetic acid etc during fermentation and also lower down the pH, making the environment acidic against pathogens. Soy products are considered as a good substrate for functional foods since fermentation by lactic acid bacteria has the potential to reduce the levels of some carbohydrates which can be responsible for gas production in the intestinal system, increase free isoflavone levels and flavor desirable changes in bacterial populations in the gastrointestinal tract, antioxidant activity, anticarcinogenic activity, antimicrobial activity and phytoestrogen activity in post menopausal women. Some researchers have reported the utilization of oligosaccharides during fermentation of soymilk using either pure or mixed culture of the following probiotic lactobacilli for the development of functional fermented soy foods: *Lactobacillus cellobiosis, L. plantarum, L. fermentum, L. fermenti, L. pentosaceus, L casei, L. rhamnosus.*

Introduction

People mainly vegetarians are searching cheapest source of proteins to meet protein requirement in their diet. Soybean, (*Glycine max*) (L) Merr. is economically the most important bean in the world, providing vegetable protein for millions of people and ingredients for hundreds of chemical products and a potential source of bioactive peptides. It also contains carbohydrates, phytochemicals, saponins, phytic acid and fiber, all thought to provide health benefits. Soybean products can be good substitutes for animal products because, unlike some other beans, soybean offers a "complete" protein profile [1]. Legumes constitute one of the richest and least expensive sources of protein in the human diet throughout the globe. The proteins are an essential component of the diet, needed for survival of humans and animals. The basic function of proteins in nutrition is to supply adequate amounts of required amino acids and helps in muscle building. Soybean is an important pulse crop for human and animal nutrition. Soybean (Glycine max) is one such important plant in human and animal nutrition. Approximately 60% of all processed food products include ingredients derived from soy. Soy products have superior nutritional characteristics in terms of high protein contents and amino acids and serve as a low cost substitute to dairy milk in developing countries and as a nutritive supplement for the lactose intolerant population [2]. Soymilk is lactose free and a good source of essential fatty acids. It contains no cholesterol and little or no saturated fat. Soymilk can be a good source of high quality protein, B vitamins, potassium, iron, dietary fiber, and bio-active components, including isoflavones. Many soymilks are fortified with calcium, vitamins A and D, riboflavin, zinc, and vitamin B12. Important bio-active components, found naturally in soybeans are being studied in relation to relieving menopausal symptoms, such as hot flashes, maintaining healthy bones, and preventing prostate, breast cancers, and colorectal cancer. The content and profile of bio-active components varies from product to product, depending upon how much soy protein is in the food and how the soy protein is processed. Soymilk is a healthy, high-quality protein source that contains all essential amino acids needed for growth. In general, soy protein products are equal in quality to animal products. In addition to the excellent nutritional value of soy protein, scientists have found that consumption of soy protein can contribute to reducing the risk of heart disease. The FDA has approved a health claim stating that "25 grams of soy protein in a daily diet low in saturated fat and cholesterol can help reduce total and LDL cholesterol that is moderately high to high." Soy milk is also free from cholesterol and rich in dietary fiber. Soybean has several health beneficial agents such as antioxidants that protect cells from free radicals and prevent cancer and premature ageing of the cells. Soybean has concentrated amounts of isoflavones that possibly prevent heart diseases, lower cholesterol levels and maintain bone health. It also contains sterols and sterolins that boost antibody production [3]. Furthermore it contains an unfavorable beany flavor and large amount of galacto oligosaccharides such as raffinose and stachyose that are indigestible by mammals and are flatulence-causing factors [4]. Flatulence is the common symptom associated with consumption of pulses but more serious accompanying consequences like abdominal pain and diarrhea overshadow its social implications, [5]. α-galactosidase is abundant in plants, microorganisms and animals. It hydrolyzes a variety of simple α-D-Galactoside as well as more complex molecules such as oligosaccharides and polysaccharides [6, 7]. Several groups have developed enzymatic processes using α-galactosidase from plant [8], bacterial [9] and fungal sources [10, 11, 12] for the hydrolysis of galacto-oligosaccharides of raffinose family. [10] have successfully demonstrated the removal of significant amount of oligosaccharides by *Aspergillus oryzae* (A. oryzae) cells entrapped in agarose. Their result indicates a possibility of using whole cells for pre-treatment of soymilk to overcome the physiological problems of flatulence. They suggest constructing a fluidized bed reactor for continuous hydrolysis of RFOs in soymilk. Recent trend in research work has opened a new horizon on oligosaccharides for their beneficial physiological effect on the microflora of the gastrointestinal tracts in recent years, and to their applications in the development of functional foods, due to their bioactivities. It has been suggested that soybean oligosaccharides can significantly reduce abnormal blood glucose, lipid level and oxidative stress in an animal model. Prebiotic oligosaccharides are defined as food ingredients which affect the physiological function of the body in a targeted way to show a positive effect, which may justify health claims in due course [13]. However, excessive ingestion of oligosaccharides may cause intestinal discomfort, flatulence and diarrhoea due to the osmotic effect, high fermentation rates and production of gas [14]. Amongst more than 400 species of bacteria present in the intestine of adults, lactobacilli are considered to be the most beneficial to human health [15]. Members of this genus can contribute to digestion, absorption of nutrients, and prevention of colonisation by pathogens and stimulation of the immune response in humans [16]. Soymilk, as well as milk, is suitable for the growth of lactic acid bacteria (*Streptococcus thermophilus, Lactobacillus acidophilus, Lactobacillus rhamnosus, Lactobacillus casei, Lactobacillus bulgaricus*) and bifidobacteria (*Bifidobacterium longum* and *Bifidobacterium infantis*) [17, 18]. Stimulative factors for growth are considered to be oligosaccharides (Prebiotics-Raffinose, Stachyose etc.) [19], amino acids and peptides [20]. Moreover, soy products are now considered to have a potential role in prevention of chronic diseases such as atherosclerosis, cancer, osteoporosis and menopausal disorder [21, 22].

Although rich in high quality protein and aglycones, soybean contains oligosaccharides such as raffinose and stachyose that are not digested by human beings and may cause flatulence or discomforts [23]. To overcome these limitations and to develop functional fermented soy products, fermentation of soymilk by lactic acid bacteria, Bifidobacter and different fungi producing α-galactosidase enzymes to exert beneficial effects on human hosts [24].

Utilization of Oligosaccharides Present in Soy Milk

The carbohydrates of soybeans, containing little starch and hexose, are largely polysaccharides with some oligosaccharides. Carbohydrates make up approximately 35% of the soybean. Galacto-oligosaccharides (raffinose, stachyose and verbascose) comprise approximately 5% of the soybean dry matter, while starch represents less than 1% [25]. An important benefit of dietary fiber to humans is increasing the water-holding capacity of stools. Increased stool bulk, softness and increased transit time may reduce diverticular disease, hemorrhoids and possibly other diseases of the lower gastrointestinal tract. Soy milk (also called soya milk, soymilk, soybean milk, or soy bean juice) and sometimes referred to as soy drink/beverage is a beverage made from soybeans. A stable emulsion of oil, water and protein, it is produced by soaking dry soybeans and grinding them with water.

Table 1. Comparative Analysis of Soymilk with Cow milk [26]

Milk type	Whole Cow's Milk	Skimmed Cow's Milk	Soya Milk
Calories	150	90	133
Calories from fat	70	0	34
Total fat (g)	8	0	4
Saturated fat	5	0	1
Cholesterol (mg)	35	<5	0
Sodium (mg)	125	125	104
Carbohydrates (g)	12	13	13
Fibre (g)	0	0	3
Sugars (g)	12	12	Not Given
Protein (g)	8	8	9
Vit.A (%)	6	10	24
Vit.C (%)	4	4	0
Calcium (%)	30	30	19
Iron (%)	0	0	9

A liquid food obtained as a result of combining:

1) Aqueous-extracted whole soybean solids and water; or,
2) Other edible-quality soy protein solids, soybean oil, and water; to provide the basic compositional levels recommended in Table 1.

The composition of soymilk depends on different factors, such as the bean composition and parameters that depend on the type of process being used, such as the maceration time, the squeezing and dilution level during the process [27]. Soymilk produced by the traditional process presents the following composition, in average: 3.4% of protein, 1.8% of lipids, 1.5% of carbohydrate and 0.4% of ash. There are several different methods for soymilk processing. The traditional method involves soaking the beans, followed by wet grinding (cold water grinding), filtering and cooking [28]. The traditional process results in a typical oriental soymilk with an intense characteristic of "beany" flavor, which is not generally accepted by the westerns consumers. Such off-flavor characteristic results from the reaction catalyzed by soybean enzyme lipoxygenase that acts almost immediately upon wet grinding [29].The use of hot water (80-100°C) to inactivate the lipoxygenase either previously or during the grinding is the basis of other methods of the soymilk processing. The referred methods avoid the reaction of the enzyme and the formation of the attribute "beany" flavor. Some of those methods involve, like the traditional one, the grinding of the beans and the discharge of the solid residue retained in the filtration. In other methods, the bean is ground with hot water and the remaining slurry is neither filtered nor centrifuged. These methods have the advantage of avoiding generating solid residue. In one of these methods, the slurry is homogenized in two stages. One negative aspect of the methods that use the whole grain to produce soymilk, without generating solid residue by filtration or centrifugation, is the presence of another flavor attribute in the product named "chalkiness". The resulting products are also, sometimes, physically unstable, i.e., phase separation occurs shortly during the product storage, depending on the pressure of homogenization. Soymilk is considered as a suitable economical substitute for cow's milk and an ideal nutritional supplement for lactose-intolerant population. It naturally has about the same amount of protein (though not the same amino acid profile) as cow's milk.

Benefits of Soy Milk

- Excellent source of protein and dietary fiber
- Has little saturated fat and no cholesterol and lactose
- Good source of iron
- Soy protein may reduce the risk of certain cancers, including breast cancer, prostate cancer, lung
- Dietary soybean increased plasma thyroxin, an advantage for areas with endemic goiter
- Has isoflavones, which deal with many health issues; most important being the prevention of various cancers, heart disease, osteoporosis, antioxidant etc.
- As it doesn't contain galactose, soy milk can safely replace breast milk in children with galactosemia

- Lacks casein and is a good source of lecithin and vitamin E
- Safe for people with lactose intolerance, or milk allergy
- It also helps in fighting the symptoms of menopause and in promoting eye health or anticatatract
- It can conveniently be used as a weaning formula instead of cow's milk for infants
- It also shows antidiabetic and antiobesity property

Fermentation of Soymilk

The consumption of soymilk has been limited due to the presence of nondigestible oligosaccharides such as raffinose and stachyose [30], "beany flavor" due to unpleasant off-flavor of *n*-hexanal and pentanal, which are naturally present in soymilk as products of the breakdown of unsaturated acids and glucose-conjugated isoflavones which are highly polar, water soluble compounds, are more difficult to be absorbed by the intestinal epithelium and have weaker biological activities than the corresponding aglycones [31]. Monogastric animals lack the ability to synthesize sufficient α-galactosidase in their intestinal system to hydrolyze α-galactosides present in soymilk. Thus, these sugars (raffinose and stachyose) pass into large intestine, where the intestinal microflora acts on them, causing flatulence and gastrointestinal disturbance. These disturbances reduce both feed efficiency in monogastric animals and general consumer acceptance of soy foods [32, 33]. Some processes have been developed to lessen the development of undesirable flavors during processing [29, 34] and attempts have been made to reduce the levels of oligosaccharides in soy products by leaching and enzyme hydrolysis [12]. However, both of these methods are expensive, and there is an additional risk of losing valuable protein in the process.

To overcome these limitations and to improve the acceptability as well as the nutritive value of soymilk, fermentation with various organisms have been attempted. Soymilk has been examined as a substrate for the *Lactobacillus* species *L. casei* [35, 36], *L. helveticus* [37], *L. fermenti*; [38], *L. fermentum* [39], *L. reuteri* [40] and *L. acidophilus* [35, 41, 42, 43, 44].

Fermentation by α-galactosidase possessing organisms causes enzymatic hydrolysis of complex sugars. α-Galactosidase (α-D-galactoside galactohydrolase, E.C.3.2.1.22) is a carbohydrase, widely distributed in microorganisms, plants and animals, which catalyses the hydrolysis of α-D-galactose from melibiose, raffinose, stachyose, verbascose and more complex polysaccharides. This would provide simple sugars as substrate for the lactic acid bacteria as well as reduce the level of oligosaccharides present. Some researchers have reported the reduction of oligosaccharides during fermentation of soymilk using either pure or mixed culture of the following bacteria: *Lactobacillus cellobiosis, L. plantarum, L. fermentum, L. delbrueckii, L. fermenti, L. pentosaceus* and *L. bulgaricus*. Soymilk is suitable for the growth of some LAB, especially bifidobacteria [18] because of its content of growth stimulating factors such as oligosaccharides, amino acids and peptides.

Gut microflora plays an essential role in soy isoflavone absorption, metabolism and bioavailability by virtue of its enzymatic activity [45, 46, 47,48]. Differences in gut microfloral profiles may contribute to the large inter-individual variation typical for isoflavone bioavailability following soy intake. The enzymes most pertinent to isoflavone bioavailability are the β-glucosidases, for initial glucoside hydrolysis and subsequent aglycone absorption and the β-glucuroniases and sulphatases, for reabsorption of the hepatic conjugates and biliary excretion. The initial hydrolysis of soy isoflavone glucosides to their respective aglycones is suggested to be the rate-limiting step in isoflavone absorption thus, β-glucosidases activity may be particularly important in relation to isoflavone bioavailability. β-Glucosidases (β-d-glucoside glucohydrolase, EC 3.2.1.21) comprise a heterogeneous group of enzymes that are able to cleave the β-glucosidic linkages of di and/ or oligo-saccharides, or other glucose conjugates. β -Glucosidases are widely distributed in plants, fungi and bacteria and play pivotal roles in many biological processes, such as degradation of cellulosic biomass, hydrolysis of glycolipids, cyanogenesis, and modification of secondary metabolites [49]. Bacterial species of bacteroides, bifidobacteria and lactobacilli have the highest activity of this enzyme. β-Glucosidases of intestinal microflora in lower bowel can hydrolyze the glucoside isoflavones to aglycones and promote their absorption. Therefore, bacteria with β-glucosidase activity are potentially important in the production of compounds with higher estrogenicity and better absorption, facilitating the bioavailability of isoflavones. Many lactobacilli and bifidobacteria possessing this enzyme have been examined with respect to their ability to convert the soy glucosides to aglycones including *L. acidophilus* [50], *L. casei* [51], *L. fermentum* [51], *Lactobacillus plantarum* and *L. bulgaricus* [52]. In fermentation studies, the technological properties of the strains to be used as starters are very important and properly selected strains must be used in order to enhance specific characteristics of the final product. For example, the taste of soymilk can be improved using some LAB strains that decrease the beany, grassy and soy flavor [41].

Other Soy Based Fermented Foods

The main objections to soybean products by some consumers are the associated intrinsic flavor which has been described as beany/grassy or astringent and phenomenon of flatulence. Indeed, the first benefit of soy fermentation is the reduction of its beany flavor and chalkiness. Furthermore, soybeans are high in phytate (phytic acid) and oxalate (oxalic acid), substances that can block the uptake of essential minerals-calcium, magnesium, iron, copper and zinc-in the intestinal tract. Fermentation can reduce antinutritional factors, objectionable flavor and flatulent sugars; namely stachyose and raffinose. This can be achieved by using microbial cultures in the form of mono-and multi-cultures; particularly using lactic acid bacteria. However, it has been reported that acceptability of soy bean products may be enhanced by modification or processing methods. For instance, in soymilk preparation, application of heat, soaking of soybean in ethanol or alkali and acid grinding are some of modifications in cold-water soy extraction. Soy is considered a good substrate for functional foods, since fermentation by probiotics has the potential to (1) reduce the levels of some carbohydrates possibly responsible for gas production in the intestinal system, (2) increase free isoflavone levels and (3) favor desirable changes in bacterial populations in the gastrointestinal tract. During fermentation and microbial growth, the functional properties of soy products like other foods are formed as follows: protein is hydrolyzed to amino acids and peptides by proteolytic enzymes, oligosaccharides are hydrolyzed to monosaccharides, phytic acid degraded to inorganic phosphates. These properties are The most well-known soy fermented products are miso, tempeh, soy sauces (shoyu), Natto and fermented tofu.

Soy Based Probiotic Yoghurt

Soy fortified yoghurt is a nutraceuticals food. Yoghurts with 5% added soy protein concentrate qualify for the FDA-approved soy health claim of "cholesterol reducing" and also contain sufficient fiber to provide 1 g of dietary fiber per serving. Soy yoghurt is a fermented soy yoghurt made with a mixed starter culture consisting of *Streptococcus thermophilus* and

Lactobacillus delbrueckii ssp. bulgaricus. In dairy and soy applications, probiotic organisms are delivered with different fermented dairy and soy products, most notable yoghurt and soy yoghurt. Incorporation of probiotic organisms such as *Lactobacillus acidophilus, Bifidobacterium* sp., and *L. casei* in fermented products provides a potential to improve the quality of the product and the health status of consumers. Dairy and soy foods may serve as the ideal system for delivery of probiotic bacteria to the human gastrointestinal tract (GIT) due to provision of a favorable environment that promotes the growth and enhances the viability of these microorganisms.

During the bacterial fermentation, major constituents including lactose and milk proteins, soy proteins, raffinose, stachyose and other soy carbohydrates are utilized for the bacterial growth, which results in the conversion of fermentable materials into a range of products such as lactic acid, acetic acid, peptides, amino acids and different vitamins. In addition to exceptional nutritional attributes, milk and soy derived products such as fermented soy yoghurt or soy contain components that possess a range of different bioactive compounds. Compared to casein, soy protein showed a greater antioxidative ability in preventing lipid oxidation. Isoflavones have been found to increase the activities of some antioxidative enzymes in the liver. Some of these bioactive compounds are considered functional, thus making dairy and soy products important part of functional foods and nutraceuticals [53].

Soy Shrikhand

Shrikhand was prepared as per the method described by [54] from different blends of soy milk and dairy milk in different ratio. The blends were homogenized in 2 stages using baby homogenizer (Gaulin, APV, USA). In the 1st stage, pressuer was maintained at 150kg/cm^2 and in the 2nd stage pressure was 200 kg/cm^2. Finally milk was heated at 72^{0}C for 20 min and cooled to room temperature (30-37^{0}C). Dahi @1% was then inoculated to all the blends and incubated at 37^{0}C for 6 h [55]. The solid material called Chakka was prepared from the dahi and 60% sugar was added. The contents were throughly mixed and then screened through fine muslin cloth. The consumer acceptance of shrikhand decreased with increase in soy milk supplementation. The highest acceptable soy milk supplementation was 60%.

Soymilk Kefir

Soy milk is prepared from soy beans. Then kefir grains (*Lb. brevis, Lb. helveticus, Lb. kefir, Leuconostoc mesenteroides, Kluyveromyces lactis, K. marxianus, Pichia fermentans*) are added @5% to the milk. Kefir grains are small, gelatinous, yellowish, and irregularly shaped masses resembling individual miniature florets of a head of cauliflower. Yeast and lactic acid bacteria coexist as symbiotic relationship for an acid-alcohol fermentation [56]. The microoganisms constituting the kefir grains produce lactic acid, antibiotics, and bacteriocide which inhibit the development of degrading and pathogenic microorganisms in kefir milk. Kefiran is reported to possess antitumor activity [57].

Soy Beverages

Soybeans (250g) were initially soaked in 1 L of water for 16 hr at 25^0C. Subsequently, 190g of soaked beans in 500 ml of water were ground with a commercial blender unit for 3 min, filtered on whatman #4 membranes and boiled for 5 min. This milk contains 4.5% protein and 2.3% fat [58].

Tofu or Bean Curd

Tofu is a food made by coagulating soy milk and then pressing the resulting curds into soft white blocks. Tofu is made by coagulating soy milk and pressing the resulting curds. Although pre-made soy milk may be used, most tofu producers begin by making their own soy milk, which is produced by soaking, grinding, boiling and straining dried (or, less commonly, fresh) soybeans [59]. Coagulation of the protein and oil (emulsion) suspended in the boiled soy milk is the most important step in the production of tofu. This process is accomplished with the aid of coagulants. Two types of coagulants (salts-calcium sulfate and acids-Glucono delta-lactone) are used commercially [60]. The third type of coagulant, enzymes, is not yet used commercially but shows potential for producing both firm and "silken" tofu. Tofu and its production technique were introduced into Korea and then Japan. Tofu contains a low amount of calories, relatively large amount of iron, and little fat. Depending on the coagulant used in manufacturing, the tofu may also be high in calcium and/or magnesium. 100 grams of firm tofu contains

15.78 grams of soy protein. The FDA granted this health claim for soy: "25 grams of soy protein a day, as part of a diet low in saturated fat and cholesterol, may reduce the risk of heart disease” [61].

Cheonggukjang

It is a fermented soybean paste used in Korean cuisine. It contains whole as well as ground soybeans. It can be made in 2 to 3 days through fermentation of boiled soybeans, adding Bacillus subtilis, which is usually contained in the air or in the dried rice plants, at 40°C without adding salt, compared with the much longer fermentation period required for doenjang, another, less pungent variety of Korean soybean paste. Like many forms of doenjang, cheonggukjang is paste-like in texture, but also includes some whole, uncrushed soybeans. Cheonggukjang may also be made by fermenting boiled soybeans in a warm place, pounding a portion of them, and adding salt and red chili powder. Cheonggukjang is generally considered to be a healthy food (particularly in the winter), as it is rich in vitamins and other nutrients, though its very strong odor is not universally enjoyed. Some people have commented that this soup gives off the aroma of wet socks.

Doenjang

This is a traditional Korean fermented soybean paste. Its name literally means "thick paste" in Korean [62]. To produce *doenjang*, dried soybeans are boiled and stone-ground into coarse bits. This paste is then formed into blocks, which are called meju. The blocks are then exposed to sunlight or warmth. When they are exposed to the sun or warmth, dried rice plants are attached to the surface of the soybean blocks. Dried rice plants are readily available in Korea and are a rich source of bacteria (*Bacillus subtilis*). The fermentation process begins at this stage. The Bacillus subtilis bacteria reproduce, consuming soybean protein and water in the meju. The unique smell of the meju is mainly the ammonia which is produced by the bacteria. One to three months later, depending on the block size, the meju are put into large opaque pottery jars with brine and left to further ferment, during which time various beneficial bacteria transform the mixture into a further vitamin-enriched substance (similar to the way milk ferments to become yogurt). Liquids and solids are separated after the fermentation process, and the liquid becomes Korean soy sauce [63]. The solid, which is *doenjang*, is very salty and quite thick, often containing (unlike most miso) some whole, uncrushed soybeans. *Doenjang* is rich in flavonoids and beneficial vitamins, minerals, and hormones which are sometimes claimed to possess anti-carcinogenic properties. In Korean traditional meals, the menu has concentrated on vegetables and rice but *doenjang*, which is made of soybeans, has a great deal of lysine, an essential amino acid that rice lacks. There are linoleic acid of 53% and linolenic acid of 8% which have an important role in normal growth of blood vessels and prevention of blood-vessel-related illness.

Miso

This is a traditional Japanese seasoning produced by fermenting rice, barley and/or soybeans, with salt and the fungus kōjikin, the most typical miso being made with soy. High in protein and rich in vitamins and minerals, miso played an important nutritional role in feudal Japan [59]. Different varieties of miso have been described as salty, sweet, earthy, fruity and savory. The traditional Chinese analogue of miso is known as dòujiàng. Miso is a part of many Japanese-style meals. It most commonly appears as the main ingredient of miso soup, which is eaten daily by much of the Japanese population. Miso is a part of many Japanese-style meals. It most commonly appears as the main ingredient of miso soup, which is eaten daily by much of the Japanese population. Miso is a source of *Lactobacillus acidophilus*. Lecithin, a kind of phospholipid caused by fermentation, which is effective in the prevention of high blood pressure, also present in miso [64].

Conclusion

Soy fermented food products have many nutritional benefits including reducing cardiovascular disease, reducing menopausal symptoms, weight loss, arthritis, brain function. It contains phytochemicals such as isoflavones, saponins, phytosterols that promote health. Addition of probiotic culture enhances functional attribute of soy yoghurt. In future technological prospects that should be considered or addressed in view of the functionality of probiotic micro-organisms include: (i) fermentation and drying technologies, and (ii) microencapsulation. Furthermore, the probiotic dairy and soy industries should work closely not only with the regulatory authorities but also with the medical profession in order to substantiate the health claims associated with these beneficial micro-organisms. Overcoming these will help ensure the acceptability of probiotic dairy and soy products by the consumer and, hence, safeguard the future of the industry. In addition, more work is needed to develop a large-scale fractionation of protein hydrolysates to obtain products enriched with biologically active peptides of specific function that could be used as nutraceutical additives in functional foods.

References

[1] H. Snyder and T. Kwon, Van Nostrand Reinhold New York (1987).

[2] J.R. Biranjan, D.R. Ingole and S.R. Sanyal, Utility of soymilk in dietary therapy of undernourished children. Indian Medical Gazette, 9, 313–316 (1983).

[3] J. Liu, S. Wang and Y. Lin, *Nutrition and Cancer*, 44, 2, 183-187 (2002).

[4] N.R. Reddy, D.K. Salunkhe and R.P. Sharma, *Journal of Food Science*, 45, 1161–1164 (1980).

[5] D.K. Salunkhe and S.S. Kadam, Hand book of world legumes: Nutritional chemistry processing technology and utilization. Boca Raton, Florida: CRC Press Inc (1989).

[6] R. Cruz, and Y.K. Park, *Journal of Food Science*, 47, 1973–1975 (1982).

[7] P. M. Dey and J.B. Pridham, *Advance Enzymology*, 36, 911–930 (1972).

[8] B.D. Shivanna and M. Ramakrishan, *Journal of Biosciences*, 9, 109–116 (1985).

[9] R. Cruz, J.C. Batistela and G. Wosiacki, *Journal of Food Sciences*, 46, 1196–1200 (1981).

[10] S. K. Khare, K. Jha and A.P. Gandhi, *Food Chemistry*, 51, 29–31 (1994).

[11] S.J. Prashanth, and V.H. Mulimani, *Process Biochemistry*, 40, 1199–1205 (2005).

[12] H. Sugimoto and N. Buren, *Journal of Food Science*, 35, 5, 655-660 (1970).

[13] M. Roberfroid, *The British Journal of Nutrition*, 80, 4, S197 (1998).

[14] S. I. Mussatto and I.M. Mancilha, *Carbohydrate Polymers*, 68, 587–597 (2007).

[15] G.R. Gibson, *British Journal of Nutrition*, 80, S209–S212 (1998).
[16] P. K. Gopai, P.A. Sullivan and J.B. Smart, *International Dairy Journal*, 11, 19–25 (2001).
[17] J. Matsuyama, H. Hirata, T. Yamagishi and K. Hayashi, Fermentation profiles and utilization of sugars of Bifidobacteria in soymilk, Nippon Shokuhin Kogyo Gakkaishi 39: 887–893 (in Japanese) (1992).
[18] C. Chou and J. Hou, *International Journal of Food Microbiology*, 56, 2-3, 113-121 (2000).
[19] T. Sakai, K. Kohyama and T. Yasui, *Carbohydrate Polymers*, 57, 451–458 (2004).
[20] A. Bezkorovainy, and R. Miller-Catchpole, Ecology of bifidobacteria, Biochemistry and physiology of bifidobacteria. CRC Press, Boca Raton, FL, pp. 29– 72 (1989).
[21] J. Anderson and S. Gilliland, *Journal of the American College of Nutrition*, 18, 1, 43-50 (1999).
[22] K.D.R. Setchell and A. Cassidy, *Journal of Nutrition*, 129, 758S-767S (1999).
[23] I. Liener, *Critical Reviews in Food Science & Nutrition* 34, 1, 31-67 (1994).
[24] Y. Wang, R. Yu and H. Yang, *Food Microbiology*, 20, 3, 333-338 (2003).
[25] L. Karr-Lilienthal, C. Kadzere and C. Grieshop, *Livestock Production Science,* 97, 1, 1-12 (2005).
[26] GCSE D&T Food Technology, *Food Technology* (http://events.aqa.org.uk/ebooking) (2009).
[27] W. Wilkens and J. Hackler, *Cereal Chem*, 46, 391-396 (1969).
[28] K. Kwok and K. Niranjan, *International Journal of Food Science & Technology*, 30, 3, 263-295 (1995).
[29] W. Wilkens, L. Mattick and D. Hand, *Food Technol*, 21, 12, 86-89 (1967).
[30] K. Leske, C. Jevne and C. Coon, *Poultry Science*, 72, 4, 664-668 (1993).
[31] K.D.R. Setchell, *Journal of Nutrition*, 130, 654S-655S (2002).
[32] J. Rackis, *Journal of the American Oil Chemists' Society*, 58, 3, 503-509 (1981).
[33] S.V. Silva and F.X. Malcata, *International Dairy Journal*, 15, 1-15 (2006).
[34] G. Mustakas, W. Albrecht, J. McGhee, L. Black, *Journal of the American Oil Chemists' Society*, 46, 11, 623-626 (1969).

[35] T. Murti, C. Bouillanne, M. Landon and M. Smazeaud, *Journal of Food Science*, 58, 1, 153-157 (1993).
[36] M. Garro, G. Valdez, G. Oliver and G. Giori, *Journal of Food Protection*, 62, 7, 808-810 (1999).
[37] A. Angeles and E. Marth, *J Milk Food Technol*, 34, 30-36 (1971).
[38] S. Chumchuere and R. Robinson, *Food Microbiology*, 16, 2, 129-137 (1999).
[39] M. Garro, G. Valdez and G. Giori, *Food Microbiology*, 21, 5, 511-518 (2004).
[40] G. Tzortzis, A. Goulas, M. Baillon, G. Gibson and R. Rastall, *Applied Microbiology and Biotechnology*, 64, 1, 106-111 (2004).
[41] B. Mital, K. Steinkraus and H. Naylor, *Journal of Food Science*, 39, 5, 1018-1022 (1974).
[42] M. Chopra and D.I. Thurnham, *Proc Nutr Soc*, 58, 663–671 (1987).
[43] L. Shelef, K. Bahnmille, M. Zemel and L. Monte, *Journal of Food Processing and Preservation,* 12, 3, 187-195 (1988).
[44] Y. Wang, R. Yu and C. Chou, *Food Microbiology*, 19, 5, 501-508 (2002).
[45] Y. Chang and M. Nair, *Journal of Natural Products,* 58, 12, 1892-1896 (1995).
[46] S. Dhananjay, S. Kulkarni, K. G. Kapanoor, V. K. Naganagouda, H. M. Veerappa, *Biotechnology and Applied Biochemistry*, 45, 51–57 (2006).
[47] D. Tsangalis, J. Ashton, A. McGill and N. Shah, *Journal of Food Science*, 67, 8, 3104-3113 (2002).
[48] D. Tsangalis, J. Ashton, A. McGill and N. Shah, *Journal of Food Science*, 68, 2, 623-631 (2003).
[49] J.R. Bordignon, N. Kazuhiko, Y. Tadashi and N. Sayuki, Hydrolysis of isoflavone and consumption oligosaccharides during lactic acid fermentation of soybean, Food and Technology Division, Japan International Research Center for Agricuture Sciences (JIRCAS), Tsukuba, Ibaraki 305-8686, Japan (http:// www. Jircas.affrc.go.jp) (2000).
[50] H. Chien, H. Huang and C. Chou, *Food Microbiology*, **23**, 8 772-778 (2006).
[51] A. Tang, N. Shah, G. Wilcox, K. Walker and L. Stojanovska, *Journal of Food Science*, 72, 9, M431-M436 (2007).
[52] Y. Pyo, T. Lee and Y. Lee, *Journal of Food Science*, 70, 3, S215-S220 (2005).

[53] D.A. Cherry, Woodwell, National ambulatory medical care survey: 2000 summary. Advanced data from vital and health statistics. No. 328 2002 National Center for Health, Statistics Hyattsville, MD (2002).
[54] S. De, Outlines of Dairy Technology. Oxford University Press, New Delhi (1980).
[55] D.B. Trenna and E.G. Stanley, *J Food Sci*, 70, M186-M189 (2005).
[56] R.J. Liu, M.J. Chen and C.W. Lin. *J. Agric. Food Chem,* 53, 2467–2474 (2005).
[57] M. Shiomi, V. Sasaki and M. Murofushi, *Jpn J Med Sci*, 35, 75-80 (1982).
[58] [58] C.P. Champagne, J. Green-Johnson, Y. Raymond, J. Barrette and N. Buckley, *Food Res Inl,* 42, 612-621 (2009).
[59] W. Shurtleff and A. Aoyagi, The book of miso: savory, high-protein seasoning. Ten Speed Press (ISBN 1-58008-336-6) (2001).
[60] G. Shun-Tang, and O.Tomotada, *Journal of Food Science*, 70, 4, 258–262, (doi:10.1111/j.1365-2621.2005.tb07170.x) (2005).
[61] W. Shurtleff and A. Aoyagi, Tofu & soymilk production: a craft and technical manual (3rd ed.), Lafayette, California: Soyfoods Center, ISBN 978-1928914044 (2000).
[62] J. Suk, "Efficacy and Nutrition of Doenjang" (in Korean). Public website (2008).
[63] P. Kun-Young, Pusan National University. "Korean food for defeating cancer" (in Korean). Hankook Ilbo (2007).
[64] J.Z. Miller, M.H.Weinberger, S.A. Daugherty, *Journal of Chronic Diseases*, 40, 245–250 (1987).

In: Oligosaccharides
Editors: L. S. Schweizer, S. J. Krebs
ISBN: 978-1-62948-328-3

Chapter IX

Impact of Low Oligosaccharide Soybean Meal on Nutrient Digestibility, Growth Performance, and Carcass Characteristics of Broiler Chickens

***W. A. Dozier, III*[*] *and K. R. Perryman*[†]**
Department of Poultry Science, Auburn University, Auburn, AL, US

Abstract

Previous research has reported that removal of raffinose and stachyose from SBM can increase the energy utilization of SBM by broilers. Several strategies have been implemented to ameliorate the adverse effects of raffinose and stachyose on nutrient utilization, which include ethanol extraction of SBM, galactosidase supplementation, and developing new varieties of low galactooligosaccharide SBM. Soybean meals stemming from novel genetic varieties of soybeans have been developed to contain either a 75% (low oligosaccharide SBM [LOSBM])

[*] E-mail addresses: bill.dozier@auburn.edu
[†] E-mail addresses krp0008@auburn.edu

or 90% (ultra-low oligosaccharide SBM [ULSBM]) reduction in GAL concentrations. Broilers fed diets containing LOSBM had higher AME_n compared with birds fed CSBM and ULSBM. Both ULSBM and LOSBM had higher digestible AA content compared with CSBM, with ULSBM having the highest AA digestibility. Broilers fed diets formulated with ULSBM had superior feed conversion from 1 to 28 and 1 to 42 days of age compared with birds consuming LOSBM and CSBM based diets. Diets formulated with LOSBM and ULSBM had supplemental fat concentrations reduced by up to 70% compared with CSBM based diets. Due to the better nutrient profile of LOSBM and ULSBM, less supplemental fat can be utilized in diet formulation without any adverse effects on growth performance or carcass characteristics.

Introduction

Dietary energy sources represent a significant portion of diet costs in the poultry industry. The majority of energy in poultry diets is provided by cereal grains and oilseed meals with supplemental fat added to meet the desired energy density of the diet. Energy utilization of corn is relatively high, but gross energy (GE) of soybean meal (SBM) is poorly utilized by poultry [1]. By increasing the energy utilization of SBM in poultry, less supplemental fat could be used in dietary formulations, translating into lower dietary costs. Due to government mandates for the use of renewable fuels, the cost of energy-providing ingredients have increased dramatically over the past decade. Rising petroleum costs and the Renewable Fuel Mandate have led to a dramatic increase in biodiesel production since 2000 [2]. The United States produced over 4.2 billion liters of biodiesel in 2011 [2], resulting in increased demand for animal fats and vegetable oils.

As a result, the price of poultry oil has increased from $0.25 to $1.00 per kg during the past 7 years, and prices may continue to increase with demand for animal fat and vegetable oils [3]. Demand and prices for fat/oil are predicted to remain high, creating further financial strain for the poultry industry. Fat is an excellent source of energy and linoleic acid for poultry [4]. Diets fed to broilers typically contain added fat from 1 to 3% with inclusion concentrations varying with the phase of production. Increasing energy utilization of primary ingredients instead of relying on energy provided by supplemental fat to meet dietary energy recommendations can translate to lower dietary costs.

Metabolizable Energy

While the gross energy of SBM is greater than corn, poultry metabolizes 28% less energy from SBM compared with corn. Gross energy of SBM is greater than in corn, but it has 42% less metabolizable energy (ME) [1, 5]. These differences in ME can be attributed to the differences in nutrient composition between corn and SBM. Annison and Choct [6] reported that carbohydrates are more complex in SBM than cereal grains. The carbohydrate fraction of corn is primarily composed of starch, while the carbohydrate fraction of SBM contains variable concentrations of non-starch polysaccharides (NSP) and poorly digested sugars [7]. Variability in the carbohydrate fraction of SBM can occur depending on soybean variety and agronomic conditions [8, 9]. Moreover, differences in the ratio of starch and sugars to dietary fiber can have a pronounced effect on the nutritive value of the carbohydrate fraction and affect nutrient digestibility of various feed ingredients [10]. Differences in the ME of SBM have also been reported between monogastric species by the National Research Council [11, 12] [Figure 1]. Energy utilization of SBM in poultry is 27% lower compared with pigs due to anatomical differences between animals. Poultry have 2 large ceca and a short colon, while swine have a single short cecum and a long colon [13]. The larger hind gut of the pig provides a greater fermentative capacity due to an increased microflora population.

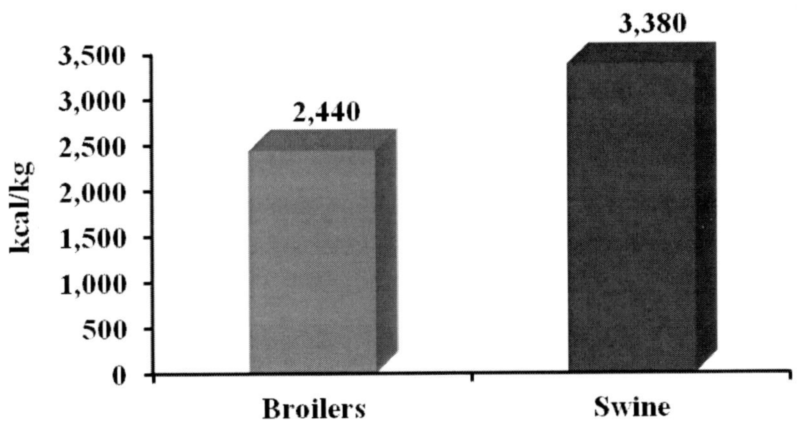

Figure 1. Metabolizable energy of soybean meal for broilers and swine. *(Adapted from NRC, 1994 [11] and 1998 [12]).*

The larger microflora population metabolizes indigestible SBM carbohydrates more efficiently. Furthermore, the greater length of the porcine alimentary tract allows for more complete SBM carbohydrate digestion compared with poultry due to increased retention time [14]. This fermentative process converts relatively indigestible carbohydrates into short chained fatty acids, which can supply a source of energy in swine [15, 16]. Poor energy utilization of SBM is related to its carbohydrate composition.

Soybean Meal Carbohydrate Composition

Soybean meal is considered the "gold standard" protein source for monogastric animals [17]. While SBM is primarily used as a protein source because of a favorable amino acid (AA) profile, SBM contains high concentrations of both protein and carbohydrates [7]. Once soybeans have been processed into meal, the resulting dehulled SBM is typically 48% crude protein, 35% carbohydrates, 10% water, 5% minerals, 1% oil, and <1% other [18]. Published research has reported high (85 to 95%) digestibility values for the protein fraction of SBM fed to chickens [19, 20], but studies are limited on evaluating the digestibility of the carbohydrate fraction of SBM in poultry. Carbohydrate digestibility of SBM has been reported at 35% in rats [21], 61% in turkeys [7], and 40% in broiler chicks [22]. A possible explanation for the low digestibility is the complex nature of SBM carbohydrates [6]. The carbohydrate fraction of SBM is composed of approximately 20 to 30% non-starch polysaccharides (NSP) [23] and 10% free sugars, including oligosaccharides [24].

Non-starch polysaccharides. Non-starch polysaccharides in SBM are a variable mixture of approximately 8% cellulose and 17% pectic polysaccharides such as type I and II rhamnogalacturonans, arabinogalactan I, and xylogalacturonan [25]. Due to the complexity of SBM carbohydrates, fermentative capacity and gastrointestinal tract adaptation are the primary factors influencing NSP digestibility in monogastric animals [18]. Pectic polysaccharides are highly complex carbohydrates, which are partially water soluble, whereas cellulose is insoluble and has limited digestibility in poultry [26, 27]. Insoluble NSP are partially resistant to microbial fermentation, while soluble NSP can be metabolized by gastrointestinal microflora producing short chain fatty acids. These short chain fatty acids are absorbed and utilized by poultry for energy [16]. However, due to the limited fermentative capacity of the chicken, the contribution of short chain fatty acids to the ME of the bird is limited to 2 to 3% of total dietary ME [28]. In swine, ME provided through microbial fermentation of complex carbohydrates can provide up to 24% of dietary ME [29], however, net efficiency of energy utilization of complex carbohydrate fermentation is low [30].

Researchers have also suggested that NSP may lead to decreased digestibility of other nutrients [31, 32]. Soluble NSP are able to bind water in the lumen of the small intestine, which increases the viscosity of the digesta, leading to changes in the physiology and microbial ecosystem of the alimentary tract [33]. A more viscous digesta transits more slowly through the gastrointestinal tract and can lead to lower O_2 concentrations, providing a favorable environment for the establishment of fermentative bacteria [34]. This finding was supported by researchers who reported an increase in microbial fermentation in the small intestine when soluble soy NSP were added to broiler diets [35]. Furthermore, viscous digesta stemming from large concentrations of dietary NSP -have been attributed to increased losses of endogenous nutrients in broilers.

Angkanaporn et al. [33] reported increased endogenous losses of protein, including digestive enzymes and AA, in roosters precision-fed diets high in soluble NSP. Viscous digesta caused by high dietary NSP concentrations have been reported to increase bile acid secretion and loss of endogenous products in the feces of rats [36]. Furthermore, it has been reported that NSP can bind bile salts, lipids, and cholesterol leading to further nutrient losses in the excreta [37]. Sequestration and increased fecal loss of enzymes, bile acids, lipids, and cholesterol could lead to major changes in the digestive and absorptive dynamics of the alimentary tract with poor overall efficiency of nutrient absorption as a consequence [18].

Free Sugars and Galactooligosaccharides. The free sugar component of the carbohydrate fraction of SBM is composed of about 5% sucrose and 6% galactooligosaccharides (**GAL**) (5% stachyose and 1% raffinose) [8, 15, 38]. Sucrose is highly digestible when consumed by poultry, while GAL have poor digestibility [39]. Raffinose and stachyose concentrations in SBM have been reported to vary depending on the location where the soybeans were grown and the specific soybean genotype [8] [Figure 2].

Limited digestibility of GAL is related to the structure of these molecules and the lack of endogenous enzymes in poultry necessary to hydrolyze and absorb these carbohydrates. Raffinose is a trisaccharide with a galactose bonded to the glucose of a sucrose molecule via an α-1-6 linkage. Stachyose is a tetrasaccharide that is analogous to raffinose with an additional galactose bonded to the galactose molecule of raffinose via an additional α-1-6 linkage

[18]. Digestibility of raffinose and stachyose is low in monogastric animals because non-ruminants do not produce α-1-6-galactosidase in the intestinal mucosa, which is necessary to hydrolyze the α-1-6-galactosidic bonds [40]. Hydrolysis of these bonds produces two free galactose molecules and a sucrose molecule. If these bonds cannot be hydrolyzed, raffinose and stachyose are unable to cross cell membranes and remain in the lumen of the gastrointestinal tract. Coon et al. [41] reported ileal digestibility values of raffinose and stachyose to be less than 1% in roosters. The poor ileal digestibility of raffinose and stachyose has led researchers to investigate the negative impact of undigested GAL present in the lumen of the small intestine on nutrient digestibility and intestinal characteristics.

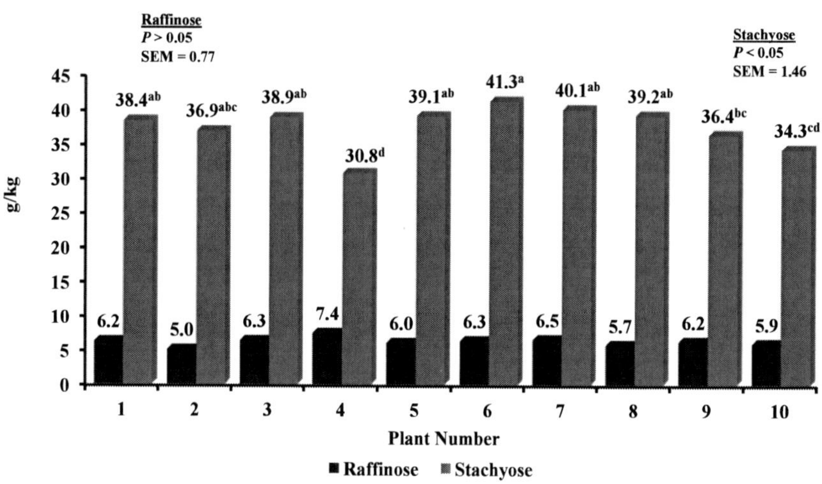

Figure 2. Raffinose and stachyose concentrations between soybean meal (SBM) samples collected from 10 processing plants located throughout the United States. (*Adapted from Greishop et al., 2003 [8]*).

Veldman et al. [42] reported a 25% reduction in crude protein (**CP**) and organic matter digestibility when pigs were fed diets containing high concentrations of raffinose and stachyose (2.8 and 15.2%, respectively). This finding was attributed to an increase in osmolality in the lumen of the small intestine. Increased osmolality of the gastrointestinal tract can be attributed to the small molecular weight of raffinose and stachyose, which results in water passing through the permeable epithelial cells into the lumen. A higher volume of water in the lumen leads to a dilution of digestive enzymes and substrates, reducing nutrient breakdown and digestibility. Smiricky et al. [43] reported reduced N and AA digestibility in growing pigs corresponding with higher concentrations of raffinose and stachyose in the diet. Furthermore, Leske et al. [44] reported a negative dose-dependent response (0.61 to 5.41% GAL) on TME_n when raffinose and stachyose were added to soy protein concentrate. Poor energy utilization of GAL in poultry has been attributed to the ability of raffinose and stachyose present in the diet to increase fluid retention, hydrogen production, and cause diarrhea [41]. Irish and Balnave [45] reported negative correlations of stachyose concentrations in the distal ileum with BW gain (r^2 = 0.90, $P < 0.05$) and feed intake (r^2 = 0.74, $P < 0.05$). Furthermore, feed conversion was positively correlated with stachyose concentration (r^2 = 0.81, $P < 0.05$). The poor performance was attributed to a rise in osmotic pressure resulting in increased fluid retention and a faster rate of passage resulting in poor nutrient digestibility.

Microflora

Due to the absence of an endogenous α-1-6-galactosidase, raffinose and stachyose are primarily metabolized by microbiota residing in the large intestine [46]. Hydrolysis and metabolism of these substrates generate short chain fatty acids, reducing the pH of the lumen, and altering the local environment [46]. As microbiota are highly sensitive to the environment of the lumen, fermentation of GAL may result in a prebiotic effect attributed to the growth of beneficial microorganisms. Research has demonstrated changes to the microbial population, indicative of a beneficial effect when indigestible polysaccharides were fed to monogastrics [47, 48, 49]. Beneficial effects have been credited to an increase in the number of *Lactobacilli* and *Bifidobacteria,* and subsequent decrease in *Clostridia* and *Enterobaceteria* due to oligosaccharide digestion [50]. Higher concentrations of short chain fatty acids have been credited in reducing colonic inflammation related to a decrease in the release of inflammatory cytokines and suppressing inflammation link gene expression *in vitro* [51]. Research has also suggested a positive correlation between the feeding of fermentable substrates and improvements in alimentary tract health, specifically through modulation of gut-associated lymphoid tissues [52]. Ito et al. [53] reported increases in both goblet cell numbers and production of luminal mucin related to increased digesta viscosity in rats fed various soluble fibers. Utilizing poultry, Faber et al. [54] demonstrated that fermentation of GAL may improve cecal health. These studies reported heavier cecal weights, increased short chain fatty acid production, and upregulation of cytokine gene expression when broilers were subjected to an *E. acervulina* challenge. In contrast to the beneficial health effects associated with GAL fermentation, these researchers described poorer feed efficiency with birds consuming diets with higher concentrations of GAL than a low GAL control. Overall, the effect of microflora and gastrointestinal adaptations attributed to the fermentation of GAL is not well understood, and it is still generally accepted that GAL can lead to intestinal disorders and poor performance in broilers [18].

Reduction of Galactooligosaccharides of Soybean Meal

Due to the relative inability of monogastric animals to digest GAL in SBM, and the potential anti-nutritional effects of GAL, researchers have developed strategies to reduce or hydrolyze the GAL of SBM. Further processing of SBM via ethanol extraction has been reported to remove over 90% of GAL content [41, 42, 44, 55]. Researchers have also utilized exogenous α-1-6 galactosidase *in vivo* or *in vitro* to hydrolyze α-1-6 galactosidic bonds present in raffinose and stachyose to increase nutrient digestibility [39, 56, 57, 58, 59, 60]. Furthermore, plant geneticists have developed new genetic varieties of soybeans that yield SBM with dramatically less GAL. While each of these methods were effective at limiting the concentration of GAL in SBM, most studies have reported inconsistent results on the effects of less GAL in SBM-based diets improving nutrient digestibility and growth performance of broilers.

Ethanol Extraction. Ethanol extraction of SBM results in a 90% reduction of raffinose and stachyose concentrations [42]. When ethanol-extracted SBM was precision fed to roosters, TME_n, DM digestibility, and apparent digestibilities of hemicellulose and cellulose were increased compared with roosters fed a control SBM [41]. These researchers attributed the increase in nutrient digestibility and energy utilization to a 50% slower rate of passage for chicks consuming ethanol-extracted SBM compared with roosters fed the control SBM. Moreover, other researchers reported that feeding roosters ethanol-extracted SBM resulted in greater TME_n concentrations compared with a control SBM [44]. When raffinose and stachyose were added to soy protein concentrate, these authors observed a decrease in TME_n [Figure 3].

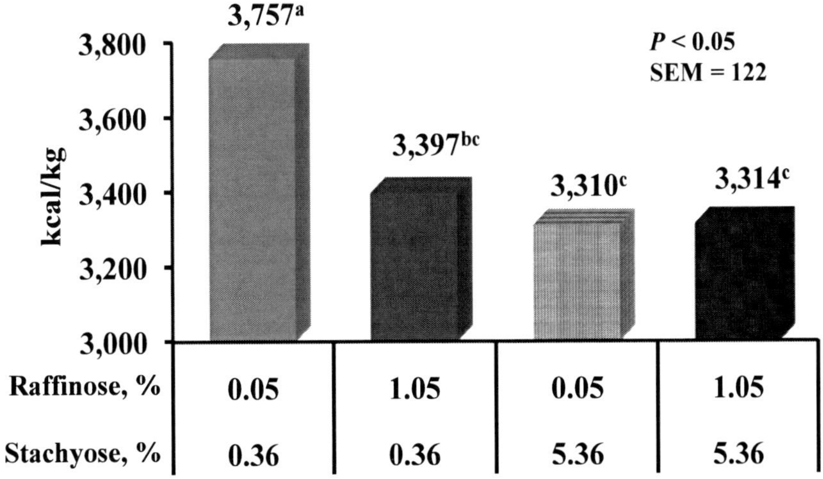

Figure 3. Effect of raffinose and stachyose supplementation on the TME_n of soy protein concentrate when precision fed to roosters. (*Adapted from Leske et al., 1993[44]*).

In addition to GAL, ethanol extraction of SBM removes a large percentage of soluble NSP. Because NSP have been reported to decrease nutrient digestibility, this could explain the positive effects ethanol extraction has on nutrient utilization [41]. Conversely, Irish et al. [56] did not observe increases in TME_n, weight gain, feed efficiency, or protein digestibility when an ethanol-extracted SBM was fed to chickens. These authors attributed the results to the ethanol extraction process altering the nutrient composition of the SBM and the low palatability of the experimental diet.

Enzyme Supplementation. Galactooligosaccharides have limited nutrient value when fed to broilers because poultry lack the enzymes required for the hydrolysis and subsequent absorption of these molecules. Research attempting to supplement diets with α-galactosidase has been met with limited success. Ghazi et al. [57] provided SBM supplemented with gradient concentrations of α-galactosidase (0, 0.025, 0.0625, 0.250 g/kg) to cockerels in a precision fed assay. These authors reported a 10.8% increase in N retention and a 15.6% increase in TME_n when feeding SBM incubated with 0.25 g/kg of α-galactosidase. Moreover, Ghazi et al. [58] found similar increases in TME_n and true N digestibility when α-galactosidase was supplemented into corn-SBM-based diets [Table 1].

Table 1. Effect of α-galactosidase supplementation on growth performance and nutrient digestibilities in growing broilers.

α-Galactosidase (g/kg)	Weight Gain (g/d)	Feed Intake (g/d)	Feed:Gain (g/g)	AME (kcal/kg)	Nitrogen Digestibility (%)
0.000	40	77	1.95	2,417	63.6
0.025	49	106	2.19	2,568	71.7
0.250	52	106	2.04	2,635	76.3
SEM	1.4	5.6	0.10	16.9	0.87
Linear Effect	$P < 0.05$	NS	NS	$P < 0.05$	$P < 0.05$

(Adapted from Ghazi et al., 2003 [58])

Graham et al. [60] reported increased TME_n when SBM was sprayed and incubated with α-galactosidase, but no differences in growth performance were observed when diets containing enzyme-incubated SBM were fed to broiler chicks from 3 to 27 days of age. Additionally, the autolysis of raffinose and stachyose of soy flakes resulted in a 94% reduction of GAL concentrations (hydrolyzed into their monosaccharide constituents) [61]. When these soy flakes were fed to chickens, no increases in TME_n, growth rate, or feed efficiency were observed compared with chickens fed water incubated soy flakes. Due to a lack of response in TME_n or growth rate, these authors hypothesized that energy utilization of SBM may not be exclusively related to GAL concentrations. Furthermore, α-galactosidase or carbohydrase cocktails containing α-galactosidase supplemented in corn-SBM-based diets have not enhanced the growth performance of broilers [39, 56, 59].

Trait Enhanced Soybean Meal. Genetic varieties of soybeans have been developed with low concentrations of raffinose and stachyose [62, 63, 64, 65, 66]. In addition to a reduction in GAL content, these enhanced soybeans contained increased concentrations of CP/AA and sucrose. The altered nutrient composition of SBM produced from low GAL soybeans may influence nutrient digestibility and growth performance of broilers. Recently, Perryman and Dozier [66] examined AME_n and apparent AA digestibility of commodity SBM (CSBM), low oligosaccharide (LOSBM), and ultra-low oligosaccharide (ULSBM) in four experiments. The LOSBM and ULSBM had 75 and 90% reductions in GAL concentrations compared with the CSBM, respectively. In experiment 1, LOSBM had a 194 kcal/kg (2,435 vs. 2,241 kcal/kg) higher AME_n compared with CSBM. In experiment 2, LOSBM displayed 141 and 134 kcal/kg more AME_n than ULSBM and CSBM, respectively. Digestible methionine, lysine, threonine, isoleucine, and valine concentrations were higher for LOSBM and ULSBM compared with CSBM in experiments 3 and 4. Digestible AA concentrations were the highest for ULSBM followed by LOSBM, and CSBM [Figure 4].

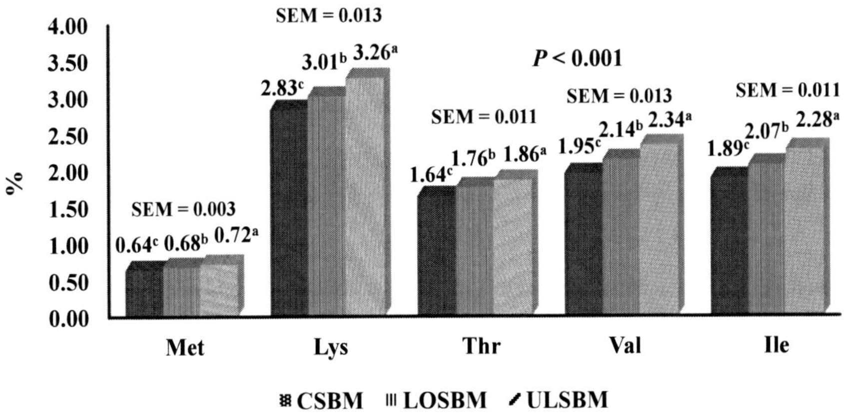

Figure 4. Total digestible AA concentrations of a commodity soybean meal (CSBM), low oligosaccharide soybean meal (LOSBM), and ultra-low oligosaccharide soybean meal (ULSBM) when fed to growing broilers. (*Adapted from Perryman and Dozier, 2012 [66]*).

Higher AA digestibilities were due to greater total AA concentrations rather than improved AA digestibility coefficients. Moreover, trypsin inhibitor concentrations were higher in ULSBM and LOSBM compared with CSBM. However, all SBM appeared to have adequate heat treatment as protein solubility index was within acceptable range [67]. Recently, Chen et al. [65] determined nutritional value of conventional and reduced-oligosaccharide SBM (87% reduction) in four chick assays. Roosters fed reduced-oligosaccharide SBM had higher AME_n on dry-matter basis (a difference of 200 kcal/kg) compared with birds consuming conventional SBM. Amino acid digestibility coefficients were similar for roosters fed two types of SBM, which infers that AA coefficients for conventional SBM can be employed when using reduced-oligosaccharide SBM. In agreement, Parsons et al. [62] fed five types of low GAL SBM to roosters and reported a 7% increase in TME_n compared with a control SBM. Conversely, Baker et al. [63] reported no differences in TME_n or AA digestibility of low GAL SBM compared with a control SBM when fed to chickens. In agreement with results determined in poultry, low GAL SBM had similar AA digestibility values compared with conventional SBM when fed to pigs [64].

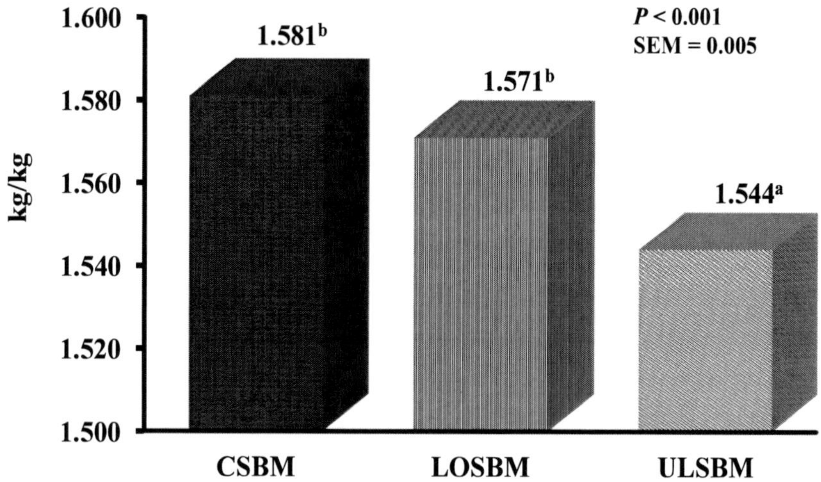

Figure 5. Feed conversion ratio for broilers fed diets formulated with either commodity soybean meal (CSBM), low oligosaccharide soybean meal (LOSBM), or ultra-low oligosaccharide soybean meal (ULSBM) from 1 to 42 days of age. (*Adapted from Perryman and Dozier, 2013[68]).*

Perryman et al. [68] determined the effects of feeding diets containing CSBM, LOSBM, and ULSBM on growth performance and meat yields of broilers during a six week production period in two experiments. Diets were formulated using AME_n and digestible AA concentrations determined from previous research [65]. Broilers receiving diets containing LOSBM grew faster and more efficiently from 1 to 14 days of age in experiment 1. In experiment 2, broilers provided diets containing ULSBM had better feed conversion than the birds fed diets containing the two other types of SBM from 1 to 28 and 1 to 42 days of age [Figure 5].

In addition, broilers fed diets containing LOSBM and ULSBM had higher total breast meat yields than birds provided diets with CSBM. Moreover, broilers fed diets formulated with ULSBM had greater carcass yield than birds given diets containing CSBM. Diets formulated with ULSBM and LOSBM had approximately 45% less supplemental fat, and broilers provided these diets exhibited no adverse effects on growth performance and meat yields when compared with broilers fed diets containing CSBM. The reduction in supplemental fat was attributable to both the higher concentrations of AME_n and digestible AA of SBM selected to have low concentrations of raffinose and stachyose. In agreement, Chen et al. [65] reported increased gain:feed and decreased feed consumption of broilers fed reduced-oligosaccharide SBM from 7 to 20 days of age compared with birds fed diets containing CSBM. Body weight gain was similar among the treatments. Conversely, Baker et al. [63] reported no observed differences in BW gain, feed intake, or feed conversion when chicks were fed diets containing low GAL SBM compared with a control SBM when fed to chickens. In agreement with results determined in poultry, low GAL SBM had similar AA digestibility values compared with conventional SBM when fed to pigs [64].

Conclusion

Genetically selecting LOSBM and ULSBM varieties of soybeans has merit in enhancing nutrient value of SBM compared with CSBM. Further research is warranted to determine the effect of fiber and sucrose concentrations on energy utilization when feeding novel genetic varieties of soybeans to broilers. In addition, low trypsin inhibitor SBM appears to have value as a protein source for poultry [69]. Low trypsin inhibitor varieties could prove advantageous if more stringent environmental regulations are implemented pertaining to hexane extraction. In the future, combinations of multiple beneficial genetic traits may enable the selection of a low oligosaccharide, low trypsin inhibitor, and high CP/AA soybeans. The agronomic implications and effects on animal performance pertaining to new genetic varieties would need to be investigated.

References

[1] F. W. Hill, D. L. Anderson, R. Renner, and L. B. Carew, Jr. *Poult. Sci.* 39,573 (1960).

[2] National Biodiesel Board, *http://www.biodiesel.org/production/production-statistics*

[3] M. Donohue and D. L. Cunningham, *J. Appl. Poult. Res.* 18, 325 (2009).

[4] S. Leeson, and J. D. Summers, *Scott's Nutrition of the Chicken*, M. L. Scott and Associates, Ithaca, N.Y. (2001).

[5] I. R. Sibbald, and S. J. Slinger, *Poult. Sci.* 6, 1612 (1962).

[6] G. Anninson, and M. Choct, *Enzymes in Animal Nutrition*, ETH-Zurich, Zurich, Switzerland. (1993)

[7] L. M. Potter, and M. Potchanakorn. *World Soybean Conf. III Proc.,* R. Shibles, ed. West View Press, Boulder, CO. (1985).
[8] C. M. Grieshop, C. T. Kadzere, G. M. Clapper, E. A. Flickinger, L. L. Bauer, R. L. Frazier, and G. C. Fahey, Jr., *J. Agric. Food Chem.* 51,7684 (2003).
[9] L. K., Karr-Lilienthal, C. T. Kadzere, C. M. Grieshop, and G. C. Fahey, Jr., *J. Livestock Prod. Sci.* 97, 1 (2005).
[10] A. Pettersson, and J. E. Lindberg, *Anim. Feed Sci. Technol.* 66, 97 (1996).
[11] NRC, *Nutrient Requirements of Poultry 9th ed.*, Natl. Acad. Press, Washington, D.C.
[12] NRC, *Nutrient Requirements of Swine. 10th rev. ed.,* Natl. Acad. Press, Washington, D.C.
[13] E. T. Moran, Jr., *Poult. Sci.* 61, 1257 (1982).
[14] M. Choct and D. J. Cadogan, *Manipulating Pig Production* VIII, Adelaide, South Australia (2001).
[15] K. E. Bach Knudsen, *Anim. Feed Sci. Technol.* 67, 319 (1991).
[16] B., Carré, B., J. Gomez, and A. M. Chagneau. *Br. Poult. Sci.* 36, 611 (1995).
[17] Cromwell, G. L. *The Farmer's Pride, KPPA News*, 11 No. 20 (1999).
[18] M. Choct, M., Y. Dersjant-Li, J. McLeish, and M. Peisker. *Asian-Aust. J. Anim. Sci.* 10,1386, (2010).
[19] C. A. Ivy, D. B. Bragg, and E. L. Stephenson. Poult. Sci. 50, 408 (1971).
[20] C. M. Parsons, Y. Zhang, and M. Araba. *Poult. Sci.* 79,1127 (1981).
[21] E. Karimzadegan, A. J. Clifford, and F. W. Hill. *J. Nutr.* 109, 2247 (1979).
[22] G. N. Lodhi, R. Renner, and D. R. Clandinin. *J. Nutr.* 99, 413 (1969).
[23] M. Choct. *Feed Mill. Int.* 7, 13 (1997).
[24] R. Macrae, R. K. Robinson, and M. J. Sadler. *Encyclopedia of Food Science, Food Technology and Nutrition,* Academic Press. London (1993).
[25] C. T. M. Fransen, *PhD Diss. Univ. of Utrecht*. Utrecht, Netherlands (1999).
[26] Carré, B., L. Derouet, and B. Leclercq. *Poult. Sci.* 69, 623 (1990).
[27] C. H. M. Smits and G. Annison. *World's Poult. Sci. J.* 52, 203. (1996).
[28] H. Jørgensen, X. Zhao, K. E. Bach Knudsen, and B. O. Eggum. *Br. J. Nutr.* 75, 379 (1996).
[29] K. E. Bach Knudsen and I. Hansen. *Br. J. Nutr.* 65, 217 (1991).

[30] J. T. Yen, J. T., J. A. Nienaber, D. A. Hill, and W. G. Pond. *J. Anim. Sci.* 69, 2001 (1991).

[31] T. Antoniou, R. R. Marquardt, and E. Cansfield. *J. Agric. Food Chem.* 28, 1240 (1981).

[32] M. Choct, and G. Annison. *Br. Poult. Sci.* 31, 811 (1990).

[33] K. Angkanaporn, K., M. Choct, W. L. Bryden, E. F. Annison, and G. Annison. *J. Sci. Food Agric.* 66, 399 (1994).

[34] D. D. Wagner, and O. P. Thomas. *Poult. Sci.* 57, 971 (1978).

[35] M. Choct, M., R. J. Hughes, J. Wang, M. R. Bedford, A. J. Morgan, and G. Annison. 1996. *Br. Poult. Sci.* 37, 609 (1996).

[36] T. Ide, T., M. Horii, K. Kawashim, and T. Yamamoto. *Br. J. Nutr.* 62, 539 (1989).

[37] G. V. Vahouny, R. Tombes, M. M. Cassidy, D. Kritchevsky, and L. L. Gallo. *Proc. Exp. Biol. Med.* 166, 12 (1981).

[38] S. Kuriyama, S. and L. B. Mendel. *J. Biol. Chem.* 31, 125 (1917).

[39] P. W. Waldroup, C. A. Keen, F. Yan, and K. Zhang. *J. Appl. Poult. Res.* 15, 48 (2006).

[40] R. Gitzelmann, R. and S. Auricchio. *Pediatrics* 36, 231 (1965).

[41] C. N. Coon, K. L. Leske, O. Akavanichan, and T. K. Cheng. *Poult. Sci.* 69, 787 (1990).

[42] A. Veldman, A., W. A. G. Veen, D. Barug, and P. A. van Paridon. 1993. Effect of α-galactosides and α-galactosidase in feed on ileal piglet digestive physiology. *J. Anim. Physiol. Anim. Nutr.* 69, 57 (1993).

[43] M. R. Smiricky, C. M. Grieshop, D. M. Albin, J. E. Wubben, V. M. Gabert, and G. C. Fahey, Jr. *J. Anim. Sci.* 80, 2433 (2002).

[44] K. L. Leske, C. J. Jevne, and C. N. Coon. *Poult. Sci.* 72, 664 (1993).

[45] G. G. Irish and D. Balnave. *Aust. J. Agric. Res.* 44, 1483 (1993).

[46] J. M. Wong, R. Souza, C. W. C. Kendall, A. Emam, and D. J. A. Jenkins. *J. Clin. Gastroenterol.* 40, 235 (2006)

[47] P. Spring, C. Wenk, K. A. Dawson, and K. E. Newman. *Poult. Sci.* 79, 205 (2000).

[48] J. W. Anderson, P. Baird, R. H. Davis Jr., S. Ferreri, M. Knudtson, A. Koraym, V. Waters, and C. L. Williams. *Nutr. Rev.* 67, 188 (2009)

[49] M. Roberfroid, G. R. Gibson, L. Hoyles, A. L. McCartney, R. Rastall, I. Rowland, D. Wolvers, B. Watzl, H. Szajewska, B. Stahl, F. Guarner, F. Respondek, K. Whelan, V. Coxam, M.-J. Davicco, L. Le'otoing, Y. Wittrant, N. M. Delzenne, P. D. Cani, A. M. Neyrinck, and A. Meheust. *Br. J. Nutr.* 104, 1 (2010)

[50] R. Nemcova, A. Bomba, S. Gancarcikova, R. Herich, and P. Guba. *Berl. Munich Tierarztl Wochenschr* 112, 225 (1999).

[51] S. Tedelind, F. Westberg, M. Kjerrulf, and A. Vidal. *World J. Gastroenterol.* 13, 2826 (2007).

[52] P. D. Schley and C. J. Field. *Br. J. Nutr.* 87, 221 (2002)

[53] H. Ito, M. Satsukawa, E. Arai, K. Sugiyama, K. Sonoyama, S. Kiriyama, and T. Morita. *J. Nutr.* 139, 1640 (2009).

[54] T. A. Faber, R. N. Dilger, A. C. Hopkings, N. P. Price, and G. C. Fahey Jr. *Poult. Sci.* 91, 3132 (2012).

[55] K. L. Leske, O. Akavanichan, T. K. Cheng, and C. N. Coon. *Poult. Sci.* 70, 892 (1991).

[56] G. G. Irish, G. W. Barbour, H. L. Classen, R. T. Tyler, and M. R. Bedford. *Poult. Sci.* 74, 1484 (1995).

[57] S. Ghazi, S., J. A. Rooke, H. Galbraith, and A. Morgan. *Br. Poult. Sci.* 38, 528 (1997).

[58] S. Ghazi, S., J. A. Rooke, and H. Galbraith. Br. Poult. Sci. 44, 410 (2003).

[59] M. T. Kidd, G. W. Morgan, Jr., C. J. Price, P. A. Welch, and E. A. Fontana. *J. Appl. Poult. Res.* 10, 65 (2001).

[60] K. K. Graham, M. S. Kerley, J. D. Firman, and G. L. Allee. *Poult. Sci.* 81, 1014 (2002).

[61] C. R. Angel, J. L. Sell, and D. R. Zimmerman. *J. Agric. Food Chem.* 36, 542 (1988).

[62] C. M. Parsons, L. M. Potter, and R. D. Brown, Jr. *Poult. Sci.* 60, 2687 (1981).

[63] K. M. Baker, P. L. Utterback, C. M. Parsons, and H. H. Stein. *Poult. Sci.* 90, 390 (2011).

[64] K. M. Baker and H. H. Stein. *J. Anim. Sci.* 87, 2282 (2009).

[65] X. Chen, C. M. Parsons, and N. Bajjalieh. *Poult. Sci.* 92, 1830 (2013).

[66] K. R. Perryman and W. A. Dozier, III. *Poult. Sci.* 91, 2556 (2012).

[67] M. Araba and N. M. Dale. *Poult. Sci.* 69, 76 (1990).

[68] K. R. Perryman, H. Olanrewaju, and W. A. Dozier, III. *Poult. Sci.* 92, 1292 (2013).

[69] T.Loeffler, M. Y. Shim, R. B. Beckstead, A. B. Batal, and G. M. Pesti. *Poult. Sci.* 92, (2013).

In: Oligosaccharides
Editors: L. S. Schweizer, S. J. Krebs
ISBN: 978-1-62948-328-3

Chapter X

Lipo-Oligosaccharides in Miller Fisher Syndrome

Kyung Yeon Lee*
Department of Pediatrics, Ulsan University Hospital,
University of Ulsan College of Medicine, Ulsan, Republic of Korea

Abstract

Miller Fisher syndrome is a clinical variant of Guillain-Barré syndrome, characterized by the acute-onset of ophthalmoplegia, ataxia, and areflexia, with a relative sparing of strength in the extremities and trunk. In 1992, the discovery of immunoglobulin G antibodies against GQ1b ganglioside in patients with Miller Fisher syndrome inaugurated a new era in research on the pathogenesis of Miller Fisher syndrome. Subsequent studies revealed that the oculomotor, trochlear, and abducens nerves have significantly higher percentages of the GQ1b than the other cranial nerves. Because of the molecular similarity between lipo-oligosaccharides on GQ1b ganglioside of the human neuronal cell surfaces and lipo-oligosaccharides on the outer membranes of certain bacteria, e.g., *Campylobacter jejuni*, anti-GQ1b antibody generated in the bacterially infected patients cross-reacts with these lipo-oligosaccharides, which can cause Miller Fisher syndrome. This antibody is now widely used as a diagnostic marker for Miller Fisher syndrome. However, a

* Corresponding author: Kyung Yeon Lee. E-mail: pdsnoopy@naver.com.

small number of anti-GQ1b-negative Miller Fisher syndrome cases have also been documented, although they presented with clinical features nearly identical to those of patients who were anti-GQ1b-positive. Despite the enormous advancement of knowledge on the pathogenesis of Miller Fisher syndrome thanks to the progress of comprehension about the structure of lipo-oligosaccharide in human gangliosides and bacterial cell membranes and the role of anti-GQ1b antibody, we still have unresolved questions about Miller Fisher syndrome.

Introduction

Miller Fisher syndrome is a clinical variant of Guillain-Barré syndrome, characterized by a clinical triad of acute-onset ophthalmoplegia, ataxia, and areflexia, with a relative sparing of strength in the extremities and trunk [1]. In 1956, Miller Fisher reported on three patients with this disorder, and postulated that it represented a distinct clinical variant of Guillain-Barré syndrome with a good prognosis, because one of his patients had demonstrated cerebrospinal fluid albuminocytologic dissociation [1]. In 1992, the discovery of immunoglobulin G antibodies against GQ1b ganglioside (anti-GQ1b antibody) in patients with Miller Fisher syndrome inaugurated a new era in research on the pathogenesis of Miller Fisher syndrome [2]. Subsequent study revealed that a monoclonal anti-GQ1b antibody immunostained the paranodal regions of the extramedullary portions of the human oculomotor, trochlear, and abducens nerves, and that the human oculomotor nerve contained a larger amount of GQ1b [3].

On the other hand, *Campylobacter jejuni* is the most frequently identified antecedent agent in Miller Fisher syndrome [4]. Monoclonal antibodies against the GQ1b ganglioside were demonstrated to react to lipo-oligosaccharide fractions from *Campylobacter jejuni* isolated in patients with Miller Fisher syndrome [5].

Subsequent investigations have demonstrated that there is a structural similarity between the lipo-oligosaccharide moiety of the GQ1b ganglioside in human neuronal cell membrane and the lipo-oligosaccharides of the outer membranes in certain bacteria, e.g., *Campylobacter jejuni*. A cross-reactive immune response to them can cause Miller Fisher syndrome. In this chapter, the structure of the lipo-oligosaccharides of the gangliosides in human neuronal cell membranes and bacterial outer membranes is examined and the roles of cross-reactive antibodies against those lipo-oligosaccharides in the pathogenesis of Miller Fisher syndrome along with the clinical features of Miller Fisher syndrome are reviewed.

History of Miller Fisher Syndrome

In 1956, Miller Fisher reported on three patients with acute ophthalmoplegia, ataxia and areflexia and a relative sparing of strength in the extremities and trunk subsequent to upper respiratory infectious symptoms [1]. They showed good recoveries without particular treatment. Cerebrospinal fluid obtained from those patients revealed normal cell counts in all three patients, and elevated protein in one patient. He postulated that it was a variant of Guillain-Barré syndrome with a good prognosis due to cerebrospinal fluid albuminocytologic dissociation in one patient in addition to areflexia. Afterthen, Smith and Walsh described it as the "syndrome of external ophthalmoplegia, ataxia and areflexia (Fisher)" in 1957 [6], Goodwin and Poser reported it as "Fisher's syndrome" in 1963 [7], and Patel et al. called it "Miller Fisher syndrome" in 1966 [8]. Since then, the disorder has been called Miller Fisher syndrome or Fisher syndrome.

Clinical Features of Miller Fisher Syndrome

Miller Fisher syndrome is known as a variant of Guillain-Barré syndrome, accounting for 5% to 10% of Guillain-Barré syndrome cases. In the series of Guillain-Barré syndrome cases documented by Ropper et al, Miller Fisher syndrome accounted for 3% [9]. In other Western populations, Miller Fisher syndrome is reported to have an incidence of approximately 1%-5% of Guillain-Barré syndrome cases [10]. However, the incidence is considerably higher in Asian countries such as Taiwan (19%) and Japan (25%) [11, 12]. According to the study by Mori et al. which reviewed the 5 reliable reports including more than 5 patients with Miller Fisher syndrome, the male-to-female ratio for the 118 patients was 2:1 (range, 1:1-5.3:1), and the mean age at Miller Fisher syndrome onset was 43 years [13]. The ratio of Miller Fisher syndrome to Guillain-Barré syndrome without Miller Fisher synrome was 1:4.1 (range 1:3.0-1:13). The incidence of Miller Fisher syndrome hospital cases was 1.8 cases per year [13].

Miller Fisher syndrome is defined as having the clinical triad of ataxia, ophthalmoplegia and areflexia, based on Fisher's original article [1]. It frequently starts with acute or subacute onset diplopia, ataxia or both [12]. Limb weakness is absent or minimal. Symptoms other than the clinical triad are (in descending order according to frequency) ptosis, bulbar palsy, pupillary abnormalities such as mydriasis, facial palsy, dysesthesia, minimal limb weakness, decreased superficial sensation, decrease in deep sensation and autonomic abnormalities, such as tachycardia and urinary symptoms [12, 13].

The median time to the nadir of symptoms after neurologic onset was 6 days (range 2 to 21 days) [12]. In 28 of the 50 patients who had received no immunotherapy, the median recovery periods between neurologic onset and the beginning of recovery from ataxia and ophthalmoplegia were 12 and 15 days, respectively. The respective median periods for the disappearance of ataxia and ophthalmoplegia were 32 and 88 days. Almost all the patients recovered from the ataxia, ophthalmoplegia, and areflexia in that order [12].

Anti-GQ1b Antibody in Miller Fisher Syndrome and Its Variants

The detection rate of anti-GQ1b antibodies in patients with Miller Fisher syndrome was estimated at between 83% and 100% [3, 14, 15]. These antibodies are now widely used as a diagnostic marker for Miller Fisher syndrome. Anti-GQ1b immunoglobulin G antibody titers peak at disease onset and then decrease with time, whereas cerebrospinal fluid albuminocytologic dissociation is present in only 25% of patients with Miller Fisher syndrome during the first week, but is present in 84% during the third week [5, 16].

Miller Fisher syndrome has been thought to be related to Guillain-Barré syndrome because they share some clinical features, such as acute or subacute onset disease with spontaneous remission, the presence of an antecedent infection, areflexia and albuminocytologic dissociation in the cerebrospinal fluid. Miller Fisher syndrome includes a variety of clinical variants such as acute ophthalmoplegia without ataxia [17-20], acute ataxia without ophthalmoplegia [21, 22], ataxic Guillain-Barré syndrome [23, 24], acute sensory ataxic neuropathy [25-27], acute oropharyngeal palsy [28], and pharyngeal-cervical-brachial weakness [29]. In addition, Miller Fisher syndrome clinically overlaps with Biekerstaff brainstem encephalitis in that both exhibit ataxia, ophthalmoplegia, albuminocytologic dissociation in cerebrospinal fluid, positive serum IgG anti-GQ1b antibodies, and antecedent *Campylobacter jejuni* infection [5, 14], although patients with Biekerstaff brainstem encephalitis show decreased consciousness and sometimes reveal normal or brisk deep tendon reflex and cerebrospinal fluid pleocytosis, which indicates the involvement of the central nervous system.

Among these disorders, Miller Fisher syndrome, Biekerstaff brainstem encephalitis, acute ophthalmoplegia without ataxia, acute sensory ataxic neuropathy, and pharyngeal-cervical-brachial weakness have been shown to be associated with the anti-GQ1b antibodies, although patients with acute sensory ataxic neuropathy more often have the anti-GD1b antibody and less frequently have anti-GQ1b antibodies compared with patients with Miller Fisher syndrome [27]. Patients with pharyngeal-cervical-brachial weakness have anti-GT1a antibody with or without GQ1b antibody [29, 30]. Thus, some authors have proposed that this spectrum of disorders should be named the "anti-GQ1b antibody syndrome" to consolidate their common serological profiles [30, 31].

Antecedent Infections in Miller Fisher Syndrome

According to one report by Mori et al. on antecedent infections in 50 patients with Miller Fisher syndrome, before the onset of Miller Fisher syndrome, respiratory symptoms occurred in 76% of the patients, gastrointestinal symptoms in 4%, and fever in 2% [12]. No infectious symptoms occurred in 18% of the patients. The median interval between onset of infectious symptoms and development of neurological symptoms was 8 days (range, 1 to 30 days). A variety of infectious agents, such as *Campylobacter jejuni* [32, 33], *Haemophilus influenzae* [34], *Mycoplasma pneumoniae* [35], Epstein-Barr virus [36], cytomegalovirus, enterovirus [37], varicella-zoster virus [38], human immunodeficiency virus [39], *Helicobacter pylori* [40], Influenza A [41], and Norovirus [42] have been reported as preceding infectious agents in Miller Fisher syndrome. Although not all of these microorganisms have been proven causative pathogens of Miller Fisher syndrome, *Campylobacter jejuni* and *Haemophilus influenzae* have been demonstrated their association with Miller Fisher syndrome [43]. According to a prospective case-controlled study by Koga et al., which examined five pathogens – *Campylobacter jejuni*, cytomegalovirus, Epstein–Barr virus, *Mycoplasma pneumoniae* and *Haemophilus influenzae* – in 73 patients with

Miller Fisher syndrome and 73 sex- and age-matched hospital controls to identify the antecedent pathogens associated with Miller Fisher syndrome, the serologic incidence of *Campylobacter jejuni* (21%) and *Haemophilus influenzae* (8%) infections was significantly higher in the patients with Miller Fisher syndrome compared to the matched control patients [43]. Interestingly, *Haemophilus influenzae* was isolated from the sputum of one of Fisher's original patients [1].

Streptococcus pyogenes, *Staphylococcus aureus*, *Coxiella burnetii*, and the mumps virus are also reported as antecedent agents in Miller Fisher syndrome, but no statistical correlations have been demonstrated [44]. In addition to antecedent infection, postvaccination Miller Fisher syndrome cases after use the combined pertussis, diphtheria, and tetanus toxoid vaccine [45] and novel influenza A (H1N1) vaccine [46] were also reported.

Lipo-Oligosaccharide of *Campylobacter Jejuni* Cell Membrane

Lipo-oligosaccharides are glycolipids found in the outer membranes of Gram-negative bacteria, such as *Campylobacter* and *Haemophilus*. The term is synonymous with the low molecular weight form of bacterial lipopolysaccharide, which is also referred to as an endotoxin [47]. Lipo-oligosaccharides play a critical role not only in maintaining the integrity and functionality of the outer membrane of the Gram-negative cell envelope, but also in the pathogenesis of certain bacterial infections because they are capable of acting as immunostimulators and immunomodulators [47]. Furthermore, the lipo-oligosaccharides of some bacterial strains display molecular mimicry of host cells, and the variety of lipo-oligosaccharides is responsible for antigenic diversity, which helps in the evasion of host immune defenses.

In the outer membrane of Gram-negative bacteria, lipo-oligosaccharides contain two main structural features: a lipid A-based outer membrane-anchoring moiety, and an oligosaccharide core [48]. The oligosaccharide core is composed of an oligosaccharide inner core, and a short oligosaccharide chain called the outer core, which faces the environment. The outer core oligosaccharide chain varies depending on the bacterial strain [47, 48]. Fraction of the lipo-oligosaccharides from several bacterial strains have been shown to be structurally similar to human host cell surface molecules.

Campylobacter jejuni, a Gram-negative bacillus, is not only one of the most common causes of human gastroenteritis in the world, but also the most well-known pathogen concerning the pathogenesis of Miller Fisher syndrome, Guillain-Barré syndrome, and their variants.

As with other Gram-negative bacteria, *Campylobacter jejuni* also has sialylated lipo-oligosaccharides in its outer membrane, which is structurally similar to the sialylated lipo-oligosaccharides of the gangliosides in human neuronal cell membranes [49].

Since the first report in 1993, several studies have demonstrated ganglioside-like structures in the lipo-oligosaccharide outer core of *Campylobacter jejuni* strains isolated from patients with Miller Fisher syndrome and Guillain-Barré syndrome [50]. Mass spectrometry and nuclear magnetic resonance analyses of individual *Campylobacter jejuni* strains have displayed the presence of GD3-like lipo-oligosaccharides in Miller Fisher syndrome-inducing *Campylobacter jejuni* strains [51]. GT1a- and GD1c-like lipo-oligosaccharides that mimic GQ1b have also also identified in *Campylobacter jejuni* strains isolated from patients with Miller Fisher syndrome [4, 52, 53]. The immunization of GQ1b/GT1a-deficient mice with a GT1a-like lipo-oligosaccharide generated monoclonal IgG antibodies that reacted with GQ1b and GT1a [43]. Recent studies have shown that not only GQ1b, but also some ganglioside complexes including GQ1b, are targets for serum antibodies in patients with Miller Fisher syndrome [54, 55]. Kaida et al. reported that seven of 12 (58%) patients with Miller Fisher syndrome were found to have IgG antibodies to ganglioside complexes containing GQ1b, of whom five had IgG antibodies to the GQ1b/GM1 complex and two had antibodies to the GQ1b/GD1a complex [54]. In fact, GD1c, GQ1b, and GT1a have in common the structure (αNeuAc(2–8)αNeuAc(2–3)βGal(1–3)GalNAc) in the terminal residues [56]. These investigations suggest that not only GQ1b-like lipo-oligasaccharides, but also GD3-, GT1a- and GD1c-like lipo-oligasaccharides and ganglioside complexes including GQ1b contain sialylated lipo-oligasaccharides that are structurally similar to that of the human GQ1b ganglioside, which can result in Miller Fisher syndrome (Figure 1).

Campylobacter jejuni exhibits considerable structural variation in its lipo-oligosaccharide outer core, which may give the bacterium opportunities to evade a host immune response. In the patients infected by the bacteria, however, it may cause diverse clinical manifestations due to a variety of antigenicity. Several genetic mechanisms responsible for this lipo-oligosaccharide variation in *Campylobacter jejuni* have been documented [56].

First, there is extensive variation in the gene content of the lipo-oligosaccharide biosynthesis gene locus.

In addition, variation in homopolymeric tracts, single-base deletions, insertions, and mutations can lead to gene inactivations or glycosyltransferases with different acceptor specificities, resulting in the expression of different lipo-oligosaccharide structures. Detailed mechanisms of biosynthesis of ganglioside-like lipo-oligosaccharide in *Campylobacter jejuni* are described next section.

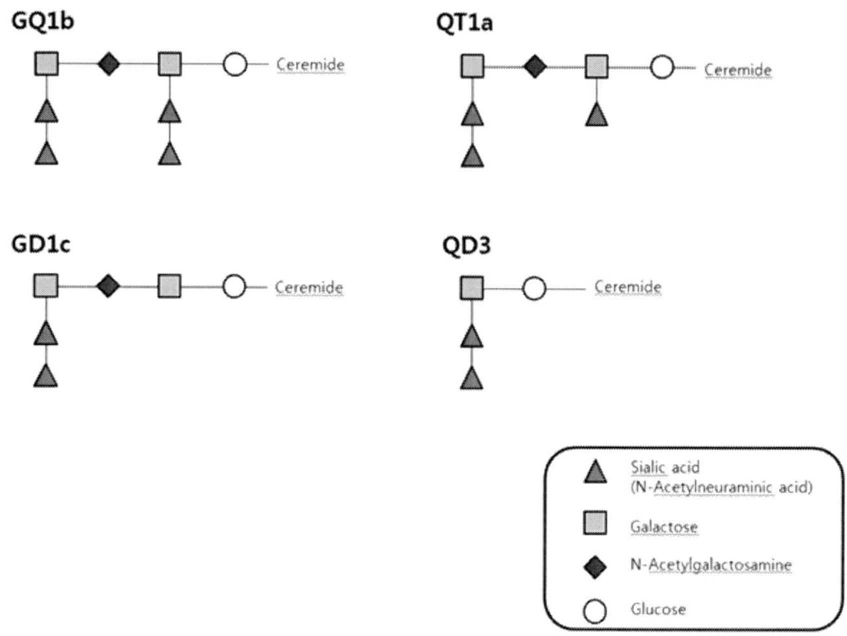

Most patients with Miller Fisher syndrome have antibody against GQ1b ganglioside, whereas a small number of patients have other kinds of antibodies, such as antibodies against GT1a-, GD1c-, or GD3-like lipo-oligosaccharides. Note that GQ1b, GT1a, GD1c, and GD3 contain in common the structure composed of "Sialic acid–Sialic acid–Galactose" in the terminal residues.

Figure 1. Gangliosides related to Miller Fisher syndrome.

Meanwhile, *Haemophilus influenzae*, which is a Gram-negative bacterium causing respiratory infections and known as an antecedent pathogen of Miller Fisher syndrome, has also been shown that its lipo-oligosaccharide includes an αNeuAc(2–8)αNeuAc(2–3)βGal branch, a component of the GQ1b ganglioside [57].

Lipo-Oligosaccharides Biosynthesis Genes in *Campylobacter Jejuni*

Various ganglioside-like structures have been identified in the lipo-oligosaccharides fraction of the *Campylobacter jejuni* cell wall [58]. This variation in lipo-oligosaccharides structure is the result of differences in the presence of lipo-oligosaccharides biosynthesis genes and in the DNA sequence polymorphism within these genes [56]. Based on the differences in gene content observed thus far, eight different classes (classes A through H) of the lipo-oligosaccharides biosynthesis gene locus have be identified [59, 60]. However, only strains of classes A, B, or C lipo-oligosaccharides locus express ganglioside-like lipo-oligosaccharides [61]. *Campylobacter jejuni* Sialyltransferase Cst-II or -III, N-acetylgalactosaminyltransferase CgtA, and galactosyltransferase CgtB are essential for the biosynthesis of ganglioside-like lipo-oligosaccharide [56].

The *cst-II* gene encodes a sialyltransferase that is necessary for the transfer of sialic acid onto the lipo-oligosaccharide core in *C. jejuni* class A and B strains [62]. *C. jejuni* class C strains depend on the *cst-III* gene for lipo-oligosaccharide sialylation [62]. *Campylobacter jejuni* isolates from Guillain-Barré syndrome (96%) and Miller Fisher syndrome (88%) carry the cst-II or -III, cgtA and cgtB genes which encode these enzymes significantly more often than do isolates from uncomplicated enteritis (72%) [63]. Cst-II sialyltransferase consists of 291 amino acids. The variation in the nucleotide sequence of the *cst-II* gene affects its activity, and the 51st determines its enzymatic activity [56]. Cst-II (Thr51) has only α-2,3-sialyltransferase activity (monofunctional) and produces GM1- and GD1a-like lipo-oligosaccharides.

In contrast, Cst-II (Asn51) has both α-2,3- and α-2,8-sialyltransferase activities (bifunctional) and synthesizes GT1a-, GD1c- or GD3-like lipo-oligosaccharides, in which each terminal trisaccharide is identical to that of GQ1b. Neuropathic *Campylobacter jejuni* strains were more frequently found to have cst-II, particularly cst-II (Thr51), than enteritic ones (82% vs. 52%) [64]. Strains with cst-II (Asn51) regularly expressed the GQ1b-like lipo-oligosaccharides, whereas those with cst-II (Thr51) had the GM1-like and GD1a-like lipo-oligosaccharides.

The presence of these bacterial lipo-oligosaccharides in patients with neuropathy corresponded to autoantibody reactivity. Patients infected with *Campylobacter jejuni* (cst-II Asn51) were more often positive for anti-GQ1b

IgG antibodies and developed Miller Fisher syndrome presenting with ophthalmoplegia and ataxia.

In contrast, patients who had *Campylobacter jejuni* (cst-II Thr51) were more frequently positive for anti-GM1 and anti-GD1a IgG antibodies and had Guillain-Barré syndrome presenting with limb weakness.

Human Ganglioside and Anti-GQ1b Antibody in Miller Fisher Syndrome

A ganglioside is a molecule composed of a glycosphingolipid (ceramide and oligosaccharide) and one or more sialidic acids (e.g. n-acetylneuraminic acid, NANA) linked on the oligosaccharide chain [65]. The name "ganglioside" was first applied by the German biochemist Ernst Klenk in 1942 to lipids newly isolated from ganglion cells of the brain [66]. Although gangliosides should be named systematically in the same way with the position of the sialic acid residue(s) indicated as for branched structures, they are more conveniently named by a short-hand nomenclature system proposed by Svennerholm in which M, D, T and Q refer to mono-, di-, tri- and tetrasialogangliosides, respectively, and the numbers 1, 2, 3, etc. refer to the order of migration of the gangliosides on thin-layer chromatography [66]. For example, the order of migration of monosialogangliosides is GM3 > GM2 > GM1. To indicate variations within the basic structures, further terms are added, e.g. GT1a, GD1c, etc. (Figure 1).

Gangliosides are membrane glycolipids that are highly enriched in the nervous system. They can amount to 6% of the weight of lipids from the brain, where they constitute 10 to 12% of the total lipid content of neuronal membranes. The brain contains as much as 20 to 500 times more gangliosides than most non-neural tissues, with three times as much in grey as in white matter. As the brain develops, there is an increase in the content of gangliosides and in their degree of sialylation [66].

The oculomotor, trochlear, and abducens nerves, along with the optic nerve, have significantly higher percentages of GQ1b than the other cranial nerves and the ventral and dorsal roots of the spinal cord [67]. Immunohistochemical studies have shown that GQ1b is highly expressed in the extramedullary regions of the human oculomotor, trochlear, and abducens nerves, whereas no accumulation of GQ1b was found in other cranial nerves, including the optic nerve [3]. Monoclonal antibodies against GQ1b and GT1a also bind the vast majority of motor end-plates of human oculomotor muscles [68]. Anti-GQ1b antibodies cause functional and structural changes at neuromuscular junctions where GQ1b is expressed in mouse nerve–muscle preparations [69].

These findings suggest that anti-GQ1b IgG antibodies attack the neuromuscular junctions of oculomotor muscles and produce ophthalmoplegia in patients with Miller Fisher syndrome.

Meanwhile, it remains controversial whether the pathophysiology of ataxia in Miller Fisher syndrome originates from peripheral proprioceptive sensory neuropathy or from cerebellar lesion.

Muscle spindles contain specialized sensory fibers responsible for the proprioceptive sense, which primarily detects changes in the length of the muscle. The neural components and intrafusal muscle fibers of muscle spindles have been labelled by monoclonal anti-GQ1b as well as anti-GD1b antibodies in humans, and the staining patterns suggest that the group Ia afferents in intrafusal muscle spindles strongly express GQ1b and GD1b [68]. These group Ia afferents are the likely targets in patients with Miller Fisher syndrome, resulting in ataxia.

Postural body sway analysis study, which measures other aspects of cerebellar and peripheral nerve functions, demonstrated that 9 of 10 patients with Miller Fisher syndrome have a dysfunction of the proprioceptive afferent system [70].

Moreover, some investigators have found peripheral nerve dysfunction in electrophysiological studies of patients with Miller Fisher syndrome, which supports a peripheral ataxia mechanism [71, 72]. However, one study showed selective staining of the molecular layer of human cerebellum by sera from patients with Miller Fisher syndrome who have anti-GQ1b antibodies [73]. Cerebellar changes on the MRIs of patients with Miller Fisher syndrome have also been reported [74].

In summary, due to the structural similarity between lipo-oligosaccharides of GQ1b gangliosides at the neuromuscular junctions of oculomotor muscles and on muscle spindles and GQ1b-like lipo-oligosaccharides in the *Campylobacter jejuni* outer membraine, in patients with *Campylobacter jejuni* infection, anti-GQ1b antibodies may attack neuromuscular junctions of oculomotor muscles and muscle spindles, and can produce the characteristic clinical features of Miller Fisher syndrome.

Anti-GQ1b-Negative Miller Fisher Syndrome

Although the anti-GQ1b antibody produced by bacteria containing a GQ1b-like lipo-oligosaccharide plays the key role in inducing Miller Fisher syndrome, a small number of anti-GQ1b-negative Miller Fisher syndrome cases also have been documented even after infection with *Campylobacter jejuni*, which is the representative bacterium related to anti-GQ1b antibody in the pathogenesis of Miller Fisher syndrome.

Several *Campylobacter jejuni* strains isolated from patients with Miller Fisher syndrome did not express GQ1b-like lipo-oligosaccharides, and instead expressed GT1a-, GD3-, or GD1c-like lipo-oligosaccharides, which contain the common structure in the terminal residues, as previously mentioned (Figure 1) [4, 60, 61, 68].

Moreover, recently, antibodies against isolated gangliosides other than these lipo-oligosaccharides, ganglioside complexes, or ganglioside-like lipo-oligosaccharides other than GQ1b-like lipo-oligosaccharides were identified in patients with Miller Fisher syndrome [13, 60, 75, 76]. Kimoto et al. reported that 22 (16%) of 139 patients with *Campylobacter jejuni*-isolated neuropathy developed Miller Fisher syndrome, and two (9%) of these 22 patients were anti-GQ1b or anti-GT1a antibody-negative, whereas four (18%) were anti-GM1 or anti-GD1a antibody-positive [60].

According to the most recent study on anti-GQ1b-negative Miller Fisher syndrome, the clinical features of the patients with anti-GQ1b-negative Miller Fisher syndrome did not differ from those of anti-GQ1b-positive patients, except for male dominance and antecedent gastrointestinal symptoms. In that study, the authors reported that 24 patients revealed negative results of anti-GQ1b ganglioside antibody among 207 Miller Fisher syndrome cases, and instead some anti-GQ1b-negative patients had IgG antibodies against GM1b (8.3%), GD1a (4.2%), GalNAc-GD1a (4.2%), or GT1a (8.3%) [76]. In addition, immunoglobulin G antibodies against heterologous ganglioside complexes and isolated GM1b, GD1c, and GalNAc-GM1b gangliosides were revealed to be serologic markers of anti-GQ1b-negative Miller Fisher syndrome [76].

Among the 24 patients with anti-GQ1b-negative Miller Fisher syndrome, three with serologic evidence of recent *Campylobacter jejuni* infection tested positive for the immunoglobulin G antibody against GD1a and GalNAc-GD1a, GM1/GT1a and GM1/GQ1b, and several ganglioside-like lipo-oligosaccharides, respectively. This study suggests that antibodies against gangliosides other than GQ1b, ganglioside complexes, or ganglioside-like lipo-oligosaccharides other than GQ1b-like lipo-oligosaccharides play a pathogenic role in the development of Miller Fisher syndrome. Recently, an anti-voltage-gated potassium channel and novel antineuronal antibodies were found in the sera of patients with Miller Fisher syndrome [77]. Although the presence and significance of a mechanism other than the anti-GQ1b antibody-mediated one remains uncertain, this kind of pioneering research can widen the boundaries of knowledge about the pathogenesis of Miller Fisher syndrome.

Unresolved Questions

The advancement of our understanding of the pathogenesis of Miller Fisher syndrome has resulted from knowledge about the structure and functions of lipo-oligosaccharides present on human gangliosides and bacterial cell membranes.

Despite the progress in comprehension of this pathogenesis, some questions remain unanswered. First, a small subset of patients with Miller Fisher syndrome reveal a negative result for the anti-GQ1b antibody, and some of those patients have other kinds of gangliosides antibodies. What causes anti-GQ1b-negative Miller Fisher syndrome, and why is the clinical manifestations of anti-GQ1b-negative patients the same as those of anti-GQ1b-positive patients?

Second, as aforementioned, the incidence of antecedent *Campylobacter jejuni* or *Haemphilus influenzae* infections is not so high in patients with Miller Fisher syndrome. Which mechanism is operating in patients with Miller Fisher syndrome who have had neither a preceding *Campylobacter jejuni* nor a preceding *Haemphilus influenzae* infection? Is there any pathogen other than *Campylobacter jejuni* or *Haemphilus influenzae* that has a GQ1b-like lipo-oligosaccharide on it? How can Miller Fisher syndrome develop in patients without prior infection?

Third, the incidence of Miller Fisher syndrome in Asian countries appears to be higher than that in Western countries. What causes regional differences in Miller Fisher syndrome prevalence?

Despite the enormous advancement of knowledge about the pathogenesis of Miller Fisher syndrome, continuous efforts and investigations are needed to answer these unresolved questions.

Conclusion

During last 20 years since the discovery of the anti-GQ1b antibody in patients with Miller Fisher syndrome, there has been a tremendous advancement in knowledge on pathogenesis of Miller Fisher syndrome. The molecular similarity between the lipo-oligosaccharides on the GQ1b ganglioside of the human neuronal cell surfaces and those on the outer membranes of certain bacteria, e.g., *Campylobacter jejuni*, may induce a cross-reactive anti-GQ1b antibody that can cause Miller Fisher syndrome.

However, anti-GQ1b-negative Miller Fisher syndrome cases also have been documented. Additional study is needed to answer the unresolved questions about Miller Fisher syndrome.

References

[1] Fisher, M. An unusual variant of acute idiopathic polyneuritis (syndrome of ophthalmoplegia, ataxia and areflexia). *N Engl. J. Med.* 1956;255:57–65.

[2] Chiba, A., Kusunoki, S., Shimizu, T., et al. Serum IgG antibody to ganglioside GQ1b is a possible marker of Miller Fisher syndrome. *Ann. Neurol.* 1992;31:677–9.

[3] Chiba, A., Kusunoki, S., Obata, H., et al. Serum anti-GQ1b IgG antibody is associated with ophthalmoplegia in Miller Fisher syndrome and Guillain-Barré syndrome: Clinical and immunohistochemical studies. *Neurology* 1993;43:1911–7.

[4] Koga, M., Gilbert, M., Li, J., et al. Antecedent infections in Fisher syndrome: A common pathogenesis of molecular mimicry. *Neurology* 2005;64:1605–11.

[5] Yuki, N. Fisher syndrome and Bickerstaff brainstem encephalitis (Fisher-Bickerstaff syndrome). *J. Neuroimmunol.* 2009;215:1–9.

[6] Smith, J. L., Walsh, F. B. Syndrome of external ophthalmoplegia, ataxia, and areflexia (Fisher); ocular manifestations in acute idiopathic polyneuritis (Guillain–Barre syndrome); report of two cases. *AMA Arch. Ophthalmol.* 1957;58:109–14.

[7] Goodwin, R. F., Poser, C. M. Ophthalmoplegia, ataxia, and areflexia. Fisher's syndrome. *JAMA* 1963;186:258–9.

[8] Patel, A., Pearce, L., Hairston, R. Miller Fisher Syndrome (variant of Landry–Guillain–Barre–Strohl syndrome –ophthalmoplegia, ataxia, areflexia). *South Med. J.* 1966;59:171–5.

[9] Ropper, A. H., Wijdicks, E. F. M., Truax, B. T. *Guillain-Barré Syndrome*: Philadelphia, FA Davis, 1991.

[10] Bogliun, G., Beghi, E. Incidence and clinical features of acute inflammatory polyradiculoneuropathy in Lombardy, Italy, 1996. *Acta Neurol. Scand.* 2004;110:100–6.

[11] Lyu, R. K., Tang, L. M., Cheng, S. Y., et al. Guillain-Barré syndrome in Taiwan: a clinical study of 167 patients. *J. Neurol. Neurosurg. Psychiatry* 1997;63:494–500.

[12] Mori, M., Kuwabara, S., Fukutake, T., et al. Clinical features and prognosis of Miller Fisher syndrome. *Neurology* 2001;56:1104–6.

[13] Mori, M., Kuwabara, S., Yuki, N. Fisher syndrome: clinical features, immunopathogenesis and management. *Expert Rev. Neurother.* 2012;12: 39–51.

[14] Ito, M., Kuwabara, S., Odaka, M., et al. Bickerstaff's brainstem encephalitis and Fisher syndrome form a continuous spectrum: clinical analysis of 581 cases. *J. Neurol.* 2008;255:674–82.

[15] Willison, H. J., Veitch, J. Immunoglobulin subclass distribution and binding characteristics of anti-GQ1b antibodies in Miller Fisher syndrome. *J. Neuroimmunol.* 1994;50:159–65.

[16] Nishimoto, Y., Odaka, M., Hirata, K., et al. Usefulness of anti-GQ1b IgG antibody testing in Fisher syndrome compared with cerebrospinal fluid examination. *J. Neuroimmunol.* 2004;148:200–5.

[17] Odaka, M., Yuki, N., Hirata, K. Anti-GQ1b antibody syndrome: clinical and immunological range. *J. Neurol. Neurosurg. Psychiatry* 2001; 70: 50–5.

[18] Yuki, N. Acute paresis of extraocular muscles associated with IgG anti-GQ1b antibody. *Ann. Neurol.* 1996;39:668–72.

[19] Yuki, N., Odaka, M., Hirata, K. Acute ophthalmoparesis (without ataxia) associated with anti-GQ1b IgG antibody: clinical features. *Ophthalmology* 2001;108:196–200.

[20] Lee, S. H., Lim, G. H., Kim, J. S., et al. Acute ophthalmoplegia (without ataxia) associated with anti-GQ1b antibody. *Neurology* 2008;71:426–9.

[21] Mori, M., Kuwabara, S., Koga, M., et al. IgG anti-GQ1b positive acute ataxia without ophthalmoplegia. *J. Neurol. Neurosurg. Psychiatry* 1999; 67:668–70.

[22] Kusunoki, S., Chiba, A., Kanazawa, I. Anti-GQ1b IgG antibody is associated with ataxia as well as ophthalmoplegia. *Muscle Nerve* 1999: 22:1071–4.

[23] Richter, R. B. The ataxic form of polyradiculoneuritis (Landry- Guillain-Barré syndrome). clinical and pathologic observations. *J. Neuropathol. Exp. Neurol. 1962*;21:171–84.

[24] Yuki, N., Susuki, K., Hirata, K. Ataxic Guillain-Barré syndrome with anti-GQ1b antibody: relation to Miller Fisher syndrome. *Neurology* 2000;54: 1851–3.

[25] Ito, M., Matsuno, K., Sakumoto, Y., et al. Ataxic Guillain-Barré syndrome and acute sensory ataxic neuropathy form a continuous spectrum. *J. Neurol. Neurosurg. Psychiatry* 2011;82:294–9.

[26] Yuki, N., Miyatani, N., Sato, S., et al. Acute relapsing sensory neuropathy associated with IgM antibody against B-series gangliosides containing a GalNAc beta 1–4(Gal3–2 aNeuAc8–aNeuAc)beta 1 configuration. *Neurology* 1992;42:686–9.

[27] Notturno, F., Caporale, C. M., Uncini, A. Acute sensory ataxic neuropathy with antibodies to GD1b and GQ1b gangliosides and prompt recovery. *Muscle Nerve* 2008;37:265–8.

[28] O'Leary, C. P., Veitch, J., Durward, W. F., et al. Acute oropharyngeal palsy is associated with antibodies to GQ1b and GT1a gangliosides. *J. Neurol. Neurosurg. Psychiatry* 1996;61:649–51.

[29] Nagashima, T., Koga, M., Odaka, M., et al. Continuous spectrum of pharyngeal–cervical–brachial variant of Guillain-Barré syndrome. *Arch. Neurol.* 2007;64:1519–23.

[30] Shahrizaila, N., Yuki, N. Biekerstaff brainstem encephalitis and Fisher syndrome: anti-GQ1b antibody syndrome. *J. Neurol. Neurosurg. Psychiatry* 2013;84:576–83.

[31] Odaka, M., Yuki, N., Hirata, K. Anti-GQ1b IgG antibody syndrome: clinical and immunological range. *J. Neurol. Neurosurg. Psychiatry* 2001; 70:50–5.

[32] Jacobs, B. C., Eudtz, H. P., van der meché, F. G., et al. Serum anti-GQ1b IgG antibodies recognize surface epitope on *Campylobacter jejuni* from patients with Miller-Fisher syndrome. *Ann. Neurol.* 1995;37: 260–4.

[33] Nisser, A., Bernheimer, H., Berger, T., et al. Serum antibodies against gangliosides and *Campylobacter jejuni* lipopolysaccharides in Miller Fisher syndrome. *Infect. Immun.* 1997;65: 4038–42.

[34] Koga, M., Yuki, N., Tai, T., et al. Miller Fisher syndrome and *Haemophilus influenzae* infection. *Neurology* 2001;57: 686–91.
[35] Merkx, H., Keyser, J., Ebinger, G. Miller Fisher syndrome associated with *Mycoplasma pneumoniae* infection: report of a case. *Clin. Neurol. Neurosurg*. 1994;96:96–9.
[36] Schnorf, H., Rathgeb, J. P., Kohler, A. Anti-GQ1b positive Miller Fisher syndrome in a patient with acute Epstein-Barr virus infection and negative *Campylobacter* serology. *Eur. Neurol.* 1998;40:177.
[37] Moustaki, M., Gika, A., Fretzayas, A., et al. Miller Fisher syndrome in associated with enterovirus infection. *J. Child Neurol.* 2012;27:521–2.
[38] Takano, H., Yuki, N. Fisher syndrome associated with chicken pox and anti-GQ1b antibody. *J. Neurol.* 1995;242:255–6.
[39] Arranz Caso, J. A., Martinez, R., Cabrera, F., et al. Miller Fisher syndrome in a patient with HIV infection. *AIDS* 1997;15:550–1.
[40] Chiba, A., Sugiyama, T., Yonekura, K., et al. An antibody to Vac A of *Helicobacter pylori* in the CSF of patients with Miller Fisher syndrome. *Neurology* 2004;63:2184–6.
[41] Hara, M., Morita, A., Ichihara, K., et al. Miller Fisher syndrome associated with influenza A infection. *Intern. Med.* 2012;51:2621–3.
[42] Shimizu, T., Tokuda, Y. Miller Fisher syndrome linked to Norovirus infection. *BMJ Case Rep*. 2012;2012.
[43] Koga, M., Gilbert, M., Li, J., et al. Antecedent infections in Fisher syndrome: a common pathogenesis of molecular mimicry. *Neurology* 2005;64:1605–11.
[44] Yuki, N. Infectious origins of, and molecular mimicry in, Guillain–Barré and Fisher syndromes. *Lancet Infect. Dis*. 2001;1:29–37.
[45] Krämer, H. H., Niemöller, U., Döring, K., et al. Postvaccination Miller Fisher syndrome after combined pertussis, diphtheria and tetanus toxoid vaccine. *J. Infect*. 2013;66:460–1.
[46] Shoamanesh, A., Chapman, K., Traboulsee, A. Postvaccination Miller Fisher syndrome. *Arch. Neurol*. 2011;68:1327–9.
[47] Moran, A. P., Prendergast, M. M., Appelmelk, B. J. Molecular mimicry of host structures by bacterial lipopolysaccharides and its contribution to disease. *FEMS Immunol. Med. Microbiol*. 1996;16:105–15.
[48] Kilár, A., Dörnyei, A., Kocsis, B. Structural characterization of bacterial lipopolysaccharides with mass spectrometry and on- and off-line separation techniques. *Mass Spectrom. Rev*. 2013;32:90–117.

[49] Louwen, R., Heikema, A., van Belkum, A., et al. The sialylated lipooligosaccharide outer core in *Campylobacter jejuni* is an important determinant for epithelial cell invasion. *Infect. Immun.* 2008;76:4431–8.

[50] Yuki, N., Taki, T., Inagaki, F., et al. A bacterium lipopolysaccharide that elicits Guillain–Barré syndrome has a GM1 ganglioside-like structure. *J. Exp. Med.* 1993;178:1771–5.

[51] Salloway, S., Mermel, L. A., Seamans, M., et al. Miller-Fisher syndrome associated with *Campylobacter jejuni* bearing lipopolysaccharide molecules that mimic human ganglioside GD3. *Infect. Immun.* 1996;64: 2945–9.

[52] Kimoto, K., Koga, M., Odaka, M., et al. Relationship of bacterial strains to clinical syndromes of *Campylobacter*-associated neuropathies. *Neurology* 2006;67:1837–43.

[53] Nam Shin, J. E., Ackloo, S., Mainkar, A. S., et al. Lipo-oligosaccharides of *Campylobacter jejuni* serotype O:10: structures of core oligosaccharide regions from a bacterial isolate from a patient with the Miller-Fisher syndrome and from the serotype reference strain. *Carbohydr. Res.* 1997; 305:223–32.

[54] Kaida, K., Kanzaki, M., Morita, D., et al. Anti-ganglioside complex antibodies in Miller Fisher syndrome. *J. Neurol. Neurosurg. Psychiatry* 2006;77:1043–6.

[55] Kanzaki, M., Kaida, K., Ueda, M., et al. Ganglioside complexes containing GQ1b as targets in Miller Fisher and Guillain-Barré syndromes. *J. Neurol. Neurosurg. Psychiatry* 2008;79:1148–52.

[56] Gilbert, M., Karwaski, M. F., Bernatchez, S., et al. The genetic bases for the variation in the lipo-oligosaccharide of the mucosal pathogen, *Campylobacter jejuni*. Biosynthesis of sialylated ganglioside mimics in the core oligosaccharide. *J. Biol. Chem.* 2002;277:327–37.

[57] Houliston, R. S., Koga, M., Li, J., et al. A *Haemophilus influenzae* strain associated with Fisher syndrome expresses a novel disialylated ganglioside mimic. *Biochemistry* 2007;46:8164–71.

[58] Moran, A. P. Structure and conserved characteristics of *Campylobacter jejuni* lipopolysaccharides. *J. Infect. Dis.* 1997;176:S115–21.

[59] Godschalk, P. C., Kuijf, M. L., Li, J., et al. Structural characterization of *Campylobacter jejuni* lipooligosaccharide outer cores associated with Guillain-Barré and Miller Fisher syndromes. *Infect. Immun.* 2007;75: 1245–54.

[60] Parker, C. T., Horn, S. T., Gilbert, M., et al. Comparison of *Campylobacter jejuni* lipooligosaccharide biosynthesis loci from a variety of sources. *J. Clin. Microbiol.* 2005;43:2771–81.

[61] Gilbert, M. P., Godschalk, C. R., Parker, C. T., et al. Genetic bases for the variation in the lipooligosaccharide outer core of *Campylobacter jejuni* and possible association of glycosyltransferase genes with post-infectious neuropathies, In: Ketley, J. M. and Konkel, M. E., editors. *Campylobacter: molecular and cellular biology,* 1st ed. Pullman, WA: Horizon Bioscience; 2005; p219–48.

[62] Louwen, R., Heikema, A., van Belkum, A., et al. The sialylated lipooligosaccharide outer core in *Campylobacter jejuni* is an important determinant for epithelial cell invasion. *Infect. Immun.* 2008;76:4431–8.

[63] Koga, M., Gilbert, M., Takahashi, M., et al. Comprehensive analysis of bacterial risk factors for the development of Guillain–Barré syndrome after *Campylobacter jejuni* enteritis. *J. Infect. Dis.* 2006;193:547–55.

[64] Koga, M., Takahashi, M., Masuda, M., et al. *Campylobacter* gene polymorphism as a determinant of clinical features of Guillain–Barré syndrome. *Neurology* 2005;65:1376–81.

[65] Sonnino, S., Mauri, L., Chigorno, V., et al. Gangliosides as components of lipid membrane domains. *Glycobiology* 2007;17:1R-13R.

[66] The American Oil Chemists' Society Lipid Library. Ganglioside, structure, occurrence, biology and analysis.

[67] Chiba, A., Kusunoki, S., Obata, H., et al. Ganglioside composition of the human cranial nerves, with special reference to pathophysiology of Miller Fisher syndrome. *Brain Res.* 1997;745:32–6.

[68] Liu, J. X., Willison, H. J., Pedrosa-Domellof, F. Immunolocalisation of GQ1b and related gangliosides in human extraocular neuromuscular junctions and muscle spindles. *Invest. Ophthalmol. Vis. Sci.* 2009;50: 3226–32.

[69] Willison, H. J., Halstead, S. K., Beveridge, E., et al. The role of complement and complement regulators in mediating motor nerve terminal injury in murine models of Guillain–Barré syndrome. *J. Neuroimmunol.* 2008;201–202:172–82.

[70] Kuwabara, S., Asahina, M., Nakajima, M., et al. Special sensory ataxia in Miller Fisher syndrome detected by postural body sway analysis. *Ann. Neurol.* 1999;45:533–6.

[71] Jamal, G. A., Ballantyne, J. P. The localization of the lesion in patients with acute ophthalmoplegia, ataxia and areflexia (Miller Fisher

syndrome). A serial multimodal neurophysiological study. *Brain* 1988; 111:95–114.

[72] Ropper, A. H., Shahani, B. Proposed mechanism of ataxia in Fisher's syndrome. *Arch. Neurol.* 1983;40:537–8.

[73] Kornberg, A. J., Pestronk, A., Blume, G. M., et al. Selective staining of the cerebellar molecular layer by serum IgG in Miller-Fisher and related syndromes. *Neurology* 1996;47:1317–20.

[74] Giroud, M., Mousson, C., Chalopin, J. M., et al. Miller-Fisher syndrome and pontine abnormalities on MRI: a case report. *J. Neurol.* 1990; 237: 489–90.

[75] Lee, K. Y. Anti-GQ1b-negative Miller Fisher syndrome after *Campylobacter jejuni* enteritis. *Pediatr. Neurol.* 2012;47:213–5.

[76] Koga, M., Gilbert, M., Takahashi, M., et al. GQ1b-seronegative Fisher syndrome: clinical features and new serological markers. *J. Neurol.* 2012; 259:1366–74.

[77] Tüzün, E., Kürtüncü, M., Lang, B., et al. Bickerstaff 's encephalitis and Miller Fisher syndrome associated with voltage-gated potassium channel and novel anti-neuronal antibodies. *Eur. J. Neurol.* 2010;17:1304–7.

In: Oligosaccharides
Editors: L. S. Schweizer, S. J. Krebs
ISBN: 978-1-62948-328-3

Chapter XI

Oligosaccharides from Ancient Beverages Fermented by Symbiotic Cultures

José Maurício Schneedorf*
Lab. Biochemistry, Biomedical Sciences Institute,
Federal University of Alfenas, Alfenas, MG, Brazil

Abstract

Symbiotic colonies of bacteria and yeasts (Scoby) are known to ferment dairy or non-dairy foods (sugar, fruits, cereals, vegetables, tubers, roots), leading to traditional or indigenous drinks distributed worldwide. These Scoby drinks remain today as a house art, as no special condition is required for their formulation. The fermented beverages have

*E-mail address: zemasfs@gmail.com

different microbial strains (sometimes probiotics) together with bioactive compounds (sometimes prebiotics), including oligosaccharides related to alleged health properties of the drinks. These oligosaccharides are known to play a role in gastrointestinal equilibrium, pathogen inhibition and immunity enhancement, among others. This chapter exploits the origin, structure, and health properties of oligosaccharides of Scoby beverages, either inherent to the carbon source used for souring or released from microbial enzyme activities (e.g., glucosyltransferase, dextransucrase, fructofuranosidase, fructosyltransferase, inulinase). The oligosaccharides found in ancient Scoby drinks, e.g., kefir, kvass, pulque, kombucha, boza, koumiss, and borde, among others, as well as non-Scoby indigenous drinks (as miki and koji) are discussed. These carbohydrates, e.g., isomaltotriose, rhamnoarabinan, fructans, and inulin, have typically less than 20 degree of polymerization. The main strains involved in their production belongs to *Lactobacillus, Leuconostoc, Lactococcus, Saccharomyces, Kluyveromyces*, among others. Complementarily, some activities of an oligosaccharide isolated from Kefir d'aqua and named AK (aqueous kefiran) on biomimetic systems and living cells are presented by means of a combined electrochemical impedance spectroscopy and cyclic voltammetry data.

Keywords: Oligosaccharides in ancient beverages, oligosaccharide, fermentation, beverages, scoby

1. Introduction

"Scoby" is an acronym for "Symbiotic Culture/Colony of Bacterial and Yeasts". In general, Scobys are viewed as a slightly sweet and carbonated acidic beverage resulting from numerous compounds that are produced by the symbiotic culture of bacteria and yeasts [1]. Indigenous fermented beverages from Scobys are noticed to be produced since the dawn of civilisation, for more than 6000 B.C. (Fertile Crescent, Middle East), using several carbon sources, as sugar, fruits, cereals, vegetables, tubers, roots and milk, among others. Beer brewed by the Babylonians and exported to Egypt around 3000 B. C. (Figure 1) was most likely the product of both alcoholic and lactic fermentations, whereas wine is thought to be first developed around 1500 B.C. [2]. Fermentation served to improve the shelf-life and safety of foods, and enabled people in the moderate and cooler regions to survive winter seasons and drought periods [3]. Fermentation has also been a means of improving sensory quality and acceptability of many raw materials to such an extent that several foods are preferred in a fermented state. Their preparation followed ancient souring methods based on a skilled worker, and have actually spread in homes, villages and smallscale industries [4].

Figure 1. Early writing proto-cuneiform recording the allocation of a fermented beverage, probably from southern Iraq, Late Prehistoric period, 3100-3000 BC. British Museum (ME 140855). It contains calculations of basic ingredients required for the production of cereal products, including fermented beverages (Licensed by Takomabibelot under Creative Commons).

Most of all is still a traditional family art, although some artisanal drinks have been distributed in industrial scale. Many of these beverages, however, are little known outside their native countries. Notwithstanding, Scobys and similar fermented drinks are distributed throughout the globe. As can be seem in Figure 2, these beverages are mainly found in Europe, former USSR, Africa,

Asia and Japan, in agreement to these older regions in the world.

There are several types of fermented drinks produced around the world, which can be classified based on the raw materials used or the type of fermentation involved in the manufacturing process. Alcoholic fermented beverages can be classified into wines and beers, while the great majority of non-alcoholic fermentations are souring, mainly lactic acid or acetic fermentations [5]. As other fermented foods, Scobys are generally prepared form starter cultures. In a general sense, a starter culture may be described as a preparation containing high numbers of viable microorganisms, which may be added to bring about desirable changes in a food substrate [3]. These mixed-culture fermentation has come about by the selection over the centuries of certain microorganisms that can grow together to produce a desirable product free from toxins and infectious organisms [6]. This was done unknowingly by establishing ecological conditions that allowed the useful organisms to complement each other's activities and to live in harmony.

As a mixed-culture fermentation, Scobys fermentations can offer a number of advantages over conventional single-culture fermentations [7]: they provide a higher product yield and growth rate, render multistep transformations that would be impossible for a single microorganism, allow a stable association between its microbiota through beneficial and even essential growth compounds produced/used by their microorganism, exclude undesired microorganism by competition, permit better utilization of substrates, can be mantained indefinitely by unskilled people with a minimum training, offer protection against contamination, and enable the utilization of cheap and impure substrates, among others. In the other hand, the microbial content in Scobys are difficult to be established, leading to difficulties for detect and control possible contaminations, and to establish the optimal conditions of balance and growth. Moreover, defining the nature of the product and the microorganisms involved for patent and regulatory purposes is more difficult than for a single strain product [6].

Microbiological characterization of traditional fermented beverages produced domestically and at semi-industrial level represent the first stage in the process for industrialization and scale-up of this type of product. Results from this initial characterization should allow the identification of microorganisms that are essential in these processes, as well as to characterize the microbial communities involved in fermentations, in order to improve beverages commercially [8].

Figure 2. Distribution of some Scoby fermented beverages discussed in this chapter.

1.0.1. Microbial

The preparation of traditional beverages are normally carried out by spontaneous fermentation involving mixed cultures of bacteria, yeast and/or fungi (Scoby cultures). The type of bacterial flora developed in each fermented beverage depends on the water activity, pH, salt concentration, temperature and the composition of the food matrix. Some microorganisms may participate in parallel, while others act in a sequential manner with a changing dominant flora during the course of the fermentation [5]. Lactic acid bacteria play an important role in the preservation, microbiological stability and production of organoleptic compounds of fermented foods [9]. Yeast consumes lactic acid, leading to a decrease in the pH value, which in turn encourages the growth of secondary pH-sensitive microbiota. Certain amino acids may be synthesised and the availability of B group vitamins may be improved. Bacterial growth may also be stimulated by the these amino acids and vitamins, sometimes produced by the yeast. These organisms show diverse metabolic capabilities (including proteolytic and lipolytic activities), which can play substantial roles in the development of tastes and aromas [8]. For instance, fruits are naturally rich in juices and sugars and are slightly acidic. The components induce growth of yeasts and are naturally used for making alcoholic beverages. Alcohol fermentation results in the production of ethanol, and yeasts are the predominant organisms (e.g. wines and beers). Plant materials containing fermentable sugars provide suitable substrates for yeast species, although only *Saccharomyces* has been typically associated with spontaneous alcoholic fermentations. Lactic acid fermentation (e.g. fermented milks and cereals) is mainly carried out by lactic acid bacteria. A second group of bacteria of importance in some beverage fermentations are the acetic acid strains. These bacteria are able to convert alcohol to acetic acid in the presence of excess oxygen [5]. In general, natural fermentation of cereals, fruits, milk or sugary solutions leads to a decrease in the level of carbohydrates as well as some non-digestible poly and oligosaccharides. Some componds formed during fermentation of raw materials are organic acids (lactic, acetic, pyruvic, lauric butyric, succinic, valeric, myristic), alcohols (ethanol, n-propanol, isobutanol, 2,3-butanediol, amy alcohol), aldehydes and ketones (acetaldehyde, formaldehyde, acetone, acetoin, diacetyl, 2-butanone), and carbonyl compounds (furfural, methional, glyoxal, 2-methyl butanal), among others [5].

Multiple reports have described the health benefits of some probiotic strains and/or prebiotic compounds also presented in Scoby beverages [10], e.g. on gastrointestinal infections, antimicrobial activity, protection against carcinogenesis and mutagenesis, prevention of injuries caused by free radicals, control of the intestinal flora, gastrointestinal resistance, improvement in lactose metabolism, reduction in serum cholesterol, immune system stimulation, improvement in inflammatory diseases and vitamin uptake [11].

Lactic and acetic acid bacteria are generally recognized as safe (GRAS) and polysaccharides isolated from them offer an alternative source of microbial polysaccharides for use in food formulations [12]. These carbohydrates have several beneficial health effects, including the decrease of blood pressure induced by hypertension, production of β-interferon, cortisol and norepinephrine, increased phagocytic activity of peritoneal and lung macrophages and increased IgA cells in these sites, antitumor activity, antimicrobial activity, and anti-inflammatory activity [13].

Some of the Scobys beverages exhibit the above effects derived from their bioactive compounds released from the plant secondary or microbial metabolism, as enzyme-linked activities, peptides and oligosaccharides. Some nondigestible but fermentable oligosaccharides are involved in the promotion of health in the host, and are known to provide improvements in nutritional status and additional health benefits.

1.0.2. Oligosaccharides

Oligosaccharides are low molecular weight carbohydrates, containing sugar moieties with degrees of polymerization (DP), usually between 3 and 20, intermediate in nature between simple sugars and polysaccharides. These carbohydrates present in diverse fermenting sources differ in their chemical structure by the number of carbohydrate monomers, the type of moieties (e.g. glucosyl-, fructosyl-,galactosyl-, xylosyl-), and the position and conformation (α or β) of hexose linkages [14]. The oligosaccharides are about 0.3–0.6 times as sweet as sucrose. In fact, the sweetness depends on chemical structure, the degree of polymerization of the oligosaccharides present and the levels of mono- and disaccharides in the mixture [15]. These low sweetness attribute, non-cariogenicity and low caloric value are exploited in food formulations as a replacement of sucrose [16]. Further, the low caloric densities make the oligosaccharides useful as bulking agents in food formulations. They are used as humectants because of their high moisture-retaining capacity without increasing water activity. Furthermore, based on their physiological properties, these carbohydrates are also grouped as digestible or non-digestible.

Most oligosaccharides are quantitatively hydrolyzed in the upper part of the gastrointestinal tract. The resulting monosaccharides are transported via the portal blood to the liver and, subsequently, to the systemic circulation. Such carbohydrates are essential for health as they serve both as substrates and regulators of major metabolic pathways. Nevertheless, some oligosaccharides present specific physicochemical properties and resist to the digestive process, reaching the caeco-colon as they have been eaten. In the caeco-colon, most of the non-digestible oligosaccharides are hydrolyzed to small oligomers and monomers, which are further metabolized by one, a few, or most of the anaerobic bacteria [15].

Natural sources of nondigestible oligosaccharides of various types presented in indigenous Scoby beverages can be found as natural components in different carbon sources. Therefore, functional oligosaccharides are found in varying concentrations in milk, honey, sugarcane juice, soybean, lentils, mustard, fruits and vegetables such as onion, asparagus, sugar beet, artichoke, chicory, leek, garlic, banana, rye, barley, yacon, wheat, tomato, salsify and bamboo shoots [17]. During manufacture of fermented beverages of Scobys, oligosaccharides are produced in variable amounts mainly depending on the strains used, so functional properties of the final products may have due not only to its probiotic properties but also to the presence of oligosaccharides. Many plant, fungal and bacterial species are capable of producing fructans (Figure 3), which can be basically distinguished by their type of linkages. Fructan synthesis is catalyzed by fructosyltransferases, which cleave the main substrate sucrose and release glucose in a first step. The energy derived from sucrose cleavage afterwards is used to transfer the remaining bound fructose moieties to the growing polymer chains with sucrose as starting precursor for polysaccharide formation [18].

Several oligosaccharides have been reported to possess prebiotic activity, although the data often arise from single studies where the test oligosaccharide is compared with a placebo in either in vitro studies or human trials. In parallel, functional oligosaccharides have gained outstanding popularity owing to their physiological benefits to the consumers [19]. The functional oligosaccharides include fructooligosaccharides, glucooligosaccharides, isomaltooligosaccharides, soybean meal oligosaccharides, mannan oligosaccharides, galactooligosaccharides, gentiooligosaccharides, isomaltulose, lactosucrose, maltooligosaccharides, xylooligosaccharides, pectin-derived acidic oligosaccharides, and cyclodextrins [17]. Notwithstanding, convincing scientific evidence for their suitability for use as prebiotics exists mainly for inulin/oligofructose (Figure 3) and for galactooligosaccharides [20]. Short-chain fructooligosaccharides (FOS) are considered as representative prebiotics due to their ability to preferentially stimulate the growth of intestinal bifidobacteria, to reduce colon cancer risk, and to enhance calcium absorption in the gut [18].

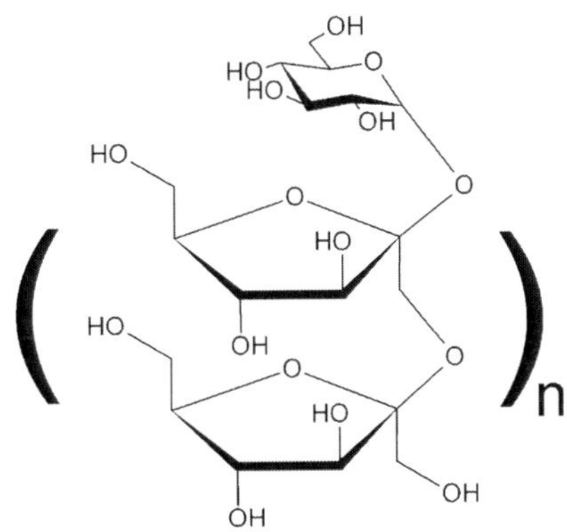

Figure 3. Structure of inulin, a linear polymer of β(2-1)-linked fructofuranoside units with one glucopyranose unit at the reducing end. Inulin, commonly presented in tubers and vegetables, is sometimes preserved in Scobys fermented beverages or hydrolyzed by bacterial inulinases.

2. Main Scobys

2.1. Kefir

Kefir is a probiotic originated from the old continent, fermented by a mixture of several lactic and acetic acid bacteria and yeasts. Kefir consists of a gelatinous and irregular grains formed by a microbial consortium embedded in a resilient polysaccharide matrix named Kefiran [13]. Kefir is generally prepared by culturing fresh or pasteurized milk with Kefir grains [21], leading to an acid, viscous, and slightly carbonated dairy beverage [22] . Kefir is also presented as its sourish product in milky or sugary suspensions (Figure 4) containing vitamins, aminoacids, peptides, carbohydrates, ethanol, and volatile compounds [23]. In this sense, the FAO/WHO (2001) has proposed a definition of Kefir based on the microbial composition of both Kefir grains (the starter culture used to produce

Kefir) and the final Kefir product [22].

Historically, Kefir grains were considered a gift from Allah among the Muslim people of the northern Caucasian mountains. The word Kefir is derived from the Turkish word "keif", which can be translated to good feeling for the sense experienced after drinking it, or their promoted health claims. Kefir grains have been passed from generation to generation among the tribes of Caucasus. From its origin in the Caucasian mountains of the former USSR, the making and consumption of Kefir has spread to many parts of the world. It has been postulated that the longevity of Caucacian peasants is partially due to their frequent consumption of Kefir [24]. Resembling small cauliflower florets in appearance, they vary in size from approximately 3 to 30 mm [25]. As beverage, Kefir is a fermented drink that is self-carbonated, has a sharp acidic taste and a yeasty flavour and contains a low percentage of alcohol [26].

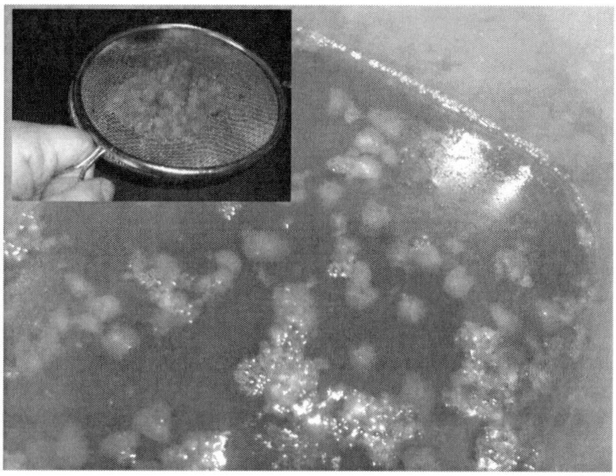

Figure 4. Kefir grains suspended in molasses solution, showing its characteristic carbonation along the floating grains.

2.1.1. Sources and Composition

Kefir is produced by diverse spectrum of microbial species present. Lactobacilli are present as the largest portion of the microbial population with lactoccoci, acetic acid bacteria and yeasts making up the remaining portion of the microorganisms present in the Kefir grains cultured in milk [27]. The end products of milky fermentation are lactic acid, acetaldehyde, acetoin, diacetyl, ethanol and carbon dioxide. Besides, vitamin B_1, B_{12}, vitamin K, folic acid, calcium and amino acid contents increase in Kefir during fermentation. The microorganisms present in Kefir are responsible for the lactic, acetic, and alcoholic fermentation of milk that yields a product with remarkable characteristics [28]. Hence, the strains commonly found in Kefir cultures, both for milky or sugary suspensions, are known to contain lactic acid bacteria (*Lactobacillus, Lactococcus, Leuconostoc, Acetobacter* and *Streptococcus spp.*), yeasts (*Kluyveromyces, Torula, Candida* and *Saccharomyces spp.*) and acetic acid bacteria, all trapped in a complex matrix of polysaccharides and proteins [25]. *S. cerevisiae* represented the largest and most commonly identified yeast isolates. The presence of *S. cerevisiae* contributes to the enhancement of the organoleptic quality of the Kefir beverage, promoting a strong and typically yeasty aroma as well as its refreshing, pungent taste. Furthermore, this yeast also reduces the concentration of lactic acid, removes the hydrogen peroxide and produces compounds that stimulate the growth of other bacteria, thus increasing the production of exopolysaccharides [26]. Generally, the country or even an inner region, the carbon source and experimental conditions of fermentation have identified, with both culture-dependent or molecular methods, several complimentary genera beyond the above mentioned. For instance, *Chyseomonas, Toruspola, Kloekera* [23], *Lachancea* and *Kazachstania* [26], and several others were found throughout the world.

The entrapped polysaccharide is named Kefiran, an exopolysaccharide produced by microorganisms present in the Kefir grains fermenting milk. Kefiran is produced by *Lactobacillus kefiranofaciens* isolated from Kefir grains. It has a repeating hexasaccharide structure with equal amounts of glucose and galactose residues in the chain sequence, and has several health promoting properties [12]. The molecular weight of Kefiran is approximately 1–4 x 10^6 Da [13].

Sugary Kefir grains are very similar to milky Kefir grains in terms of their structure, associated microorganisms and products formed during the fermentation process, albeit without the characteristic cauliflower look of them. Kefir d′aqua, sugary Kefir, or water Kefir, is generally a home made fermented beverage based on a sucrose solution with or without fruit extracts. Different from the milky bacteria-encapsulated polysaccharide Kefiran, an oligosaccharide have been isolated from an aqueous fraction of Kefir grains [13]. It is composed of repeating units of glucose, rhamnose, galactose and arabinose (ratio 4.0:2.4:1.0:2.6), leading to an oligomer with a molecular weight about 3500 Da [29].

2.1.2. Activities

Beneficial effects of Kefir are supported by an increasing body of scientific evidence [30]. The benefits of consuming it in the diet are numerous, as it is reported to possess antibacterial, anti-inflamatory, immunological, pro-digestive, antitumoral, antioxidant and hypocholesterolemic effects, among others [27, 24]. Therefore, Kefir is well known to resist to a large spectrum of pathological strains [13], and it seems to be recognized as safe, although its culture contamination has been reported as a source of health impairments. Antibiotic activity of both Kefir and purified aqueous Kefiran (AK) has been evidenced from disk diffusion method and susceptibility tests data against some well known pathogenic bacteria (*S. pyogenes, S. salivarius, S. aureus, P. aeruginosa, S. tiphymurium, E. coli, L. monocytogenes, Helicobacter pylori* and *C. albicans*) [21, 13].

Kefiran derived from rice culture is highly soluble in water and exerts hypocholesterolemic and hypotensive effects. It is also known to lower blood cholesterol and reduce blood pressure in spontaneously hypertensive stroke-prone rats [31]. In addition, Kefiran reduced atherosclerosis in rabbits fed a high-cholesterol, as well as accelerated sterol excretion and decreased histamine excretion in rats fed a high-cholesterol diet or administered orotic acid.

In parallel, both Kefir and Kefiran, major milk-based, have been used to obtain technically and commercially feasible biotechnological products, as starter cultures by casein immobilization in cheese production, food-grade additive of milk gels for fermented products, industrial scale-up of alcoholic fermentation of whey, for batch alcoholic fermentation, for exploiting waste residues from the citrus industry, among others [13]. Kefiran also improves viscosity and viscoelastic properties of acid milk gels and itself is also able to form gels at low temperature with interesting viscoelastic properties. Furthermore, Kefiran can also form edible transparent films at low concentrations [12].

The protective effect of Kefiran was demonstrated in several models, as in a hemoglobin release assay with human red blood cells against *Bacillus cereus*. This strain and its extracellular factors are able to produce different biological effects on cultured human enterocytes (Caco-2 cells) and erythrocytes [30]. After treatment with Kefiran, an antagonism of the activity of *B. cereus* hemolysins have been found. In another assay using mouse foot edema model, the AK isolated from water Kefir presented a mean inhibitory activity of 30 % against carrageenan after the stimulus, suggesting that the water Kefiran exerted an anti-inflammatory activity [29]. In a cicatrizing test with rats following inoculation of *Staphylococcus aureus*, animals treated topically with a Kefir oligosaccharide cream presented a faster reduction of the infected-induced wound, a well developed granulation of the epithelium together with neovascularization areas, and less trauma after treatments, as compared to the negative control group [32]. Futhermore, Kefiran may be effective against allergic and inflammatory disorders if its pharmacodynamic efficacy is improved [31]. Although the anti-inflammatory responses of sugary Kefir and its derivatives are poorly related in the literature, Kefir has proved to exert a beneficial effect on acute inflammatory responses, additonally improving the immune status of treated animals [13].

Orally administrated Kefiran is reported to directly inhibit mast cell activation, induce early change in the balance of cell populations in intestine through the interaction of Kefiran with intestinal mucosa, and to reduce atherosclerosis by oligosaccharides digested from Kefiran absorbed in blood [31]. However, the pathway by which Kefiran is absorbed and metabolized in the body is not elucidated completely. In this respect, there is a lack of knowledge concerning the immune mechanisms induced by exopolysaccharides produced by lactic acid bacteria, which would allow a better understanding of the functional effects described to them [33].

In this chapter, the direct action of AK was evaluated by electroanalytical techniques with biomimetic membranes (supported bilayer membrane - sBLM and liposomes) and living cells, aiming to get some information about the action of AK on membrane structure and cell metabolism.

2.1.3. Action of Aqueous Kefiran (AK) in Biomembranes as Studied by Electrochemical Impedance Spectroscopy and Cyclic Voltammetry

Electrochemical impedance spectroscopy (EIS) is a powerful and sensitive characterization tool for studying the charge transfer processes occurring at interfaces, namely electrode-solution interfaces. In this sense, EIS has been used successfully as a tool for the measurement of electric changes of bulk and interfacial properties in batteries, corrosion, semiconductors, thin-films and analyte determinations, among others. Besides these, identification of membranes, biosensor characterization and fabrication, enzyme–substrate interactions after antigen–antibody, receptor–ligand, or DNA–DNA interactions, biomembranes and living cells can be successively monitored by EIS [34]).

Basically, EIS measures the impedance (resistance) of a system as a function of frequency of a small alternate current (AC) perturbation, with the response being a current that differs in amplitude and phase with the applied voltage. The data are compared with an equivalent electrical circuit for obtaining electrical parameters of impedance [35]. A Randles circuit is the simplest equivalent circuit that describes an electrochemical cell where a single-step Faradaic process in the presence of diffusion may occur. It combines three components, namely electrolyte resistance between working and reference electrode (Re), the double-layer capacitance (Cdl) and the Faradaic impedance due to the charge-transfer process at the working electrode-electrolyte interface. This impedance is normally subdivided into a charge-transfer resistance (Rct) and the Warburg impedance (Zw), which reflects the influence of the mass transport of the electroactive species on the total impedance of the electrochemical cell. In this sense, EIS is a technique which infers more information on the electrochemical interfacial properties of surface-modified electrodes by providing electrical parameters, as can be viewed in Figure 6.

The electrical components can be identified by studying the variation in impedance response of an electrochemical cell over a wide frequency range (normally 100 kHz–0.1 Hz) in the presence of a redox probe, e.g. potassium ferro/ferrycianide. A typical electrochemical impedance spectrum is presented in the form of a Nyquist plot (Zi vs. Zr, Figure 6). The Nyquist plot includes a semicircle region lying on the Zr axis observed at higher frequencies corresponding to the electron-transfer-limited process, followed by a linear part with a 45° angle degree at lower frequencies, representing the diffusion-limited electron transfer process (Zw) [36].

Methods

EIS and cyclic voltammetry (CV) were both performed on a potentiostat-galvanostat instrument PG-39MCSV provided with an impedance module IP3929 (Omni Metra Instr. Cient. Ltda, Nova Friburgo, RJ, Brazil). Impedance measurements were performed from 100 kHz to 100 mHz with a signal amplitude of 5 mV. The experiments were carried out with a three-electrode system composed by an Ag/AgCl (KCl-saturated) as reference electrode and a platinum wire (0.5 mm diameter) as the counter electrode. The working electrode comprised an Au disc of 3 mm diameter melted in a glass tube, and previously cleaned and activated with 0.5 mol/L H_2SO_4. EIS runs were performed in the presence of 5 mmol/L $Fe(CN)_6{}^{3-/4-}$ as a redox probe contained in 0.1 mol/L sodium phosphate buffer pH 7.4 as support electrolyte, at a final volume of 0.6 mL, and at a system potential of E^0 of 230 mV.

Supported bilayer lipid membrane (sBLM), constructed liposomes and peritoneal macrophages were used as target for AK action. sBLM was formed by dissolving L-α-phosphatidylcholine (PC) and cholesterol (CH) in chloroform at a final concentration of 20 and 7 mg/mL, respectively. Then, 30 μL of this solution was dropped on the working electrode surface and rapidily transferred into a 0.1 mol/L KCl solution for 20 min, in which the supported lipid layer was formed spontaneously [37]. For liposome preparation, a chloroform containing PC at 10 mmol/L solution was evaporated in test tubes to dryness with argon, following evacuation for 60 min for total remotion of the solvent. Subsequently, the film was hydrated in sodium phosphate buffer 0.1 mol/L pH 7.4 after complete dispersion at room temperature. Then, the multilamellar vesicles produced were ultrasonicated in water bath for10 min. In the experiments using living cells, male swiss mice were injected intraperitoneally with 1 ml of 0.5% sodium caseinate in 0.85% saline solution. Five days later, peritoneal macrophages were withdrawn with the phosphate buffer. The cells were suspended in 0.5 mL Leibovitz L-15 culture medium and incubated with cover slips in 5% CO_2 at 37 ^oC for 24 h before assays.

After electrode cleaning, 30 μL of a stock solution of poly(methyl methacrylate)-block-poly[(2-dimethylethylamino)ethyl methacrylate](PMMA-*b*-PDMAEMA) 10 mg/mL in acetone was pippeted on the electrode surface, following evaporation for 4 h at room temperature. The cationic copolymer was used for enhance cell adsorption on the untreated Au electrode surface. Then, 100 μL of macrophage suspension (1x10^6 cell/mL) was pippeted on the electrode, and the set was tightly fastened at the bottom of a modified Eppendorf tube, with its copper wire outside the chamber for providing electrical contact. This electrochemical cell was filled with 500 μL of L-15 medium and incubated at 37 ^oC up to 2 h before the assays.

Aqueous Kefiran (AK) was isolated from Kefir samples cultured in the lab following a modified procedure that has all ready been described [13]. AK was incubated with biomembrane systems and cells during 5-45 min, and usually at 0.1% (w/v).

Results

The Nyquist diagrams, corresponding to the gold electrode before and after surface/bulk perturbation with sBLM, liposomes, and murine macrophages are presented in Figures 5, 6 and 7. When an electrode surface is modified by the presence of some material, the electron transfer kinetics of $Fe(CN)_6^{3-/4-}$ is perturbed. In this sense, cyclic voltammetry (CV) can be a valuable tool for testing the kinetic barrier of an interface. The extent of kinetic limitation of the electron transfer process increases with the increasing thickness and decreasing defect density of the barrier, the biomembranes in this case. As shown inset Figure 5, CV curve changes drastically from a electrochemically reversible one (curve *a*) to a capacitive featured one (curve *b*) after Au electrode has been covered by sBLM. Hence, at the formal potential of 230 mV, the Faradaic current at the modified sBLM electrode, which is originated from $Fe(CN)_6^{4-}$ oxidation (forward scan) or $Fe(CN)_6^{3-}$ reduction (backward scan), is several orders of magnitude smaller than that at bare Au electrode. This means the electron transfer rate is greatly reduced after the Au surface has been modified by sBLM, suggesting its coverage on the electrode.

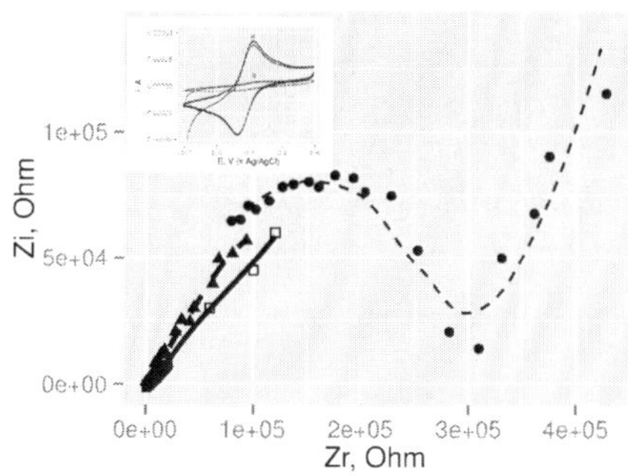

Figure 5. Nyquist plot from electrochemical impedance spectroscopy of Au nude electrode (□), electrode modified with sBLM (•), and after interaction for 45 min with 0.1 % AK solution (▲) in 5 mmol/L $Fe(CN)_6^{4-/3-}$ solution containing 0.1 mol/L sodium phosphate buffer, pH 7.4. Inset: Cyclic voltammetry of charge transfer of ferricyanide ions on the electrode surface. (a) at bare Au electrode, (b) Au electrode modified with supported bilayer lipid membrane in 5.0 mmol/L $K_3[Fe(CN)_6/K_4[Fe(CN)_6]$ (1:1) mixture containing 0.1 mol/L sodium phosphate buffer, pH 7.4.

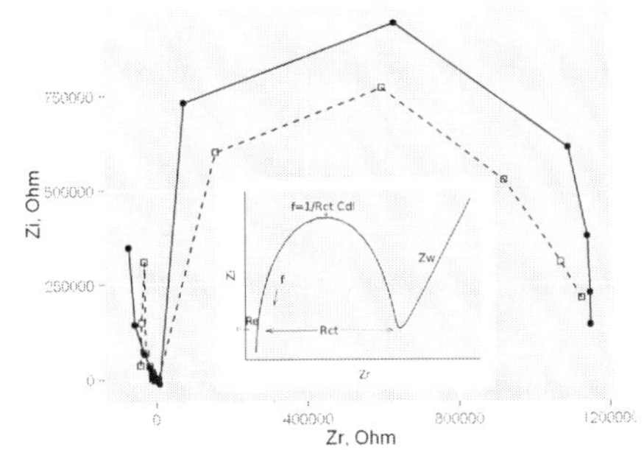

Figure 6. Nyquist plots for the impedance measurement of liposome preparation (•) and 30 min-AK treated liposome (□) in 5.0 mmol/L $K_3[Fe(CN)_6]/K_4[Fe(CN)_6]$ (1:1) mixture containing 0.1 mol/L sodium phosphate buffer, pH 7.4. Inset: electrical parameters of impedance as obtained from the Nyquist plot of EIS spectrum. Re - electrolyte resistance between working and reference electrode; Cdl - double-layer capacitance; Rct - charge-transfer resistance; Zw - Warburg impedance component; f - angular frequency.

Figure 5 also shows the Nyquist plot of the EIS spectra of bare electrode, the electrode modified with sBLM and the sBLM electrode after incubating with AK. It can be devised that each of the three impedance spectra includes a linear portion, representing the electron transfer (nude or AK-treated sBLM electrode) or diffusion process (sBLM electrode) at the electrode surface. The gold electrode with sBLM, however, includes a semicircle immediately followed by the Warburg line, whose slope is equal to unity. This arc characterizes the electron transfer and Warburg line mass transfer previously discussed [38]. After treating with AK solution, however, this semicircle disappears completely from the spectra, suggesting a disruption of membrane layers from the electrode induced by AK. As suggested from a previous work [11], the mechanism of interaction of AK could involve hydrogen bonding between the carbohydrate and the phosphate head group of the phospholipid. Three ways have been proposed to explain the interaction that can lead to membrane disruption in target, depending on the AK concentration range: low (carpet-like), median (pore formation) and high concentration (detergent-like effect).

As AK is able to cause defects/pores in sBLM structures, it is interesting to test it on ghost cells or liposomes, since a direct action was detected from the oligosaccharide activity on the sBLM. Thus, multilamelar vesicles were constructed with this goal, and the EIS spectrum is shown at Figure 6. An impedance arc similar to that presented from sBLM surface coverage can be viewed at higher frequencies (continuous line). It can be noted that the presence of AK in the liposome suspension have induced a decrease in the size of the semicircle after 30 min incubation. It indicates a reduction of the insulating system lead by multilamelar liposomes, and suggests a decrease in the level of reconstituted vesicles AK-induced, in agreement with the results from AK interaction with sBLM data.

Keeping in mind the action of AK as an anti-inflammatory agent in rat paw edema and induced granulomatous tissue models, and a disrupting agent observed with membrane systems described above, we follow to test it towards the adsorptive behaviour of peritoneal macrophages. The challenge of AK on cell adsorption is sketched in Figure 7. A linear trend is presented for the nude electrode, suggesting that the electron transfer process was mainly controlled by diffusion. When the macrofage cells were coated on the electrode surface (100 μL of $1\text{x}10^6$ cells/mL for 120 min), the electron transfer of the electrochemical probe was modified, hindering the diffusion process at higher frequencies. Thus, this deviation from linearity highlight an effect of the cell suspensions near the electrode surface. After treatment with AK, however, the diffusion of $Fe(CN)_6{}^{3-/4-}$ redox probe increased, similarly to that presented by the untreated electrode. This finding suggests an increase of the charge transfer process at higher frequencies, and therefore a dettachment of cells from the electrode surface AK-induced, allowing the $Fe(CN)_6{}^{3-/4-}$ redox probe to reach the surface. These combined results with liposomes, supported bilayer membrane and macrophage suspension suggest a direct action of the oligosaccharide kefiran isolated from sugary cultures of Kefir (AK) on the structure of the membrane of living cells.

2.1.4. Preparation

For preparing non-aqueous Kefir, milk can be heated to 90° for 10 min in a water bath, cooled to inoculation temperature (25° C) and 5% Kefir grains added. The inoculated milk can be incubated at room temperature for 20 h. At the end of the incubation, the grains can be separated from the Kefir drink by filtration through a plastic sieve, washed and maintained at refrigerator temperature until the next culture passage (subculture) for 24- or 48-h intervals [39]. As yeasts numbers increase during storage, correspondingly significant increases

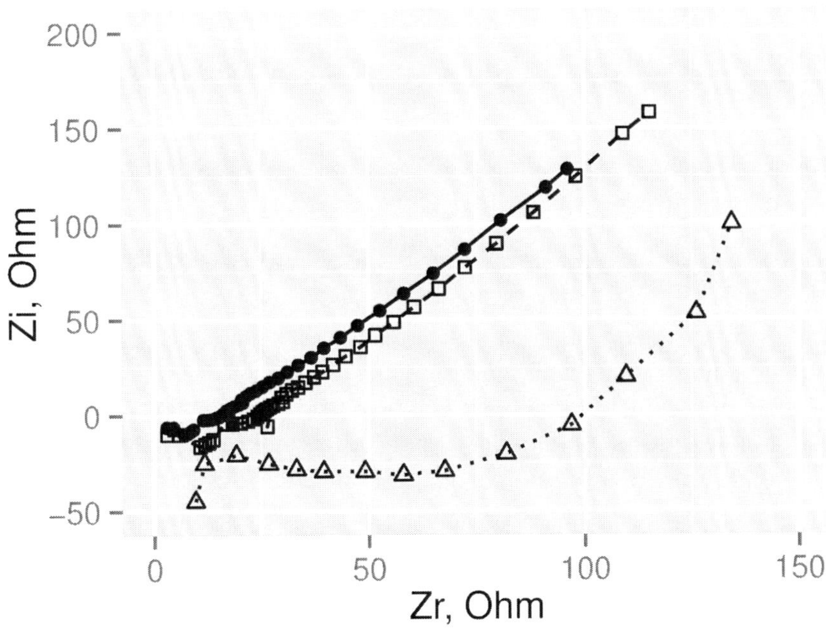

Figure 7. Nyquist plot from EIS spectra of Au nude electrode (•), macrofage suspended (1x10^6 cells/mL) (△) on the electrode surface, and adsorbed cells after treatment with 0.1%AK solution after 45 min incubation (□), in 5 mmol/L $Fe(CN)_6{}^{4-/3-}$ solution containing 0.1 mol/L sodium phosphate buffer, pH 7.4.

in ethanol and CO_2 levels can be expected between 4 and 8 weeks of storage, due to the main end products of yeast fermentation, ethanol and CO_2. Ethanol amounts reaches the expected range as the alcohol content seldom exceeds 0.1% in modern commercial Kefir production [25].

For sugary Kefir, starter grains storage frozen can be thawed and continuously grown in 50-200 g/L of commercial molasses for 15 days. The medium can be changed at 24- or 48-hour intervals. The grains can be separated from the Kefir drink by filtration through a plastic sieve, washed with sterile water, and maintained at room temperature between culture passages [32] . Kefir grains are retrieved by sieving and re-used for new fermentations. However, for the commercial production of large volumes of Kefir, direct fermentation with Kefir grains is impractical as this would demand a correspondingly large quantity of grains. In addition, the vigorous fermentation gives a product that contains a high level of CO_2 [25]. Commercial production of Kefir thus often involves several fermentation steps, including the making of a mother culture and a bulk starter. After fermentation, Kefir is slightly carbonated with a sharp acidic and slight yeasty flavour. When consuming Kefir, a prickling sensation on the tongue is typical, due to CO_2 produced by the yeast microflora [25].

2.2. Kombucha

Kombucha are another type of health beverages that are generally consumed as teas produced by fermenting the 'Manchurian mushroom' [40], a symbiosis of yeasts and acetic bacteria [41]. It is also the most common English name for a gelatinous consortium of several yeast strains and acetic acid bacteria [42], usually cultivated in a solution composed mostly of black or green tea and white sugar. The sweetened tea exposed to Kombucha fermentation gives a pleasantly sour beverage. The etymology of 'kombucha' is also obscure. One theory says it is derived from the Japanese words for seaweed, "kombu", and tea, "chao". Another story says that 'Kombu' was the name of a Korean physician who successfully treated a Japanese emperor with the bacterial-yeast consortium in 220 BC for curing his digestive troubles [1]. It is known under a number of trivial names, such as "red tea fungus", " champignon de longue vie", "ling zhi", "kocha kinoko", "chainii grib", "chainii kvass" and many others [43]. The product is a slightly sweet and carbonated acidic beverage resulting from numerous compounds that are produced by the symbiotic culture [1]. It is currently consumed worldwide, but historically, the beverage has been consumed in Asia for over two millennia [44], also in China, Russia, and Germany [45].

2.2.1. Sources and Composition

Whilst there have been no yeast species associated universally with Kombucha fermentation, osmotolerant, fermentative and acid-producing species are among the most common isolates [46]. The Kombucha tea fermented for about 14 days, and is composed of two portions: a floating cellulose pellicle layer and the sour liquid broth [47]. The jelly-like membrane floats in the nutrient solution of tea and sugar exposed to oxygen (Figure 8). At the right temperature, it multiplies continuously. At first it spreads over the entire surface of the tea, and then thickens. [42].

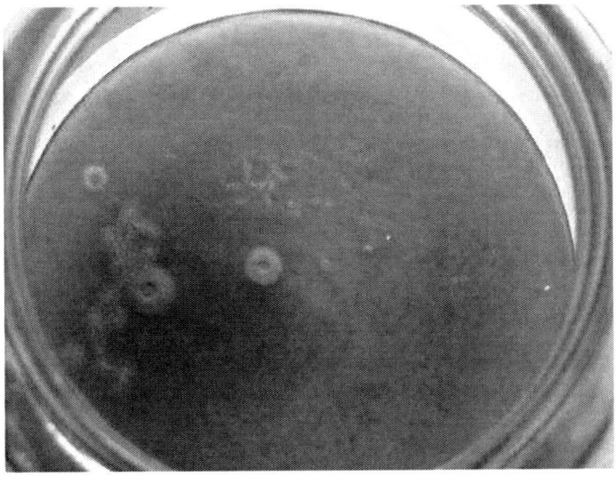

Figure 8. Kombucha tea showing the characteristic jelly-like pellicle on the surface. (Licensed by Nirinjan Singh and Anahata Balance under Creative Commons).

Only a few key bacteria are involved in Kombucha fermentation, mainly acetic bacteria such as *Acetobacter xylinoides, Acetobacter aceti, Acetobacter pasteurianus* and *Acetobacter xylinum* [41]. *A. xylinum* leads to the formation of the cellulose pellicles floating at the surface of the growth medium [1], in which the embedded cells benefit from the close contact with the atmospheric oxygen. As the cell mass of bacteria and yeasts became attached to the cellulose film, Kombucha can be considered as a fungus-like mixture of cellulose and microorganisms [43]. Another strains as *Lactobacillus* and *Bacterium gluconicum* may also be encountered. Furthermore, a broad spectrum of yeasts has been reported including species of *Zygosaccharomyces, Schizosaccharomyces, Saccharomyces, Saccharomycodes, Candida, Torulopsis, Brettanomyces/Dekkera,* [43] *Candida, Kloeckera, Pichia, Torulospora* and *Zygosaccharomyces* [46]. Kombucha yeasts hydrolyze sucrose into glucose and fructose, producing ethanol, while acetic acid bacteria convert glucose to gluconic acid and fructose to acetic acid [48].

Glucose is used by the yeasts to yield ethanol and carbon dioxide. *Acetobacter* initially oxides ethanol to acetaldehyde and then to acetic acid. The secondary biochemical activity of *Acetobacter* is the oxidation of glucose to gluconic acid. Glucose is also used by acetic-acid bacteria to synthesize microbial cellulose. Fructose remains part of the ferment broth and is utilized by the microorganisms to a lesser degree [45]. Acetic-acid concentrations may rise to levels as high as 30 g/L if the tea is allowed to ferment for up to 30 days. The usual concentration of acetic acid consumed in Kombucha is 10 g/L. Gluconic acid is also present in substantial quantities, about 20 g/L.

The diversity of yeasts reportedly associated with Kombucha may be related to the material sampled and the stage of production [46]. Jerusalem artichoke tubers extract is one of them (*Helianthus tuberosus*). Tubers are suitable for the dietetic nutrition because of the low monosaccharide content and presence of some polyfructan ingredients which act as prebiotic [49].

The basic reserve carbohydrate in the roots and tubers of Jerusalem artichoke is polyfructan inulin that consists of a homologous series of linear molecules of fructose with a degree of polymerization between 3 and 50. The content of fructose varies from 75-98% of dry matter, depending of the growth and storage treatment. During fermentation, different fructo-oligosaccharides and fructose are produced as a result of polyfructan hydrolysis caused by the activity of inulinases. In the enzymatic inulin hydrolysis, a synergic action of endo- and exo-inulinases originated from plants or microorganisms has been reported [49].

In addition to ethanol and acetic acid, a great number of other compounds, depending on the fermenting substrate, emerge as a result of numerous reactions [48], as several organic acids and some vitamins [50]. For instance, the Kombucha fermentation processes on pure sucrose and molasses differ significantly, which is related to the higher quantity of invert sugar, biotin and amino nitrogen in the molasses [50]. Several nitrogen-containing compounds, such as amino acids, methylxanthine alkaloids (caffeine, theophylline and theobromine), proteins [1], catechin gallates, adenine, gallic acids, tannins, gallotannin, aminophylline, polyphenols and flavonols are also found in the beverage. During black tea manufacture, catechin groups also undergoes oxidations and other chemical reactions forming the dimeric theaflavins and oligomeric thearubigins [51]. The combination of tea contents besides a yellow volatile oil that is solid at ordinary temperatures and has strong aromatic odor and taste [51], results in a typical astringency, brightness, mouth feel (thickness) and color of the black tea beverage [43].

2.2.2. Activities

Fungus tea has been reported to have medicinal effects against metabolic diseases, arthritis, indigestion and various types of cancer [47], regulation of gastric, intestinal and glandular activities, relief of joint rheumatism, gout and haemorrhoids, positive influence on cholesterol level, arteriosclerosis, blood cleansing, diabetes, nervousness [44] and toxin excretion (e.g. paracetamol-induced hepatotoxicity and chromate (VI)-induced oxidative stress [47]). The detoxifying property of Kombucha is presumably due to the capacity of glucuronic acid to bind to toxin molecules and to increase their excretion from the organism by the kidneys or the intestines, as gluconic acid is present in substantial quantities, about (2%) [44].

The beneficial effects of kombucha tea is attributed to the presence of acetic acid (around 1%), tea polyphenols, gluconic acid, glucuronic acid, lactic acid, vitamins, aminoacids, antibiotics and a variety of micronutrients produced during fermentation [47]. Tea polyphenols which includes epicatechin, epicatechin gallate, epigallocatechin, epigallocatechin gallate and theaflavin have been reported to possess various biological activities [47]. The action of Kombucha intake on the nervous system could be associated with its content of the B complex of vitamins [44].

Kombucha has been intensively consumed during a long time worldwide for its prophylactic and therapeutic properties [41]. Important metabolites are organic acids. Active ingredients of Kombucha tea that may exert beneficial effects have confirmed that Kombucha tea has antimicrobial activity against a spectrum of Gram-positive and Gram-negative organisms, the effect being attributed to its acetic acid content [48]. As far as lactic acid is concerned, it appears in Kombucha beverage in its most potent form. It was found that L-lactic acid present in dietetic Kombucha beverages was important for the functioning of the digestive system. Also, lactic acid may be beneficial to blood circulation and contribute to balancing acids and bases in the human body.

Kombucha has the ability to inhibit some enzymatic systems related to dietetic catabolism of lipids and sugars, as porcine pancreatic α-amylase. The inhibition potency increased during Kombucha fermentation and after two weeks, its IC_{50} reaches a value as low as the concentration of one thousand-diluted Kombucha beverage [1]. The effective inhibitors could be the phenolics, the concentration of which in the one thousand-diluted sample was approximately 1 mg/L. The gallate-containing monomers could be the effective inhibitors from

Kombucha beverage. For instance, a 200-mL cup of this beverage should contain more than 100 mg of phenolics, which could inhibit, to some extent, pancreatic α-amylase in the small intestine, playing a key role in the digestion of carbohydrates, and with possible effects on starch digestion and net absorbed glucose [1]. Furthermore, in alloxan-induced diabetes in experimental rats, the administration of Kombucha was also able to reduce pancreatic lipase activity [52]. This decrease is related to the hydrolysis of non-absorbable dietary triglycerides into absorbable monoglycerides and free fatty acids, which, in turn, leads to the decrease of plasma cholesterol and neutral lipid levels in blood. Hence, the beneficial effect of dietary Kombucha can be also attributed to its potent hypoglycemic and antilipidemic properties, as well as antioxidant potential [52].

Many claimed beneficial effects of kombucha such as alleviation of inflammation and arthritis, cancer prevention and immunity enhancement may be associated to its antioxidant activities. Kombucha tea prepared from green tea, black tea and tea waste material have significative antioxidant activities [51]. Antioxidant activity is dependent on the structure of the free-radical scavenging compounds, the substituents present on the rings of flavonoids and the degree of polymerization. Antioxidation properties of Kombucha was evidenced by incubating murine hepatocytes with hydroperoxidant, a compound that can be metabolized to free radical intermediates [42]. This model mimics several aspects related to liver pathology characterized by increased lipid peroxidation and cytotoxicity due to oxidative stress. Treatment of Kombucha tea in combination with the toxin kept the antioxidant status quite similar to that of normal hepatocytes, prevented the disruption of mitochondrial membrane potential and blocked the activation of mitochondria dependent apoptotic signaling pathways [42].

Many previous studies proved that Kombucha of fermented black tea exerts antibacterial activity against a broad range of bacteria, and around 7-12 fermentation days, influencing the gastrointestinal microbial flora of humans. Particularly, Kombucha was proved to exert an antimicrobial activity against *Helicobacter pylori, Salmonella typhimurium, Staphylococcus aureus, Agrobacterium tumefaciens, Bacillus cereus, Shigella sonnei, Salmonella enteritidis, Escherichia coli* [41], *S. typhi, S. typhimurium, Shigella flexneri, S. dysenteriae, Clostridium botulinum* and *Vibrio ssp* [45]. The antimicrobial activity of Kombucha against *Helicobacter pylori* (primary cause of gastritis and peptic ulcer disease), *Escherichia coli, Staphylococcus aureus*, and *Agrobacterium tumefaciens*, made with a low tea usage level (4.4 g/L), was attributable to the acetic-acid content [45]. Small amounts of acetic acid as little as 1 g/L are known to inhibit pathogenic and spore-forming bacteria, too.

2.2.3. Preparation

Kombucha was usually prepared leaving the tea at room temperature for up to 7-10 days, although the roles of fermentation time were not seriously considered [42]. In the other hand, it has been shown that the composition of different Kombucha preparations is greatly affected by the individual tea fungus culture used. This probably results from the variability of the normal microflora present in different tea fungus samples [47]. The use of different souring sources can also affect the final product. A traditional carbon source for Kombucha fermentation is sucrose. Application of any other sugar (lactose, glucose or fructose) may have a distinct influence on the formation of ethanol and lactic acid, but have only a minor effect on the flavour of the fermented tea [48]. Black and green teas, manufactured from *Camellia sinensis (L) O. Kuntze*, are the usual and the best substrates known for the preparation of Kombucha drinks [41].

Production of the tea is achieved by infusing tea leaves in freshly boiled water and sweetening it with about 100 g/L sucrose or honey. Once the tea has cooled to room temperature, the gelatinous surface growth/mat from a previous batch is added to the sweetened tea. After about a 7-10-day incubation at room temperature, the mat is transferred to a new fermentation and the Kombucha beverage is ready. The final product is a slightly carbonated, acidic beverage comprised of sugars, organic acids, tea components, vitamins, and minerals, resembling cider [45]. Another method of preparation [42] uses eight grams of black tea added to 300 mL water and allowed to boil for 5 min. The suspension is filtered through a sterile sieve and cooled to room temperature. After, the volume is adjusted to 1.5 L with water and ten percent (w/v) sucrose is added to it. The cooled tea is then poured into 3 L recipient that has been previously cleaned. Follow, it is then inoculated with freshly grown Kombucha mat that had been cultured in the same medium for 14 days and 10% (v/v) of previously fermented liquid tea broth. The recipient was covered with clean cheese cloth and fixed with rubber bands. The fermentation can be carried out under room temperature for 21 days [42].

Alternatively, Kombucha can be also prepared using molasses. The fermentation system with sucrose can be prepared using 2 L of boiled tap water, 140 g of sucrose (7 % solution) and 3 g of black tea. The tea is heated for 5 min at 100

^{o}C, and the leaves removed by filtration [44]. As Kombucha cultures grow, they form a flat macroscopic colony on the surface of the tea solution, taking on the shape of the receptacle. After a week to ten days, the original 'mother' cultures produce a secondary layer of growth of some called 'baby' cultures [40].

2.3. Koumis

Koumiss (or Kumis, Kumys) is a lactic acid beverage produced traditionally by the nomadic populations in the steppes areas of Central Asia (Xinjiang, Qinghai and Inner Mongolia regions or provinces of China) [53], some regions of Russia, and SouthWest Colombia [54], and it is formed by fermentation of milk by indigenous organisms [55]. It is a slightly alcoholic fermented milk beverage (1-2%), originally obtained by using natural mix starters (lactic acid bacteria and yeast). It is usually made from mares or camels milk fermented by its normal microbiota [56]. In Colombia, Koumiss is a fermented cow milk, widely consumed in rural and urban areas. Koumiss is creamy and sparkling beverage, with a sight degree of sourness [56]. It is consumed within 3 days, adding sugar cane and cinnamon before serving. The remaining beverage is used as an inoculum for the following day [54]. Currently, Koumiss is manufactured at an industrial level in those countries using pure cultures of yeasts and lactic acid bacteria [56].

2.3.1. Sources and Composition

During lactic acid fermentation, mares' milk showed different behaviour from that of bovine milk at the macroscopic and microscopic levels. This was confirmed by a lower acidification rate, different gelation points, demineralization at higher pH and different coagulum structure [57]. These differences could be attributed to inherent factors in mares' milk, such as the mineral and/or the casein composition of its colloidal phase, the micellar structure, and the probable presence of antimicrobial defences. Several characteristics of mares milk, such as a high level of polyunsaturated fatty acids and a low cholesterol content, seem to support the interest in increasingly using mares milk for human consumption [58]. Koumiss generally contains about 2 % alcohol, 0.5-1.5% lactic acid, 24 % milk sugar and 2 % fat [56]. During the production stages of Koumiss, it was found 2 to 10 times as much free amino acids as the original milk used for souring the beverage, but with low levels of aspartic acid, serine and phenylalanine content [59]. The drink contains many bacteria and components derived from the bacteria as well as various enzymes [60]. Hence, Koumiss contains as a rule the lactose-fermenting yeast species *Kluyveromyces marxianus* and *K. lactis*, some galactose-fermenting yeast species as well as mostly different yeast species with pellicle producing tendency [61]. *K. marxianus* is known to produce some galacto-oligosaccharides enzymatically, leading to proliferation of lactic acid bacteria and bifidobacteria in the intestines to promote human health [62]. These oligosaccharides are produced through the transgalactosylation of various β-galactosidases which, in its turn, catalyze the transgalactosyl reaction as well as the hydrolysis of lactose. Moreover, galacto-oligosaccharides are known to stabilize cellular membrane and to protect *K. lactis* in cryopreservation studies [63], suggesting an microecological interaction between the main strains in Koumis. In the other hand, *K. marxianus* is also able to produce an extracelluar glycoprotein inulinase, which catalyzes the hydrolysis of the polyfructoside inulin outside the cell wall [64].

Yeast ferment lactose to alcohol, which contributes to the Koumiss products typical flavor and taste. The composition of indigenous yeast species in Koumiss differed among the locations from which it is sampled. For instance, ten different yeast species were found in samples from the Qing Hai region, twelve species from the Xin Jiang region and eleven species from the Inner Mongolia region [65]. *K. marxianus, Kazachstania unispora* and *S. cerevisiae* were the commom strains found in all these regions. In the other hand, *Lactobacillus* species are widespread, namely *L. delbrueckii, L. acidophilus, L. casei, L. rhamnosus, L. paracasei* and *L. Kefir* [56]. Depending on the lactic acid content, Koumiss can exist into so-called "strong" (*Lactobacillus bulgaricus, Lactobacillus rhamnosus*), "moderate" (*L. acidophilus, L. plantarum, L. casei, L. fermentum*), and "light" Koumiss [56]. "Strong" Koumiss have pH to 4.5–3.9 at the end of the process, and the conversion ratio of lactose into lactic acid is about 80-90 %, whereas "moderate" Koumiss, that have the best fragrance and taste, averages 50 % of lactic conversion. "Light" Koumiss (*Streptococcus thermophilus* and *Str. cremorisis*) is a slightly acidified product (pH 4.5–5.0) [66]. Colombian Koumiss, in the other hand, presents high levels of *enterococci* strains, that could play an important role in the characteristics of this product, probably through proteolysis, lipolysis and citrate breakdown, hence contributing to their typical taste and flavour [54].

2.3.2. Activities

Koumiss seems to present some beneficial influence on enhancing immunity [56], while exhibits some antibacterial activity, too. In mice and chickens, supplemented Koumiss was found out to make for a notable increase in the number of animal specimens exhibiting a strong immune response to the antigens they had been previously immunized with [61], revealing an anti-immunodepressive effect for the beverage [67]. In another work, Koumiss was able to suppress increase in cholesterol and neutral lipid levels in the blood of rats fed by substituting 5 % Koumiss diet [60]. An anti-*Listeria* bacteriocin produced by *L. plantarum* was isolated from a Koumiss sample [56], producing a large inhibition zone against *L. monocytogenes, E. coli* and *Shigella spp.* Called Pediocin LB-B1, the peptide is heat resistant and stable between pH 2–10 [53]. Interesting, Koumiss also present some strains with known antihypertensive activities, as *Lactobacillus delbrueckii subsp. bulgaricus, Lactobacillus helveticus, Lactobacillus acidophilus, Lactobacillus casei, Lactobacillus jensenii, Lactococcus lactis, Lactococcus cremoris, Leuconostoc mesenteroides spp cremoris, Enterococcus faecalis* and *Enterococcus faecium*, that have been proved to reduce blood pressure [66]. Nothwithstanding, *enterococci* can be also opportunistic pathogens in humans, allowing adherence to host tissue, invasion and resistance to host defense, besides its bioactive pontential as hypotensive agent [54].

2.3.3. Preparation

Traditionally, Kumis is a home-made beverage produced by spontaneous fermentation of raw whole milk for 2 or 3 days depending on room temperature and milk producers. It is manufactured by the mixture of old koumiss with fresh raw mares' milk. Fermentation takes place in 3-8 h and is carried out by a mixture of indigenous yeasts and lactic acid bacteria [57]. The product of this fermentation is a low alcoholic (1-2 %), creamy and sparkling beverage, with a slight degree of sourness. It is stored at about 4-10 ^{o}C and consumed within 3 days, adding sugar cane and cinnamon before serving. The remaining beverage is used as an inoculum for the following day [54]. Alternatively, the milk previously miling of mares gradualiy 20-30 days after foaling (six times a day) can be placed in wooden containers that contain 10-20 % of the previous day's milk already in full fermentation [55]. The contents of each container are then frequently stirred and covered by a wooden lid. It is kept at room temperature though this is not constant, given the addition of warmed fresh milk. During the day the Koumiss is drawn off from the container for immediate consumption [55]. Chemical, microbiological and organoleptic characteristics of the samples on souring can vary along 10-15 on storage at 4 ^oC [58].

2.4. Kvass

Kvass (Figure 9) is a traditional Russian cereal fermented beverage produced from malt, rye xour, stale rye bread, and sucrose, and is consumed mainly in Eastern Europe. Kvass contains not more than 1.5 % of alcohol by volume, but if it stands for longer time, the concentration can become 2.5 % or higher. In this sense, Kvass is considered one of the few examples of traditional, non-alcoholic cereal-based beverages [68]. It has a goldenbrown color, the pleasant flavor of rye bread, low sweetness, and has sufficient carbonate to give kvass a sparkle on the palate. Kvass has been very popular in Russia and the former Soviet Union, and its consumption has increased due to massive advertising stressing its benefits for health [69]. Unlike beer, the Kvass is generally considered to be a nonalcoholic beverage and is drunk by children of all ages without any limit.

Figure 9. A cup of Kvass (Licensed by Liftarn under Creative Commons).

2.4.1. Sources and Composition

The microflora of Kvass fermentation is consistently composed of lactic acid bacteria and *Saccharomyces cerevisiae*, but the composition on species level is expected to be quite variable due to divergences in fermentation techniques and feedstock. In this sense, it is characterized by a combination of lactic fermentation with alcoholic fermentation, similar to Boza, a Turkish low/non-alcoholic fermented beverage discussed below. Microbial associations in Kvass fermentations are poorly investigated, although *Lactococcus lactis var. diacetilactis* was proposed as a starter culture in combination with *S. cerevisiae* to improve aroma and organoleptic properties of Kvass [68]. The pH of Kvass is around 3.5. The final product contains carbohydrates, proteins and amino acids, lactic and acetic acids and vitamins originating from the raw materials or as a result of microbial activity. The predominant carbohydrates are maltose, maltotriose, glucose, and fructose. Its ethanol content is about 1 % or less but it is considered to be spoiled if ethanol accumulates to higher levels [68]. Ethanol, lactate, and acetate have been found at concentrations about 190-270, 3-12, and 1-7 mM, respectively. Fructose (30-110 mM), glucose (22-48 mM), and maltose (12-32 mM) are also found in different levels. Maltotriose and maltodextrins were present in much smaller concentrations [68]. The addition of sucrose to Kvass wort enables the formation of poly- and oligosaccharides during fermentation by bacterial glycosyltransferases. Dextransucrase activity is frequently found in strains of the genus *Leuconostoc* [69]. Dextransucranase catalyzes the formation of dextran when in medium containing sucrose as sole substrate (Figure 10). However, in medium containing an acceptor (maltose, fructose, glucose, etc.) as substrate, besides sucrose as the second substrate, the enzyme produces prebiotic oligosaccharides [16]. Likewise, isomalto-oligosaccharides are formed from sucrose by *L. mesenteroides* during Kvass fermentation, and by *Weisella cibaria* during growth in gluten-free sourdough [69]. In the past decade, more than a dozen dextransucranases from *Leuconostoc spp.* and related genera were characterized on the genetical and biochemical level [68]. The enzymes generally catalyse the synthesis of diand oligosaccharides in addition to polymer formation; the oligosaccharides formed include leucrose, panose, isomaltose, and the corresponding higher oligosaccharides.

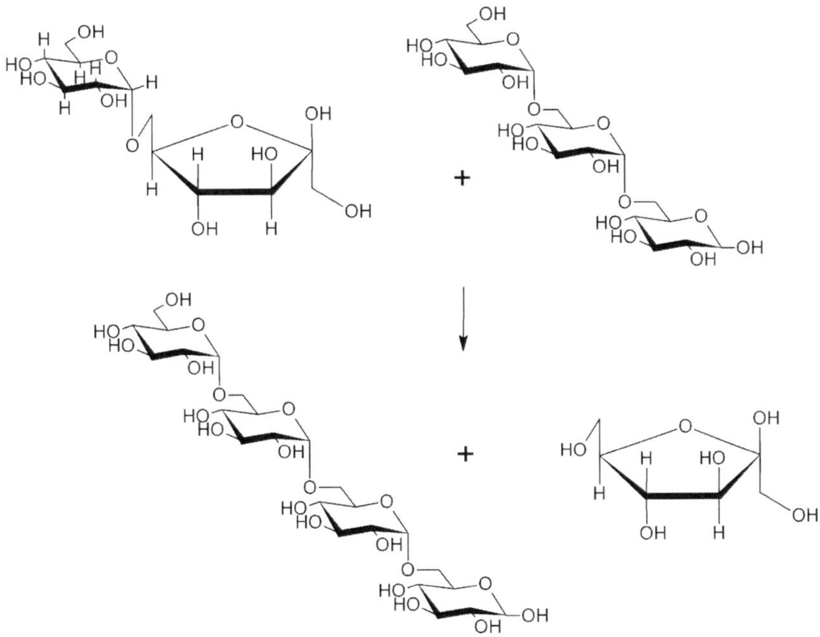

Figure 10. Formation of dextran from sucrose and oligosaccharides by dextransucrase found in some Scoby beverages.

2.4.2. Activities

Isomaltotriose and higher isomalto-oligosaccharides (Figure 11) are only partially digested in the small intestine, and therefore they are currently applied in food as alternative, low caloric sweetener and as prebiotic. The amount of isomaltotriose in Kvass was well below the dose of 10 g isomalto-oligosaccharides that were reported to exert a bifidogenic effect in humans. However, the current knowledge on oligosaccharide synthesis by glucosyltransferases enables a directed optimisation of oligosaccharide formation in Kvass. Other non-digestible poly- and oligosaccharides present in Kvass may further contribute to a prebiotic effect [68].

2.4.3. Preparation

Two main Kvass making techniques exist using either stale sourdough bread or malt as main raw materials. In Kvass fermentations from stale sourdough bread,

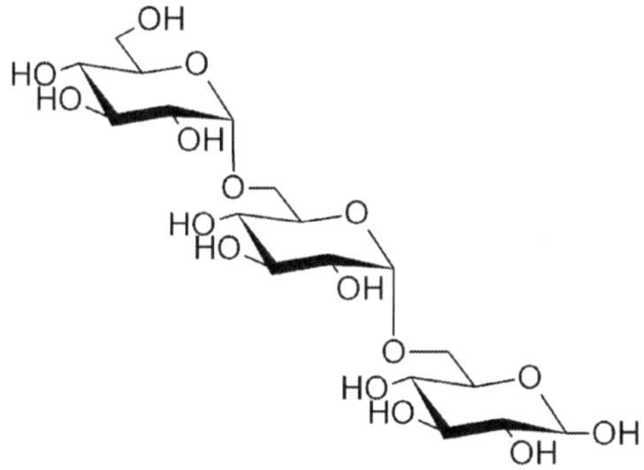

Figure 11. Structure of isomaltotriose, an oligosaccharide found in Kvass fermentation.

all sugars needed for yeast fermentation are derived from the bread-making process, in the second technique, gelatinized starch is cleaved by malt enzymes. Prior to fermentation, the Kvass batter is diluted in boiling water and clarifyed by sedimentation. Sucrose is added to the Kvass wort and fermentation is initiated by addition of bakers yeast or a previous batch of Kvass. Fermentation is terminated by chilling of the product to 4 ^{o}C before nutrients are exhausted [68]. Kvass is usually sold unfiltered and contains yeast. Commercial bottled Kvass, especially its cheaper varieties, is produced from prefabricated ingredients, but contains alcohol nonetheless. Noticeably, the labels of the bottled and canned Kvass sometimes bear no information on the alcohol content. As other mildly alcoholic beverages in Russia and the former Soviet Union (*tahn, kumys kefir*) Kvass is sold without indication of the alcohol contents on the label [69].

2.5. Borde

Borde is one of the most popular indigenous Ethiopian fermented beverages. Some of the known beverages in the country are *tella, munkoyo, shamita, cheka, korefe, keribo, bukire, merissa*, among others. Tella is alcoholic, while the rest are considered to be low or non-alcoholic beverages [70]. Ethiopian Borde is produced from a variety of locally available cereal ingredients, such as unmalted maize, barley, wheat, finger milleet and/or sorghum, and use traditional techniques. Borde is consumed while actively fermenting due to its poor keeping quality. It is a very popular and low-cost meal replacement in southern and western Ethiopia and some other parts of the country, and it is widely consumed by both adults and children (up to 3 L/day) [70]. The Borde preparation involves rudimentary equipment, and has four major phases marked by the introduction of fresh ingredients into the fermentation pot at different times. The final product become opaque, effervescent, whitish-grey to brown colour and with thick consistency and sweet-sour taste. As Borde is produced through an overnight fermentation, production of acid and alcohol during fermentation is limited. The pH of mixtures of Borde ingredients at the start of fermentation was between 4.28 and 4.31.

2.5.1. Sources and Composition

The natural fermentation of Borde is unpredictable, although the spontaneous fermentation typically results in a low pH beverage, high counts of aerobic mesophiles, lactic acid bacteria and yeasts. As Borde preparation involves several stages with cooking at about 90 ^{o}C, a source of microorganism from added malt results in a competitive activities of different microorganisms, whereby strains best adapted and with the highest growth rate will dominate during particular stages of the process. The most important strains in Borde fermentation appeared to belong to *Lactobacillus* species, together with *Leuconostoc, Pediococcus* and *Streptococcus*. Although *Pediococcus* can be considered as the most dominant cocci [71], more than 118 *Lactobacillus* species have been isolated from Borde souring [72]. The most *Lactobacillus* strains seems to belong to *L. viridescens, L. revis, L. curvatus, L. collinoides, L. pontis, L. sanfransisco* species [72]. During bacteria isolation procedures, it is also found *Enterococcus falvenscens, P. acidilactici, P. pentosaceus*. However, *W. confusa, L. viridescens, L. brevis, P. pentosaceus* and *L. curvatus* appeared to be the most important lactic acid population in Borde fermentation. *L. curvatus* is able to reduce some non-digestible oligosaccharides present in foods, as soy raffinose and mellibiose [73], an α-D-galactopyranosyl-type sugars. These galactooligosaccharides are utilized by intestinal strains of bifidobacteria and lactobacilli present in the gut, being hydrolized by α-D-galactoside galactohydrolase. Diverse other oligosaccharides are also found in Borde (Figure 12), as melibiose, trehalose, gentiobiose, and togatose [71].

2.5.2. Activities

Studies on traditional fermented foods of Ethiopia have shown that lactic acid bacteria have inhibitory effect on various food-borne pathogens [72]. *Lacto-*

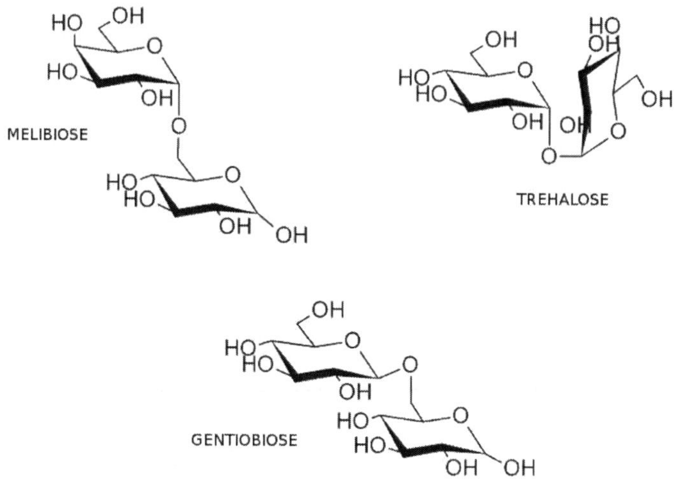

Figure 12. Some disaccharides produced form Borde fermentation. Melibiose can be formed through raffinose hydrolysis, producing a reducing disaccharide with an α-1,6 linkage between galactose and glucose (D-Gal-α(1-6)-D-Glc). Trehalose is formed by two glucose residues linked by an α-, α-1,1 glycosidic bond, while gentiobiose, by a β(1-6) linkage.

bacillus has demonstrated a significant role in suppressing the growth of various test strains in fermenting Borde, as *Staphylococcus aureus, Shigella flexneri, Salmonella spp.* and *Escherichia coli O157:H7* [72]. Althought these inhibitions could be viewed markedly at ambient temperature storage, it is thought to be due only to the activity of *Lactobacillus*.

2.5.3. Preparation

Borde is traditionally fermented from maize or wheat. The souring beverage can be prepared by an experienced brewer from a mixture of cereal ingredients following the traditional recipe and four phases of its fermentation. The common technology for their preparation is time-consuming, unhygienic and complex, which involves grinding, souring, roasting, steam cooking, boiling, cooling, mashing, wet-milling and wet-sieving operations [71]. This may create a variable bio-physico-chemical environment which select for involved microflora that overcome technological hurdles in the natural fermentation of Borde. The hydrolysis of added malt by amylolytic enzymes in the 2nd phase thus create an environment suitable for proliferation of microorganisms during traditional (four phases) fermentation of Borde. Briefly, the maize flour is fermented for 24 to 48 h (Phase I), succeeding blending of 80 % of the required malt previously fermented for 24 h (Phase II), addition of steamed dough balls, blending the mix with a thick mash, and left it for 18 h fermentation (Phase III). Finally, the Phase IV comprises additon of thick porridge from 48 h fermented maize dough and sorghum flour, homogenisation to thick slurry, sieving, filtration, and pouring back into a jar, following a final fermentation for 4 h.

2.6. Boza

Boza (or Bouza) is a low-alcohol beverage produced from the fermentation of barley, oats, millet, maize, wheat or rice. Boza is known since the times of the pharaohs, and nowadays consumed mainly in Egypt, Turkey and in some Eastern Europe countries (e.g. Albania and Romania) [74]. Turkish Boza is made by mixed yeasts and lactic acid fermentation of cooked maize, wheat and rice flours, whereas Bulgarian Boza is prepared by spontaneous fermentation of those cereals diluted with sweetened water to 10 %. Boza is a light yellow, thick, highly viscous and sour drink [74], and have a pleasant sweet-sour, bread-like taste. Boza′s origin dates back to the ancient populations from Anatolia and Mesopotamia, enjoyed its main spreading under the Ottomans, and became one of the principal distributed trades in that period. Its name originates from the Persian work ”buze”, meaning millet [75].

2.6.1. Sources and Composition

The low pH (3.9-4.0) and the high acidity of Boza indicate a fermentation by lactic acid bacteria, while the alcohol is due to yeast fermentation [74]. The protein content of Boza ranges from 1.5 to 2.0 % and due to the alcoholic fermentation involved in its formation, a significant contribution of vitamin B can be expected. Furthemore, Boza is a nutritious beverage because of its lactic acid, protein, carbohydrate, fibre and vitamin content. Alcohol content of Turkish Boza samples was found to be in the range of 0.03-0.39 % (w/v) [76]. Microflora identification in Bulgarian samples of Boza shows that it mainly consists of yeasts and a diverse population of lactic acid bacteria [77]. However, relatively few yeast species, and no filamentous fungi, have been isolated [78]. Lactic acid bacteria thus far described include *Lactobacillus acidophilus, Lactobacillus brevis, Lactobacillus coprophilus, Lactobacillus coryniformis, Lactobacillus fermentum, Lactobacillus plantarum, Lactobacillus rhamnosus, Lactobacillus sanfrancisco, Lactococcus lactis subsp. lactis, Leuconostoc mesenteroides, L. mesenteroides subsp. dextranicum, Leuconostoc raffinolactis, Pediococcus pentosaceus, Oenococcus oeni, Weissella confusa* and *Weissella paramesenteriodes* [78]. Among these strains, *L. brevis* has been shown to produce an α-glucosidase able to hydrolize polysaccharides and dextrin residues present in natural fermentation of cereals [79]. Another strain present in Boza, *L. plantarum* has been reported to present an oligosaccharide-producing enzyme, β-galactosidase (β-D-galactoside galactohydrolase). This enzyme can catalyze two basic reactions, hydrolysis of structurally related galactosides and transglycosylation reactions, resulting in a mixture of galacto-oligosaccharides [80].

2.6.2. Activities

Many of the lactic acid bacteria produce antimicrobial compounds, including bacteriocins, increasing the shelf life of the product. In this sense, previous studies have shown that Boza is a good source of bacteriocin-producing lactic acid bacteria [78]. By definition, bacteriocins produced by lactic acid bacteria are proteinaceous antibacterial compounds that exhibit bactericidal or bacteriostatic activity against genetically closely related bacteria. A bacteriocin produced by *Lactococcus lactis subsp. lactis* isolated from Boza was partially characterized and presents health benefits [75]. Mesentericin ST99, produced by *Pediococcus pentosaceus* ST18 and *Leuconostoc mesenteroides subsp. dextranicum* ST99 isolated from Boza "Belogratchik" were also identified. Pediocin ST18 and mesentericin ST99 revealed good anti-*Listeria* activity [81], as determined by colonies producing inhibition zones (agar-spot test and diffusion method). Another lactic acid bacteria, as *Lactobacillus plantarum, Lactobacillus pentosus, Lactobacillus rhamnosus*, and *Lactobacillus paracasei*, produced bacteriocins active against *Lactobacillus casei, Escherichia coli, Pseudomonas aeruginosa* and *Enterococcus faecalis* [81].

In the other side, fermented foods and beverages are more susceptible to formation of biogenic amines, since several microorganisms are involved in the fermentation process and the raw materials used contain considerable amounts of proteins [76]. Furthermore, It is known that cereals are good materials production of mycotoxins especially for ochratoxin-A. It is produced by single *Penicillium* species, *P. verrucosum*, by *Aspergillus ochraceus* and certain *Aspergillus* species, and by *A. carbonarius* [74]. Another harzadous compound found in Boza fermentation is biogenic amines. This compounds are basic nitrogenous organic compounds having a recognised activity and which may occur naturally in foods and beverages. They are generated either as a result of endogenous amino acid decarboxylase activity in raw food material or by the growth of decarboxylase-positive microorganisms under conditions favourable to enzyme activity [76]. Biogenic amines have important metabolic roles in living cells. Polyamines and putrescine are essential for growth, and other amines, like histamine, tyramine and serotonin, are involved in nervous system functions and the control of blood pressure. Although they are needed for many critical functions in humans and animals, high concentrations of these amines can have toxicological effects. Consumption of foods containing high amounts of biogenic amines may cause problems such as headaches, nausea, hypotension, hypertension, cardiac palpitation, etc. Furthermore, biogenic amines can react with nitrite and produce nitrosamines, many of which are known to be carcinogenic, mutagenic, teratogenic and embryopathic. Tyramine was the prevailing biogenic amine in all Boza samples [76]. However, the concentrations of biogenic amines detected in Turkish Boza were below the maximum permissible limits.

2.6.3. Preparation

As other Scoby beverages, interactions between microorganisms are not controlled during the fermentation process in Boza, which leads to variations in quality and stability of the product [82]. In this sense It is necessary to use starter cultures to avoid variations in the quality of this type of traditional fermented product. Different raw materials at varying concentrations and different fermentation processes are normally used in Boza production and consequently, its quality can vary a lot across [83]. For the preparation of Boza, selected and cleaned cereal or combination of two or more cereals is milled to the size of semolina and cooked in an open or steam-jacketed boiler after water addition [76]. The cooked material is transferred into suitable marble vessels for cooling.

The cooled material is strained to remove bran, hull and other foreign materials. After sugar (sucrose) addition (Bulgarian Boza), the broth is fermented by adding previously fermented Boza, sourdough or yoghurt as starter culture. Fermentation is carried out at 30 ^{o}C for 24 h. After fermentation, Boza is cooled to refrigeration temperature, bottled in plastic containers and should be consumed within 35 days. In Turkey Boza is usually consumed in winter and served with cinnamon and roasted chickpeas [76]. Alternatively, Boza can be prepared by coarsely grinding wheat grains, placing a portion of them in a wooden basin and kneading them with water into a dough. The dough is cut into thick loaves, which are lightly baked. The remainder of the grains is moistened with water, germinated, dried, ground and mixed with the loaves of bread, which are soaked in water [74]. While cooked cereal is strained to remove most of the solids, sugar is added to taste and inoculated with a starter culture.

2.7. Pulque

Pulque is a traditional alcoholic fermented beverage currently produced and consumed mainly in the central states of Mexico (Figure 13). The fermentation process starts in the maguey, where naturally occurring microorganisms present in the aguamiel ferment part of the available carbohydrates [84]. However, Mexicans showed a real preference for beer, with more than 54 % of the total ethanol consumption as compared to 11% for Pulque [85]. Pulque consumption has dramatically decreased since the Sixties. It is produced from the sap known as aguamiel, which is extracted from several species of maguey (*Agave americana, A. atrovirens, A. ferox, A. mapisaga, A. salmiana*) [86]. Agaves, which belong to the family of *Agavaceae*, are monocotyledoneous and monocarpic plants.

They have played an important role for Mexican inhabitants in the past and they still have important ecological and economic roles in this part of the world. They protect the soil against both water and wind erosion [87]. The Agave plants are also cultivated for their capacities to grow under arid or semi arid conditions and stop land turning to desert. In this sense, the tradition of the Tlachique (aguamiel harvest) as well as the sowing and use of maguey-Pulquero has been preserved as part of the Mexican heritage [85].

2.7.1. Sources and Composition

Fermentation occurs in Pulque thanks to microorganisms naturally present in the aguamiel or those incorporated during the transport or even preparation. It

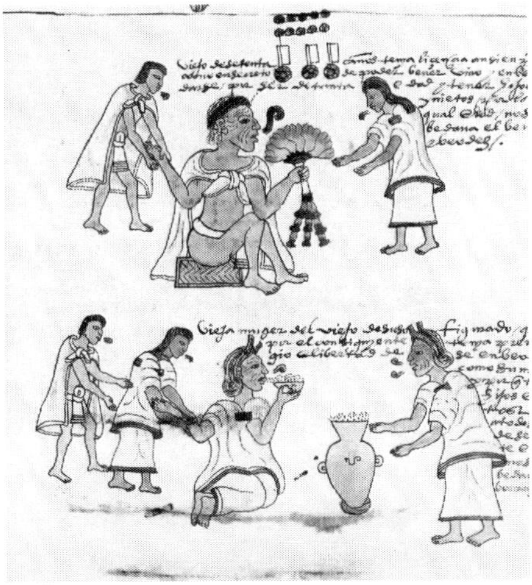

Figure 13. An illustration from Codex Mendoza depicting elderly Aztecs smoking and drinking pulque. (Licensed by Billycuts under Creative Commons).

can also be accelerated by the addition of a portion of Pulque previously produced. Thus, the beverage is the result of the action of a large bacterial diversity [85]. Microorganisms frequently identified in Pulque samples comprise several yeast and bacterial species including homo- and hetero-fermentative lactic acid bacteria, the alcohol producing bacteria *Zymomonas mobilis* and the dextran producing bacteria *Leuconostoc mesenteroides* [84]. The development of viscosity due to exopolysaccharide synthesis has been the main criteria to determine the degree of fermentation (fresh or mature Pulque) [84]. Important industrial microorganisms isolated from various Pulque samples, comprise several yeasts, such as the ethanol producing *Saccharomyces cerevisiae* and species of *Kluyveromyces* producers of inulinase [88]. Isolated bacterial species include the alcohol producing *Zymomonas mobilis* and the dextran producing lactic acid bacteria *Leuconostoc mesenteroides*. Among them, *S. cerevisiae, L. mesenteroides* and *Z. mobilis* have been proposed as essential microorganisms for the Pulque fermentation process.

Aguamiel contains around 11.5 wt % of dry matter, which was composed mainly of sugars (75 wt %). Fructose, and sucrose content represented >65 wt % of the dry matter, but the concentration of sucrose was always significantly lower than that of the two monosacharides [85]. Among sugars, about 10 wt %

are fructo-oligosaccharides (FOS), which are known to be important in the food industry for their prebiotic properties. Other components include 0.3 wt % of free amino acids (AA, with most essential amino acids and four neurotransmitters: GABA, GLY, GLX and ASX), 3 wt % of proteins, and 3 wt % of ashes [85]. In addition, its chemical composition has been also partially elucidated, highlighting the importance of the amino acid content, as well as the presence of ascorbic acid and iron in aguamiel. The concentrations in free AA in the different samples of aguamiel was estimated as 0.26 % of dry matter, corresponding to 0.3 g /L. All of the essential AAs are present in all aguamiel samples except for methionine, which was present at only a very low level (<0.01 % of dry basis) in the sample collected after 4.5 months of production. GABA is distributed throughout the biological world even if its level in plant tissues is low (ranging from 0.03 to 2.00 μmol/g fresh weight). GABA synthesis is induced by diverse environmental stresses including cold, heat, salt, and mild or transient environmental factors, such as wind and rain. GABA is known to be the main inhibitory neurotransmitter in the sympathetic nervous system. It also plays an important role in cardiovascular function. [84]. In Agave plants, GABA concentration is estimated to be 0.25 μmol/mL, corresponding to 26 mg/L of aguamiel [85].

Pectins, also produced from Agave plant, are a polysaccharide family of covalent linked galacturonic acids mainly, including homogalacturonan, rhamnogalacturonan I (Rg-I), rhamnogalacturonan II (Rg-II), xylogalacturonan and apiogalacturonan, which constitute the major part of the plant cell wall [87]. Rg-II is a small structural complex polysaccharide that contains 12 different glycosyl-residues connected via 20 different linkages. The concentration of Rg-II in Agave Pulquero juice is around 270 mg/L which corresponds to a concentration of 293 mg of Rg-II per kg of fresh Agave stem. This value is very similar to the concentration found in apple juice (350 mg/L) with the same enzymatic process [85]. Its backbone is composed of a 1,4-linked α-D-galacturonosyl residues in which four structurally well-defined oligosaccharide side chains are attached. The hydrolysate of Agave Pulquero Rg-II have revealed a series of oligosaccharides co-eluted with standard of 1,4-linked α-D-galacturonides obtained by treatment with an endopolygalacturonase. Agave Pulquero Rg-II has an average degree of polymerization between 8 and 9. In summary, a Rg-II like structure can be isolated in juices obtained by enzymatic liquefaction of Agave Pulquero stem. The glycosyl residue and glycosyl-linkage compositions of Agave Rg-II and wine Rg-II are very similar. It confirms that Agave Rg-II corresponds to the theoretical model described for sycamore and wine Rg-II structure. [87] Agave tequilana fructans are polymers of fructose with one internal or external glucosyl residue. The degree of polymerization ranges from 3 to 29 fructose units [89]. The proportion of free and polymerized fructose as well as the branching degree of Agave pulquero changes as a function of plant age.

Fructanase production by *K. marxianus* was considered to be highly inducible by inulin. This yeast is normally found in different fermenting musts of Agave beverages. These findings explain the vigorous growth performance of *K. marxianus* with fructans as carbon source. The fructans found in aguamiel have been proposed as thermoprotective prebiotics in the spray-drying of *Bifidobacterium bifidum*. Some yeasts had the capacity to grow on only one type of fructan. Yeasts like *Candida boidinii* and *Rhodosporidium fluviale* have the metabolic capability for the selective degradation of Agave tequilana fructans. In the other side, *Schizosaccharomyces pombe, Cryptococcus albidus*, and *Hanseniaspora osmophila* can produce enzymes only able to degrade the fructans with elongations, which possibly are generated by other fructanases produced by the Mezcal microbiota during Mezcal fermentation. Complimentarily, an inulinase hyperproducing strain of *Kluyveromyces* have been also found in the aguamiel [85].

Dextransucrases, which belong to family of the glycoside hydrolases, are cell-associated or soluble extracellular enzymes produced by the soil bacteria belonging to the *Leuconostoc* genus. They catalyze the synthesis of high-molecular-weight glucans (D-glucose polymers) from sucrose. If efficient acceptors like maltose or isomaltose are added to the reaction medium, dextransucrases catalyze the synthesis of low-molecular-weight oligosaccharides and, to a minor extent, high-molecular-weight glucan polymers. Different kinds of glucans with different sizes and structures, depending on the dextransucrase-producing strain, are synthesized [90].

2.7.2. Activities

The main chemical components of interest in Pulque are the polysaccharides (approximately 10 % by wt in the dilute sap), and saponins. Saponins and sapogenins are widely distributed in Agave species, The most common types are gitogen, hecogenin, tigogenin, manogenin, diosgenin and smilagenin. Saponins can produce a variety of effects, among them hemolysis and detergency. Their abundant use in folk medicine for skin infections may be explained by the antibiotic and fungistatic activity generally exhibited by saponins. Other investigations show that some saponins exhibit antiviral activity, probably due to their effect on surface tension. Hederagenin is effective against *Trichomonas* and shows anti-inflammatory activity against carrageenin-induced edema [91]. Even if the spurning actually present in the Agave species do not show all of these activities, saponins could make a significant contribution to an antiseptic effect. Another antimicrobial effect of Pulque has also been devied from the observation that salt increases the effectiveness of the maguey syrup for the pyogenic bacteria, *S. aureus*. This demonstrates that the Aztec healers probably included salt in their preparation of maguey syrup because the effectiveness of the remedy was increased [91] (Figure 13).

2.7.3. Preparation

Pulque production consist of three types of fermentations: acidic, alcoholic, and viscous, making this traditional beverage an interesting environment were microorganisms or genes with potential biotechnological applications. Among these, some encoding sugar transporters, hydrolytic enzymes, exopolysaccaride, and isolated strains for lactic acid or ethanol production [86]. The collection is done every 12 h after scraping the stem of the agave (Figure 14), and the production lasts from 4 to 6 months [85].

Figure 14. Pulque extraction in Mexico. The pulque is an alcoholic native beverage extracted from agave or Maguey (Licensed by Elias Jorge under Creative Commons).

In pulque production, freshly collected aguamiel is transported in wood barrels or in bags made from young goat skins and transferred into large barrels where fermentation takes place [86]. The process is accelerated by the addition of a portion of previously produced Pulque and the fermentation time varies from few hours to overnight, depending on whether the sap is collected at daybreak or dusk. The development of viscosity due to exopolysaccharide synthesis and the alcohol content are the main parameters used to determine the degree of fermentation. The process is static and performed under non-aseptic conditions; therefore, the populations of microorganisms involved in the fermentation are those naturally occurring during aguamiel accumulation in the maguey and those incorporated during collection, transport, inoculation and manipulation [88]. It has been proposed that the fermentation process starts in the maguey, where naturally occurring microorganisms present in the aguamiel ferment part of the available carbohydrates. The process is accelerated by the addition of the seed (a portion of previously produced Pulque). Fermentation time varies from a few hours to overnight, depending if the sap is collected at daybreak or at dusk. The development of viscosity due to exopolysaccharide synthesis has been the main criteria to determine the degree of fermentation (fresh or mature Pulque). The final product is placed in wood barrels and distributed daily for sale and consumption, without the addition of any preservatives. The entire process is performed under non-aseptic conditions, therefore the mixture of micro organisms involved in the fermentation process are those naturally occurring in aguamiel and those incorporated during its collection, transport, inoculation and manipulation [86].

2.8. Other Scobys

Some fermented beverage made with Scobys can be also prepared from extracts of several fruits and vegetables by natural fermentation. For instance, isolated yeasts as *Zygosaccharomyces psedrouxii* and *Pichia anomala* have been mixed with lactic acid bacteria (*Leuconostoc spp.* to produce a riched oligosaccharides beverage containing a fructopyranosyl residues [56].

Another Scoby beverage fermented by gelatinous grains is Tibi (or Tibetan mushroom) [92]. The grains originate from Mexico where they occur in the form of hard granules on the disk-shaped leaves of the *Opuntia* plant. Similarly to Kefir, the drink leads to an acidic, mildly alcoholic drink prepared by fermentation of a sucrose solution [93]. Figs, dates, raisins, or ginger roots provide the growth factors. The grains are a symbiotic association of *Lactobacilius brevis, Streptococcus lactis*, and *Saccharomyces cerevisiae*. Furthermore, the grains can be used again after one or two days of fermentation at room temperature, by immersing them in a fresh medium. The Tibi grains are translucent, waxy, and tough in consistency. Moreover, the grains contain a matrix of dextran consisting mainly of a backbone of glucopyranosyl residues. It has been identified an anti-inflammatory activity of Tibi fermentation in rat models of induced granuloma and rat paw edema [94].

The Tibi fermentation is also similar to the Ginger Beer plant produced in England [95] (Figure 15). It occurs as jelly-like, semi-tranparent, yellowish-white masses, aggregated into brain-like clumps, or forming deposits at the bottom of the fermentations. It also presents resemblances to Kefir grains, with which, however, it is by no means identical. It consists essentially of a symbiotic association of a specific *Saccharomycete* and a *Schizomycete*, morphologically comparable to a Lichen.

Figure 15. An illustration showing a drugstore from the eighteen century presenting a Ginger Beer preparation - "soda water & ginger beer" (Licensed by Carlos Adanero under Creative Commons).

However, Ginger Beer can also comprises other species of yeasts, bacteria and mould-fungi, casually associated with these. Other works reported the presence of two different microbes in every sample, and seemed vital to the production of Ginger Beer, *Saccharomyces pyriformis* and *Brevibacterium vermiforme*. Together, they produce the essential ingredients of traditional ginger beer: carbon dioxide and alcohol. Several other Scobys are related from different countries, although Japanese fermented drinks can be considered one of the most described. So Koji or Kome-koji is a source for several preparation in Japan, as *amazake*, *junmai-shu*, *ki-koji-kin* (*nihonshu* or original sake), all of them low-alcoholic Scoby drinks. Another Japanese Scoby beverage is Miki [96]. Different from Koji, Miki uses cooked rice and mashed raw sweet potate tubers as the saccharifying agent, and without a fungal starter, resulting in an alcoholic beverage contained 5-6% ethanol (v/v). Oligosaccharides, including maltotetraose and maltopentaose, were detected in the Miki made with higher concentrations of raw sweet potato mash.

3. Conclusion

Scobys are fermented beverages prepared since immemorial times. Some Scobys are preserved along generations due to ritualistic practices and the indigenous knowledge associated with their health benefits and sense experience. Some oligosaccharide as well as enzyme-linked oligosaccharide production are present in several Scoby preparation, depending on the source utilized for fermentation. These oligosaccharide are known to lead to some beneficial effects in humans, involving the main role of gut microbiota. However, an oligosaccharide isolated from sugary Kefir (AK) seems to be able to break up both biomimetic cell membranes and macrophages adsorption, disclosing a different path for prebiotic actions. In essence, Scobys preserve ancient recipes for their preparation even in present times, perhaps due to the convergence of the activities and sensorial pleasure related to their consumption.

Acknowledgments

The author gratefully acknowledge financial support from Minas Gerais State Research Foundation - FAPEMIG (Grant no. CBB APQ 01920/11), the kindness of Prof. Fabio Herbst Florenzano (University of São Paulo, EEL-USP, Brazil) for providing with PMMA-block-PDMAEMA, and the valuable discussions with Prof. Mariano Sabino Silva and Prof. Lavio Pareschi on the impedance assays.

References

[1] Kallel L, Desseaux V, Hamdi M, Stocker P, Ajandouz EH. Insights into the fermentation biochemistry of Kombucha teas and potential impacts of Kombucha drinking on starch digestion. *Food Research International.* 2012.

[2] Farnworth ER. *Handbook of fermented functional foods.* Taylor & Francis; 2008.

[3] Holzapfel W. Use of starter cultures in fermentation on a household scale. *Food control.* 1997;8(5):241–258.

[4] Blandino A, Al-Aseeri ME, Pandiella SS, Cantero D, Webb C. Cereal-based fermented foods and beverages. *Food Research International.* 2003;36(6):527 – 543.

[5] Blandino A, Al-Aseeri M, Pandiella S, Cantero D, Webb C. Cereal-based fermented foods and beverages. *Food research international.* 2003;36(6):527–543.

[6] Hesseltine C. Microbiology of oriental fermented foods. *Annual Reviews in Microbiology.* 1983;37(1):575–601.

[7] Panel on the Applications of Biotechnology to Traditional Fermented Foods NRC. *Applications of Biotechnology in Traditional Fermented Foods.* The National Academies Press; 1992.

[8] Okada H, Fukushi E, Yamamori A, Kawazoe N, Onodera S, Kawabata J, et al. Isolation and structural confirmation of the oligosaccharides containing α-d-fructofuranoside linkages isolated from fermented beverage of plant extracts. *Carbohydrate Research.* 2011;346(16):2633–2637.

[9] Todorov SD, Dicks LMT. Pediocin ST18, an anti-listerial bacteriocin produced by Pediococcus pentosaceus {ST18} isolated from boza, a traditional cereal beverage from Bulgaria. *Process Biochemistry.* 2005;40(1):365 – 370.

[10] Rivera-Espinoza Y, Gallardo-Navarro Y. Non-dairy probiotic products. *Food microbiology.* 2010;27(1):1–11.

[11] Barbosa A, Santos P, Lucho A, Schneedorf J. Kefiran can disrupt the cell membrane through induced pore formation. *Journal of Electroanalytical Chemistry*. 2011;653(1):61–66.

[12] Piermaria JA, Pinotti A, Garcia MA, Abraham AG. Films based on kefiran, an exopolysaccharide obtained from kefir grain: Development and characterization. *Food Hydrocolloids*. 2009;23(3):684–690.

[13] Schneedorf JM. Kefir DAqua and Its Probiotic Properties. In: *Rigobelo EC*, editor. *Probiotic in Animals*. In Tech; 2012. p. 53–76.

[14] Delzenne NM. Oligosaccharides: state of the art. *Proceedings of the Nutrition Society*. 2003 1;62:177–182. Borde.

[15] Mussatto SI, Mancilha IM. Non-digestible oligosaccharides: A review. *Carbohydrate polymers*. 2007;68(3):587–597.

[16] Vergara CMdAC, Honorato TL, Maia GA, Rodrigues S. Prebiotic effect of fermented cashew apple *Anacardium occidentale* juice. *LWT-Food Science and Technology*. 2010;43(1):141–145.

[17] Patel S, Goyal A. Functional oligosaccharides: production, properties and applications. *World Journal of Microbiology and Biotechnology*. 2011;27(5):1119–1128.

[18] Jakob F, Pfaff A, Novoa-Carballal R, Rübsam H, Becker T, Vogel RF. Structural analysis of fructans produced by acetic acid bacteria reveals a relation to hydrocolloid function. *Carbohydrate polymers*. 2012.

[19] Rycroft C, Jones M, Gibson G, Rastall R. A comparative in vitro evaluation of the fermentation properties of prebiotic oligosaccharides. *Journal of Applied Microbiology*. 2001;91(5):878–887.

[20] Martínez-Villaluenga C, Cardelle-Cobas A, Corzo N, Olano A. Study of galactooligosaccharide composition in commercial fermented milks. *Journal of Food Composition and Analysis*. 2008;21(7):540–544.

[21] Bekar O, Yilmaz Y, Gulten M. Kefir improves the efficacy and tolerability of triple therapy in eradicating Helicobacter pylori. *Journal of medicinal food*. 2011;14(4):344–347.

[22] Piermaria JA, de la Canal ML, Abraham AG. Gelling properties of kefiran, a food-grade polysaccharide obtained from kefir grain. *Food Hydrocolloids*. 2008;22(8):1520–1527.

[23] Bergmann RO, Pereira MA, Veiga SM, Schneedorf JM, Oliveira NS, Fiorini JE. Microbial profile of a kefir sample preparations: grains in natura and lyophilized and fermented suspension. *Ciência e Tecnologia de Alimentos*. 2010;30(4):1022–1026.

[24] Ho JN, Choi JW, Lim WC, Kim MK, Lee IY, Cho HY. Kefir inhibits 3T3-L1 adipocyte differentiation through down-regulation of adipogenic transcription factor expression. *Journal of the Science of Food and Agriculture*. 2012.

[25] Grønnevik H, Falstad M, Narvhus JA. Microbiological and chemical properties of Norwegian kefir during storage. *International Dairy Journal*. 2011;21(9):601–606.

[26] Magalhães KT, Pereira GdM, Dias DR, Schwan RF. Microbial communities and chemical changes during fermentation of sugary Brazilian kefir. *World Journal of Microbiology and Biotechnology*. 2010;26(7):1241–1250.

[27] Kesenkaş, Harun and Dinkçi N, Seçink, Kink Ö, Gönç S, Ergönül PG, Kavas G. Physicochemical, microbiological and sensory characteristics of Soymilk Kefir. *African Journal of Microbiology Research*. 2011;5(22):3737–3746.

[28] Londero A, Hamet MF, De Antoni GL, Garrote GL, Abraham AG. Kefir grains as a starter for whey fermentation at different temperatures: chemical and microbiological characterisation. *Journal of Dairy Research*. 2012;1(1):1–10.

[29] Moreira M, Santos M, Pereira I, Ferraz V, Barbosa L, Schneedorf JM. Atividade antiinflamatória de carboidrato produzido por fermentação aquosa de grãos de quefir. *Química Nova*. 2008;31(7):1738–1742.

[30] Medrano M, Pérez PF, Abraham AG. Kefiran antagonizes cytopathic effects of *Bacillus cereus* extracellular factors. *International journal of food microbiology*. 2008;122(1):1–7.

[31] Kwon OK, Ahn KS, Lee MY, Kim SY, Park BY, Kim MK, et al. Inhibitory effect of kefiran on ovalbumin-induced lung inflammation in a murine model of asthma. *Archives of pharmacal research*. 2008;31(12):1590–1596.

[32] Moreira M, Santos MD, Zolini G, Wouters A, Carvalho J, Schneedorf JM. Anti-inflammatory and cicatrizing activities of a carbohydrate fraction isolated from sugary kefir. *Journal of medicinal food*. 2008;11(2):356–361.

[33] Vinderola G, Perdigón G, Duarte J, Farnworth E, Matar C. Effects of the oral administration of the exopolysaccharide produced by *Lactobacillus kefiranofaciens* on the gut mucosal immunity. *Cytokine*. 2006;36(5):254–260.

[34] Fernández-Sánchez C, McNeil CJ, Rawson K. Electrochemical impedance spectroscopy studies of polymer degradation: application to biosensor development. *TrAC Trends in Analytical Chemistry*. 2005;24(1):37–48.

[35] Sezgintürk MK, Uygun ZO. An impedimetric vascular endothelial growth factor biosensor-based PAMAM/cysteamine-modified gold electrode for monitoring of tumor growth. *Analytical Biochemistry*. 2012;423(2):277–285.

[36] Pasqualotto E, Ferrario A, Scaramuzza M, De Toni A, Maschietto M. Monitoring electropermeabilization of adherent mammalian cells through electrochemical impedance spectroscopy. *Procedia Chemistry*. 2012;6:79–88.

[37] Jiang D, Diao P, Tong R, Gu D, Zhong B. Ca2+ induced $Fe(CN)_6{}^{3-/4-}$ electron transfer at Pt supported BLM electrode. *Bioelectrochemistry and bioenergetics*. 1998;44(2):285–288.

[38] Reybier K, Ribaut C, Coste A, Launay J, Fabre PL, Nepveu F. Characterization of oxidative stress in Leishmaniasis-infected or LPS-stimulated macrophages using electrochemical impedance spectroscopy. *Biosensors and Bioelectronics*. 2010;25(12):2566–2572.

[39] Güven A, Gülmez M. The effect of kefir on the activities of GSH-Px, GST, CAT, GSH and LPO levels in carbon tetrachloride-induced mice tissues. *Journal of Veterinary Medicine*, Series B. 2003;50(8):412–416.

[40] Jarrell J, Cal T, Bennett J. The Kombucha Consortia of yeasts and bacteria. *Mycologist*. 2000;14(4):166–170.

[41] Battikh H, Bakhrouf A, Ammar E. Antimicrobial effect of Kombucha analogues. *Lebensmittel-Wissenschaft+ Technologie*. 2012;47(1):71–77.

[42] Bhattacharya S, Gachhui R, Sil PC. Hepatoprotective properties of kombucha tea against TBHP-induced oxidative stress via suppression of mitochondria dependent apoptosis. *Pathophysiology*. 2011;18(3):221–234.

[43] Malbaša RV, Lončar ES, Vitas JS, Čanadanović-Brunet JM. Influence of starter cultures on the antioxidant activity of kombucha beverage. *Food Chemistry*. 2011;127(4):1727–1731.

[44] El-Nagar AKS. *Studies on cellulose production by tea fungus Kombucha*. Zagazig University; 2007.

[45] Greenwalt C, Ledford R, Steinkraus K. Determination and Characterization of the Antimicrobial Activity of the Fermented Tea¡ i¿ Kombucha¡/i¿. *LWT-Food Science and Technology*. 1998;31(3):291–296.

[46] Teoh AL, Heard G, Cox J. Yeast ecology of Kombucha fermentation. *International journal of food microbiology*. 2004;95(2):119–126.

[47] Jayabalan R, Marimuthu S, Swaminathan K. Changes in content of organic acids and tea polyphenols during kombucha tea fermentation. *Food chemistry*. 2007;102(1):392–398.

[48] Malbaša R, Lončar E, Djurić M. Comparison of the products of Kombucha fermentation on sucrose and molasses. *Food Chemistry*. 2008;106(3):1039–1045.

[49] Lončar ES, Malbaša RV, Kolarov LA. Kombucha fermentation on raw extracts of different cultivars of Jerusalem artichoke. *Acta periodica technologica*. 2007;38(38):37–44.

[50] Malbaša R, Lončar E, Djurić M, Došenović I. Effect of sucrose concentration on the products of Kombucha fermentation on molasses. *Food Chemistry*. 2008;108(3):926–932.

[51] Jayabalan R, Subathradevi P, Marimuthu S, Sathishkumar M, Swaminathan K. Changes in free-radical scavenging ability of kombucha tea during fermentation. *Food Chemistry*. 2008;109(1):227–234.

[52] Aloulou A, Hamden K, Elloumi D, Ali MB, Hargafi K, Jaouadi B, et al. Hypoglycemic and antilipidemic properties of kombucha tea in alloxan-induced diabetic rats. *BMC complementary and alternative medicine*. 2012;12(1):63.

[53] Xie Y, An H, Hao Y, Qin Q, Huang Y, Luo Y, et al. Characterization of an anti-*Listeria* bacteriocin produced by *Lactobacillus plantarum* LB-B1 isolated from koumiss, a traditionally fermented dairy product from China. *Food control*. 2011;22(7):1027–1031.

[54] Chaves-López C, Serio A, Martuscelli M, Paparella A, Osorio-Cadavid E, Suzzi G. Microbiological characteristics of Kumis, a traditional fermented Colombian milk, with particular emphasis on enterococci population. *Food microbiology*. 2011;28(5):1041–1047.

[55] Montanari G, Zambonelli C, Grazia L, Kamesheva GK, Shigaeva MK. *Saccharomyces unisporus* as the principal alcoholic fermentation microorganism of traditional koumiss. *Journal of dairy research*. 1996;63(02):327–331.

[56] Danova S, Petrov K, Pavlov P, Petrova P. Isolation and characterization of *Lactobacillus* strains involved in koumiss fermentation. *International journal of dairy technology*. 2005;58(2):100–105.

[57] Bornaz S, Guizani N, Sammari J, Allouch W, Sahli A, Attia H. Physicochemical properties of fermented Arabian mares' milk. *International Dairy Journal*. 2010;20(7):500–505.

[58] Kücükcetin A, Yaygin H, Hinrichs J, Kulozik U. Adaptation of bovine milk towards mares milk composition by means of membrane technology for koumiss manufacture. *International dairy journal*. 2003;13(12):945–951.

[59] Sajhiev A, et al. Free amino acids in natural kumiss and kumiss from reconstituted milk. *Voprosy Pitaniya*. 1970;29(4):42–46.

[60] Ishii S, Samejima K. Feeding Rats with Kumiss Suppresses the Serum Cholesterol and Trigluceride Levels. Milk Science. 2001;50(3):113–118.

[61] Fedechko I, Hrytsko R, Herasun B. *The anti-immunodepressive action of kumiss made from cow's milk.* Likars' Ka Sprava/Ministerstvo Okhorony Zdorov'ia Ukraïny. 1995;9-12:104.

[62] Cheng CC, Yu MC, Cheng TC, Sheu DC, Duan KJ, Tai WL. Production of high-content galacto-oligosaccharide by enzyme catalysis and fermentation with Kluyveromyces marxianus. *Biotechnology letters.* 2006;28(11):793–797.

[63] Elizabeth Tymczyszyn E, Gerbino E, Illanes A, Gómez-Zavaglia A. Galacto-oligosaccharides as protective molecules in the preservation of *Lactobacillus delbrueckii subsp.bulgaricus. Cryobiology.* 2011;62(2):123–129.

[64] Kushi R, Monti R, Contiero J. Production, purification and characterization of an extracellular inulinase from *Kluyveromyces marxianus var. bulgaricus. Journal of Industrial Microbiology and Biotechnology.* 2000;25(2):63–69.

[65] Mu Z, Yang X, Yuan H. Detection and identification of wild yeast in Koumiss. *Food microbiology.* 2012.

[66] Chaves-Lopez C, Tofalo R, Serio A, Paparella A, Sacchetti G, Suzzi G. Yeasts from Colombian Kumis as source of peptides with Angiotensin I converting enzyme (ACE) inhibitory activity in milk. *International Journal of Food Microbiology.* 2012.

[67] Seiler H. A review: Yeasts in kefir and kumiss. *Milchwissenschaft.* 2003;58(7-8):392–396.

[68] Dlusskaya E, Jänsch A, Schwab C, Gänzle MG. Microbial and chemical analysis of a kvass fermentation. *European Food Research and Technology.* 2008;227(1):261–266.

[69] Jargin SV. Kvass: a possible contributor to chronic alcoholism in the former Soviet Unionalcohol content should be indicated on labels and in advertising. *Alcohol and alcoholism.* 2009;44(5):529–529.

[70] Abegaz K, Beyene F, Langsrud T, Narvhus J. Indigenous processing methods and raw materials of borde, an Ethiopian traditional fermented beverage. *Journal of Food Technology in Africa*. 2002;7 (2):59–64.

[71] Abegaz K. Isolation, characterization and identification of lactic acid bacteria involved in traditional fermentation of borde, an Ethiopian cereal beverage. *African Journal of Biotechnology*. 2007;6 (12):1469–78.

[72] Tadesse G, Ashenafi M, Ephraim E. Survival of *E. coli* O157:H7, *Staphylococcus aureus, Shigella flexneri* and *Salmonella spp.* in fermenting 'Borde, a traditional Ethiopian beverage. *Food Control*. 2005;16(2):189–196.

[73] Yoon MY, Hwang HJ. Reduction of soybean oligosaccharides and properties of α-d-galactosidase from *Lactobacillus curvatus* R08 and *Leuconostoc mesenteriodes* JK55. *Food microbiology*. 2008;25(6):815–823.

[74] Uysal UD, Onc EM, Berikten D, Yilmaz N, Tuncel NB, Kivanc M, et al. Time and temperature dependent microbiological and mycotoxin (ochratoxin-A) levels in boza. *International Journal of Food Microbiology*. 2009;130(1):43–48.

[75] Todorov SD. Diversity of bacteriocinogenic lactic acid bacteria isolated from boza, a cereal-based fermented beverage from Bulgaria. *Food Control*. 2010;21(7):1011 – 1021.

[76] Yein S, ren A. Biogenic amine content of boza: A traditional cereal-based, fermented Turkish beverage. *Food Chemistry*. 2008;111(4):983 – 987.

[77] Gotcheva V, Pandiella SS, Angelov A, Roshkova ZG, Webb C. Microflora identification of the Bulgarian cereal-based fermented beverage boza. *Process Biochemistry*. 2000;36(12):127 – 130.

[78] Botes A, Todorov SD, Von Mollendorff JW, Botha A, Dicks LM. Identification of lactic acid bacteria and yeast from boza. *Process biochemistry*. 2007;42(2):267–270.

[79] De Cort S, Kumara HS, Verachtert H. Localization and characterization of α-glucosidase activity in Lactobacillus brevis. *Applied and environmental microbiology*. 1994;60(9):3074–3078.

[80] Iqbal S, Nguyen TH, Nguyen TT, Maischberger T, Haltrich D. beta-Galactosidase from *Lactobacillus plantarum* WCFS1: biochemical characterization and formation of prebiotic galacto-oligosaccharides. *Carbohydrate research*. 2010;345(10):1408–1416.

[81] Todorov SD, Dicks LMT. Screening for bacteriocin-producing lactic acid bacteria from boza, a traditional cereal beverage from Bulgaria. *Process Biochemistry*. 2006;41(1):11–19.

[82] Zorba M, Hancioglu O, Genc M, Karapinar M, Ova G. The use of starter cultures in the fermentation of boza, a traditional Turkish beverage. *Process Biochemistry*. 2003;38(10):1405–1411.

[83] Gen M, Zorba M, Ova G. Determination of rheological properties of boza by using physical and sensory analysis. *Journal of Food Engineering*. 2002;52(1):95 – 98.

[84] Escalante A, Rodrguez ME, Martnez A, Lpez-Mungua A, Bolvar F, Gosset G. Characterization of bacterial diversity in Pulque, a traditional Mexican alcoholic fermented beverage, as determined by 16S rDNA analysis. {*FEMS*} *Microbiology Letters*. 2004;235(2):273 – 279.

[85] Ortiz-Basurto RI, Pourcelly G, Doco T, Williams P, Dornier M, Belleville MP. Analysis of the main components of the aguamiel produced by the maguey-pulquero (*Agave mapisaga*) throughout the harvest period. *Journal of agricultural and food chemistry*. 2008;56(10):3682–3687.

[86] Escalante A, Elena Rodríguez M, Martínez A, López-Munguía A, Bolívar F, Gosset G. Characterization of bacterial diversity in Pulque, a traditional Mexican alcoholic fermented beverage, as determined by 16S rDNA analysis. *FEMS microbiology letters*. 2004;235(2):273–279.

[87] Ortiz-Basurto RI, Williams P, Belleville MP, Doco T. Presence of rhamnogalacturonan II in the juices produced by enzymatic liquefaction of Agave pulquero stem (*Agave mapisaga*). *Carbohydrate Polymers*. 2009;77(4):870–875.

[88] Escalante A, Giles-Gómez M, Hernández G, Córdova-Aguilar MS, López-Munguía A, Gosset G, et al. Analysis of bacterial community during

the fermentation of pulque, a traditional Mexican alcoholic beverage, using a polyphasic approach. *International journal of food microbiology*. 2008;124(2):126–134.

[89] Arrizon J, Morel S, Gschaedler A, Monsan P. Fructanase and fructosyltransferase activity of non-*Saccharomyces* yeasts isolated from fermenting musts of Mezcal. *Bioresource Technology*. 2012;110:560–565.

[90] Vidal RF, Martínez A, Moulis C, Escalier P, Morel S, Remaud-Simeon M, et al. A novel dextransucrase is produced by *Leuconostoc citreum* strain B/110-1-2: an isolate used for the industrial production of dextran and dextran derivatives. *Journal of industrial microbiology & biotechnology*. 2011;38(9):1499–1506.

[91] Davidson JR, De Montellano BRO. The antibacterial properties of an Aztec wound remedy. *Journal of ethnopharmacology*. 1983;8(2):149–161.

[92] Moinas M, Horisberger M, Bauer H. The structural organization of the Tibi grain as revealed by light, scanning and transmission microscopy. *Archives of Microbiology*. 1980;128:157–161.

[93] Horisberger M. Structure of the dextran of the tibi grain. *Carbohydrate Research*. 1969;10(3):379–385.

[94] Diniz R, Garla L, Schneedorf J, Carvalho J. Study of anti-inflammatory activity of Tibetan mushroom, a symbiotic culture of bacteria and fungi encapsulated into a polysaccharide matrix. *Pharmacological research*. 2003;47(1):49–52.

[95] Ward HM. The'Ginger-Beer Plant,'and the Organisms Composing It: A Contribution to the Study of Fermentation-Yeasts and Bacteria. *Proceedings of the Royal Society of London*. 1891;50(302-307):261–265.

[96] Teramoto Y, Hano T, Ueda S. Production and characteristics of traditional alcoholic beverage made with sweet potato as the saccharifying agent. *Journal of the Institute of Brewing*. 1998;104(6):339–341.

Index

A

absorption spectra, 189, 191
absorption spectroscopy, 231
access, xii, 55, 92, 116, 144, 146, 147, 218
accounting, 269
acetaminophen, 226
acetic acid, xiv, 29, 234, 242
acetone, 90
acetonitrile, 181
acetylation, 178
ACF, 25
acid, xii, xiii, xiv, 5, 10, 22, 24, 26, 27, 29, 33, 37, 40, 42, 43, 46, 62, 71, 75, 89, 92, 95, 96, 99, 103, 105, 108, 109, 112, 120, 121, 144, 145, 149, 153, 155, 156, 160, 169, 170, 171, 173, 175, 178, 179, 181, 182, 191, 197, 198, 201, 202, 205, 207, 208, 217, 220, 225, 234, 241, 243, 245, 255, 257, 261, 274, 276
acidity, 58
acrylate, 7
acute idiopathic polyneuritis, 280
adaptation(s), 254, 258
ADC, 180
additives, 246
adenocarcinoma, 26
adenoma, 27, 38, 76
adhesion, xi, 22, 27, 86, 100, 101, 107, 111, 115, 116, 119, 123, 124, 126, 131, 135
adhesive properties, xi, 85, 119
adipose tissue, 76
adolescents, x, 15, 16, 46, 49
adsorption, xiii, 173, 177, 180
adults, x, 16, 20, 27, 29, 49, 52, 53, 72, 75, 76, 100, 105, 236
advancement(s), xv, 13, 268, 279, 280
adverse effects, xiv, 13, 112, 251, 262
aerosols, 126
aetiology, 103
AFB1, 216
affirming, 14
Africa, 47
African-American, 147, 148, 155, 164
age, xiv, 22, 53, 102, 107, 127, 128, 131, 132, 133, 145, 146, 147, 148, 165, 252, 260, 262, 263, 269, 272
agriculture, 3, 82
AIDS, 283
albumin, 145, 151
alcohols, 61, 67, 90, 214, 224
algae, x, 2
allergic asthma, 21, 47
allergic rhinitis, 132
allergy, 21, 32, 107, 132, 239
alters, 33, 112
AME, 260
American Heart Association, 40
amine(s), 9, 95, 103, 224
amino acid(s), 59, 81, 213, 224, 234, 238, 240, 241, 242, 245, 254, 275
ammonia, 103, 244

ammonium, 180
amylase, 5, 67, 75, 77, 88, 91, 92
anaerobic bacteria, 24
anchoring, 272
antibiotic, 20, 22, 35, 90, 92, 102, 124, 129, 132
antibiotic resistance, 124
antibody, xv, 122, 235, 267, 268, 270, 271, 274, 278, 279, 280, 281, 282, 283
antigenicity, 273
anti-inflammatory drugs, 226
antioxidant, xiii, xiv, 22, 174, 179, 185, 204, 205, 234, 238
antioxidative potential, 180
antitumor, 26, 243
apoptosis, 6, 24, 26, 27, 32, 33, 36, 37, 38, 39, 43, 103, 175
appetite, 60, 74
aqueous solutions, 212
arabinogalactan, 182, 254
arrest, 26, 27, 28, 39
artery, 108
arthritis, 52, 245
ascending colon, 65
ascorbic acid, 62
Asia, 45, 47
Asian countries, 269, 279
asparagus, 65
aspirate, 164, 169
assessment, 12, 46, 91, 92, 94
asthma, 21, 132
astringent, 241
asymptomatic, 124, 165
ataxia, xv, 267, 268, 269, 270, 271, 276, 277, 280, 281, 282, 285, 286
atherosclerosis, 236
atmospheric pressure, 176
atoms, 2
atopic dermatitis, 107
atopic eczema, 107, 132
attachment, 22, 117, 118, 120, 122, 126, 129, 134, 136, 140
autoimmune diseases, 140
autolysis, 260
awareness, x, 2

B

bacillus, 273
Bacillus subtilis, 244
bacteremia, 122
bacteria, x, xiv, xv, 2, 5, 6, 7, 9, 10, 11, 14, 19, 21, 22, 23, 24, 25, 30, 31, 36, 39, 40, 42, 43, 45, 59, 65, 67, 75, 87, 88, 91, 100, 101, 102, 117, 119, 120, 121, 122, 126, 129, 140, 175, 205, 233, 236, 240, 241, 242, 243, 244, 255, 267, 268, 272, 273, 278, 280
bacterial fermentation, 6, 23, 103, 138, 242
bacterial infection, 118, 124, 272
bacterial pathogens, 127, 129
bacterial strains, 7, 87, 100, 101, 102, 103, 272, 284
bacteriocins, 102
bacterium, 124, 273, 274, 278, 284
base, 149, 274
behaviors, 50
beneficial effect, 4, 5, 16, 17, 19, 21, 105, 107, 112, 119, 129, 130, 175, 236, 257
beneficial microbes, 61
benefits, ix, 1, 3, 23, 25, 29, 31, 43, 57, 58, 80, 102, 107, 111, 112, 113, 116, 234, 245
benzene, 221
beryllium, 225
beverages, ix, xv
bile, 26, 27, 43, 255
bile acids, 26, 43, 255
bioavailability, 15, 104, 212, 240
biochemical processes, 8
biochemistry, 37, 44, 45
biodegradation, 174
biodiesel, 252, 263
biological activities, 93, 239
biological activity, 9, 119
biological fluids, 96, 222, 228, 229
biological markers, 27
biological processes, 240
biological responses, 93
biological roles, ix
biological systems, 96

biomarkers, 145, 169
biomass, 240
biosynthesis, 274, 275, 285
biotechnology, 37, 46, 80
biotin, 3
birds, xiv, 251, 258, 261, 262
blends, 242
blood, 20, 35, 45, 46, 48, 77, 103, 122, 124, 127, 166, 222, 236, 245
blood circulation, 127
blood group, 45
blood stream, 124
blood vessels, 245
blood-brain barrier, 124
body fat, 51
body mass index (BMI), 51, 53, 55, 76
body weight, 17
bonds, 2, 5, 87, 130, 212, 256, 258
bone(s), 15, 16, 18, 43, 45, 48, 80, 104, 235
bone form, 16
bone mass, 15, 104
bone mineral content, 16, 80
bone resorption, 16, 48
bowel, 5, 10, 13, 25, 33, 36, 38, 41, 46, 60, 68, 73, 74, 104, 240
brain, 245, 276
brainstem, 270, 271, 280, 281, 282
branching, 34, 146, 147, 165
breakdown, 239, 256
breast cancer, ix, xii, 144, 145, 146, 147, 148, 155, 159, 160, 161, 162, 163, 164, 165, 166, 167, 168, 169, 170, 171, 235, 238
breast milk, 12, 19, 105, 128, 132, 238
breastfeeding, 116, 130, 131, 133, 135, 136, 141
breathing, 52
brevis, 243
bronchial epithelial cells, 126
bronchus, 122
building blocks, 223
by-products, ix, 204, 207

C

cadmium, 221, 230
caecum, 48
calcium, 4, 13, 15, 16, 17, 34, 38, 43, 45, 46, 48, 103, 112, 175, 235, 241, 243
calibration, 154, 180, 181, 189, 213, 216, 228
caloric intake, 50, 61, 68
calorie, 56, 67
cancer, xii, 24, 26, 31, 33, 36, 38, 42, 52, 60, 64, 78, 103, 144, 145, 146, 147, 155, 159, 160, 161, 164, 165, 166, 167, 168, 169, 171, 235, 249
cancer cells, 26, 33
cancer progression, 145, 168
cancer therapy, 145, 147, 164
capillary, 70, 95, 178, 181, 212, 222, 223, 224, 227
capsule, 229
carbohydrate(s), ix, x, xi, xiii, xv, 2, 5, 10, 13, 23, 24, 25, 49, 50, 59, 61, 62, 64, 67, 70, 71, 72, 73, 76, 77, 81, 85, 87, 91, 94, 96, 103, 105, 109, 112, 115, 116, 118, 120, 125, 129, 139, 140, 145, 177, 178, 182, 183, 189, 191, 207, 233, 234, 236, 237, 241, 242, 253, 254, 255
carbon, xv, 10, 87, 97, 110
carboxyl, 95
carcinogenesis, 26, 33, 42
carcinoma, 97, 110, 170
cardiovascular disease(s), 4, 17, 40, 50, 103, 245
cardiovascular risk, 103
cascades, 27
case study, 111, 129
casein, 239, 242
catabolism, 103
cation, 226
causal relationship, 63
CD8+, 21
CDC, 51, 74
cecum, 9, 65, 253
cell culture, 25, 35, 37
cell cycle, 24, 26, 27, 39

cell death, 101
cell differentiation, 24
cell invasion, 284, 285
cell line(s), 24, 32, 36, 97, 103, 110, 111, 126
cell membranes, xv, 22, 256, 268, 273, 279
cell surface, xi, xv, 23, 101, 115, 117, 118, 120, 126, 127, 128, 267, 272, 280
cellular immunity, 24
cellulose, xiii, 62, 175, 177, 209, 254, 258
census, 53
central nervous system, 271
ceramide, 276
cerebellum, 277
cerebrospinal fluid, 268, 269, 270, 281
challenges, 72
cheese, 29, 43
chemical, 62, 64, 70, 71, 93, 119, 133, 177, 204, 212, 217, 234
chemical properties, 212
chemical reactions, 212
chemical structures, 119, 134
chemotherapy, 164, 168
chicken, 22, 254
chicken pox, 283
childhood, 41, 56, 116, 139
children, x, 19, 20, 49, 55, 106, 107, 122, 124, 127, 128, 140, 238, 246
chitosan, 9, 39, 45
chloroform, 182
cholera, 100, 101
cholesterol, xiii, 6, 29, 32, 61, 63, 65, 72, 80, 103, 175, 233, 234, 238, 241, 244, 255
choline, 48
chromatography, xiii, 8, 45, 70, 71, 95, 96, 133, 151, 153, 169, 173, 180, 183, 185, 186, 188, 189, 195, 198, 207, 276
chronic diseases, 18, 31, 61, 80, 236
circulation, x, xi, 85, 86, 87, 93, 97, 99, 103, 107, 126
classes, 30, 72, 86, 88, 275
cleavage, 87, 90, 97
clinical disorders, 16
clinical oncology, 38, 171
clinical syndrome, 284
clinical trials, 20, 106
closure, x, 50
CO2, 24, 29, 102
coding, 22
colon, ix, 2, 3, 5, 6, 7, 8, 9, 10, 14, 17, 19, 21, 23, 24, 25, 26, 27, 31, 32, 35, 36, 37, 38, 39, 40, 42, 43, 48, 78, 82, 103, 104, 110, 175, 253
colon cancer, 6, 25, 26, 27, 32, 35, 38, 39, 42, 43, 48, 103, 175
colon carcinogenesis, 25, 32
colonization, xi, 19, 22, 23, 41, 57, 65, 115, 118, 120, 122, 127, 128, 137, 236
colorectal cancer, 9, 68, 76, 80, 235
colostrum, 100, 106, 111, 117, 134, 136, 138
combustion, 174
commercial, 27, 30, 33, 78, 97, 100, 130, 131, 133, 219, 243
commodity, 260, 261, 262
community, 4, 37, 123
compatibility, 212
competitive advantage, 10
compilation, ix
complement, 285
complex carbohydrates, 63, 254
complexity, 10, 90, 94, 106, 130, 254
composition, ix, 2, 3, 4, 8, 10, 13, 14, 15, 38, 57, 58, 61, 62, 73, 81, 82, 91, 104, 107, 117, 119, 121, 131, 154, 157, 158, 165, 186, 189, 195, 197, 206, 229, 237, 253, 254, 259, 260, 285
compounds, ix, x, xiii, xv, 2, 9, 19, 26, 43, 45, 49, 50, 52, 57, 61, 64, 65, 66, 70, 95, 98, 123, 174, 175, 180, 183, 189, 204, 205, 207, 209, 212, 227, 239, 240, 242
comprehension, xv, 268, 279
computational modeling, 224
condensation, 95
configuration, 2, 5, 282
conflict, 90
consciousness, 271
consensus, 31, 46
consent, 147

constipation, 12, 14, 34, 104
constituents, 19, 157, 242, 260
consumers, x, 2, 4, 30, 100, 238, 241, 242
consumption, x, xiii, 3, 14, 22, 27, 28, 32, 41, 50, 52, 56, 62, 68, 73, 77, 78, 79, 94, 100, 104, 107, 137, 140, 233, 235, 239, 248, 262
control group, 8, 16, 19, 20, 26, 127, 132, 146, 164
controversial, 277
convention, 228
cooking, 51, 64, 66, 71, 72, 82, 237
cooling, 64, 68, 70, 82
copper, 221, 230, 231, 241
correlation, 146, 147, 155, 164, 165, 168, 184, 186
correlation coefficient, 155
correlations, 257, 272
cosmetic(s), 3, 22, 227
cost, 52, 214, 234, 252
covalent bond, 212
cranial nerve, xv, 267, 276, 285
critical period, 19
crop(s), 15, 50, 51, 52, 55, 56, 62, 234
crown, 224
crystallinity, 67, 68
CSF, 283
cues, 56
cultivars, 70, 74, 81, 176
culture, xiv, 8, 18, 24, 27, 40, 44, 67, 80, 91, 97, 102, 110, 123, 234, 240, 241, 245
culture conditions, 98
cure, 124
cyanide, 222
cycles, 70, 82
cyclodextrins, ix, xiii, 209, 210, 214, 219, 220, 221, 222, 223, 224, 228, 229, 230
cystic fibrosis, 124
cytokines, 9, 20, 22, 23, 45, 257
cytomegalovirus, 271
cytometry, 26, 27
cytotoxicity, 26

D

daily living, 55
deaths, x, 49, 52
decay, 99
decoration, 96
defence, 129
deficiency, 17, 45, 105
degradation, 5, 23, 58, 65, 68, 87, 92, 103, 130, 206, 210, 240
demonstrations, 70
Denmark, 149
deposition, 14
depression, 52
depth, 211
derivatives, ix, 122, 125, 175, 198, 219, 222
dermatitis, 107
desorption, 45, 70, 97, 140, 151
detectable, 66
detection, ix, 12, 70, 90, 95, 96, 97, 99, 127, 150, 153, 213, 214, 215, 216, 217, 219, 220, 221, 228, 270
developed countries, xii, 116, 119
developing countries, 54, 56, 116, 234
deviation, 183, 186
diabetes, 4, 16, 52, 76, 82
dialysis, 96, 167
diarrhea, 9, 60, 111, 138, 235, 257
diet, x, xiv, 2, 3, 4, 6, 8, 16, 17, 18, 19, 21, 27, 29, 31, 35, 37, 40, 41, 49, 50, 56, 63, 72, 77, 80, 82, 86, 93, 100, 102, 103, 105, 234, 244, 252, 256, 259
diet composition, 50, 56
dietary fiber, x, 2, 29, 43, 63, 68, 235, 237, 238, 241, 253
dietary supplementation, 25
diffusion, 18, 93, 98, 99
digestibility, ix, xiv, 65, 67, 74, 75, 76, 77, 81, 206, 252, 253, 254, 255, 256, 258, 259, 260, 261, 263
digestion, xiii, 30, 37, 45, 74, 76, 79, 82, 87, 89, 90, 91, 92, 103, 107, 108, 209, 236, 253, 257
digestive enzymes, 2, 4, 23, 67, 70, 91, 255, 256

diplopia, 270
directionality, 98
disability, 52
discomfort, 14, 100, 236
disease progression, 146, 164
diseases, x, 3, 18, 37, 47, 49, 50, 51, 52, 60, 81, 116, 117, 119, 127, 130, 133, 135, 139, 169, 235, 237
disorder, 236, 268, 269
displacement, 56, 67
dissociation, 268, 269, 270
distribution, 10, 12, 26, 53, 76, 97, 281
diversity, 106, 135, 272
DNA, 6, 74, 275
DNA damage, 27
dominance, 278
donors, 146, 155, 164
dosage, 228
double blind study, 15
dough, 29, 36
drawing, 91
drug discovery, 109
drug metabolism, 170
drug therapy, 146
drugs, xii, 133, 144, 145, 146, 167, 170, 213, 222
dry matter, 62, 236
drying, 245

E

ecology, 35, 40, 57, 80
ecosystem, 10, 25, 255
eczema, 107
editors, 285
effluent, 90
electrical resistance, 98
electrochemical impedance, xvi
electron, 181
electrophoresis, 47, 70, 109, 212, 222, 223, 224, 227
electrophoretic separation, 224
elephants, 137
elucidation, 222
emission, 152, 153, 167, 213, 214, 215, 221
enantiomers, 216, 222, 224, 228
encephalitis, 270, 271, 280, 281, 282, 286
endocrinology, 40
endothelial cells, 39, 135
energy, x, xiv, 6, 10, 24, 33, 49, 50, 51, 52, 55, 56, 57, 58, 62, 67, 73, 77, 78, 81, 87, 94, 98, 103, 108, 181, 251, 252, 253, 254, 257, 258, 260, 263
energy density, 252
energy expenditure, 55
energy supply, 55
England, 150
enlargement, 18, 218
enteritis, 275, 285, 286
enterovirus, 271, 283
entrapment, 68
environment(s), xiv, 6, 8, 51, 64, 65, 81, 87, 88, 174, 234, 242, 255, 257, 272
environmental protection, 212
environmental regulations, 263
enzymatic activity, 89, 240, 275
enzyme inhibitors, 68
enzyme(s), xv, 5, 6, 10, 24, 26, 30, 33, 36, 58, 61, 65, 67, 68, 70, 71, 75, 87, 88, 89, 90, 91, 92, 155, 208, 236, 238, 239, 240, 242, 243, 255, 259, 260, 275
epidemic, x, 49, 51, 55, 56
epidemiology, 43, 57, 136
epithelia, 23, 105, 118, 124, 126, 129
epithelial cells, 6, 22, 23, 27, 28, 39, 101, 106, 120, 121, 122, 126, 128, 134, 139, 140, 175, 256
epithelium, 6, 14, 22, 57, 93, 97, 98, 100, 103, 123, 126, 140, 239
Epstein-Barr virus, 271, 283
equilibrium, xv, 8
equipment, 94, 97
ESI, xiii, 97, 174, 181, 189, 193, 194, 199
essential fatty acids, x, 49, 50, 234
ester, 22, 62
ethanol, xiv, 59, 151, 154, 155, 241, 251, 258, 259
ethnic groups, 53, 165
ethyl acetate, 177, 182
etiology, 50

EU, 32
eukaryotic, 123
eukaryotic cell, 123
Europe, 37
evacuation, 104
evaporation, 177
evidence, x, xi, 4, 15, 85, 86, 87, 88, 89, 92, 97, 98, 100, 101, 103, 110, 112, 116, 123, 136, 165, 278
evolution, 57
excitation, 152, 153, 213, 214, 215, 217, 221
exclusion, xiii, 11, 98, 173, 180, 188, 189, 195
excretion, 16, 17, 27, 32, 57, 82, 99, 108, 109, 139, 240
exploitation, 34
exposure, 28
extraction, xiv, 174, 177, 182, 206, 208, 223, 241, 251, 258, 259, 263
extracts, 21, 31, 89, 208
extraocular muscles, 281
extrusion, 29, 41

F

facial palsy, 270
families, xi, 86
famine, 52
fat, xiv, 13, 29, 30, 48, 51, 62, 63, 73, 80, 235, 237, 243, 252, 262
fatty acids, 2, 6, 23, 24, 33, 35, 40, 41, 45, 58, 102, 103, 254, 257
FDA, 235, 241, 244
feces, 12, 27, 47, 59, 135, 255
female rat, 269
femur, 17
fermentable carbohydrates, ix, 2, 17
fermentable fiber, ix, 1
fermentation, ix, xiv, 2, 4, 7, 8, 14, 15, 19, 23, 24, 25, 38, 41, 48, 59, 65, 66, 73, 74, 78, 87, 90, 92, 102, 103, 233, 236, 239, 240, 241, 243, 244, 245, 248, 254, 255, 257
ferritin, 117
fever, 271
fiber(s), ix, xiii, 1, 2, 28, 29, 30, 36, 61, 62, 63, 68, 75, 76, 174, 178, 204, 205, 234, 237, 241, 257, 263, 277
fibrosis, 124
filtration, xiii, 96, 173, 176, 177, 178, 182, 185, 195, 238
financial, 252
financial support, 205
first generation, 4
flame, 70, 221, 231
flatulence, xiv, 14, 100, 233, 235, 236, 239, 241
flavonoids, 62, 245
flavo(u)r, xiii, xiv, 30, 233, 234, 235, 238, 239, 241
flight, 70, 97, 134
flora, 32, 33, 40, 41, 50, 77, 107
flour, 29, 30, 37, 71
fluid, 21, 29, 41, 88, 91, 164, 169, 219, 220, 257, 269, 271
fluidized bed, 235
fluorescence, ix, xii, xiii, 144, 149, 152, 153, 161, 167, 169, 170, 209, 212, 213, 214, 215, 216, 217, 219, 220, 222, 224, 225, 227, 228, 230
fluoxetine, 219, 229
fluvoxamine, 229
folate, 3, 63
food, ix, x, 2, 4, 16, 18, 19, 21, 22, 28, 29, 31, 32, 39, 42, 43, 45, 46, 48, 49, 50, 51, 52, 55, 56, 58, 62, 63, 65, 66, 67, 68, 70, 72, 73, 75, 77, 80, 81, 88, 100, 107, 108, 116, 133, 137, 175, 223, 234, 237, 241, 243, 244, 245, 249
food additive, 4, 73
food additives, 4
food industry, 100
food intake, 16
food products, 4, 18, 28, 29, 56, 88, 234, 245
food safety, 46
food security, 72
force, 17

formation, 2, 40, 42, 68, 77, 81, 82, 87, 135, 197, 198, 212, 213, 214, 216, 217, 219, 228, 238
formula, xii, 12, 19, 20, 31, 32, 33, 36, 40, 46, 100, 105, 106, 112, 113, 116, 117, 118, 119, 127, 128, 130, 131, 132, 134, 135, 138, 139, 239
fractures, 104
fragments, 98, 197, 198
free oligosaccharides, xi, 20, 101, 115, 122
free radicals, 205, 235
fructose, 64, 89, 130, 131
fruits, xv, 78
functional food, xi, xiv, 4, 12, 14, 18, 19, 33, 48, 72, 86, 133, 234, 236, 241, 242, 246
functional oligosaccharides, x, 2, 3, 110, 294
funding, 134
fungi, x, 2, 101, 236, 240
fungus, 245

G

galactosidase supplementation, xiv, 251, 260
gallium, 225
ganglion, 276
gastroenteritis, 100, 106, 273
gastrointestinal tract, ix, xiv, 4, 19, 22, 25, 31, 37, 40, 57, 58, 67, 80, 93, 108, 119, 136, 137, 234, 235, 237, 241, 242, 254, 255, 256
GC-FID, 70
gel, 47, 96, 185, 189, 195, 225
gene expression, 6, 33, 48, 257
gene promoter, 38
general colonic health, ix, 1
genes, 22, 23, 37, 39, 101, 275, 285
genetic diversity, x, 50
genetic traits, 263
genetics, 51, 103
genotype, 70, 71, 81, 255
genus, 58, 236
gland, 105
globalization, 3
glucagon, 60
glucoamylase, 5
gluconeogenesis, 6
glucose, xiii, 39, 46, 51, 60, 61, 62, 64, 65, 67, 74, 77, 89, 90, 105, 119, 124, 130, 131, 174, 180, 183, 197, 198, 210, 236, 239, 240, 255
glucose-xylose, xiii, 174
glucosidases, 46, 240
glucoside, 198, 240
glutathione, 38
glycans, 105, 117, 120, 137, 138, 165
glycerol, 62
glycogen, 51
glycol, 149
glycolytic metabolism, xi, 85
glycoprotein, ix, xii, 144, 145, 152, 154, 169, 170, 171, 314
glycoproteins, 22, 101, 126
glycosylation, ix, xii, 144, 145, 147, 160, 165, 166, 168, 169, 171
goiter, 238
grades, 160
granules, 83
graph, 213, 216
green revolution, 55, 56
Green Revolution, 75
growth, ix, x, xi, xiv, 2, 6, 7, 9, 10, 11, 12, 13, 14, 15, 18, 23, 24, 25, 29, 36, 38, 39, 86, 102, 103, 105, 116, 123, 136, 175, 233, 235, 240, 241, 242, 245, 252, 257, 258, 260, 262
growth arrest, 39
growth factor, 23, 136
growth rate, 9, 260
guidelines, 44, 75
gut microbiota, ix, 2, 4, 6, 32, 38, 57, 58, 59, 73, 75, 130, 332

H

harmful effects, 174
healing, 22

health, ix, x, xi, xv, 1, 2, 3, 4, 5, 6, 7, 10, 12, 18, 22, 23, 25, 28, 29, 31, 33, 34, 35, 38, 39, 40, 42, 43, 44, 46, 49, 52, 57, 58, 61, 64, 65, 68, 72, 76, 78, 80, 86, 87, 88, 100, 102, 107, 111, 112, 116, 130, 137, 170, 175, 206, 234, 238, 239, 241, 244, 245, 249, 257
health care, x, 44, 49
health effects, 61, 72, 257
health status, 242
heart disease, 52, 64, 235, 238, 244
height, 51, 53
Helicobacter pylori, 101, 118, 175, 271, 283
helium, 178
hemicellulose, 62, 258
hemorrhoids, 237
heterogeneity, 145, 171
hexane, 263
high blood pressure, 245
history, 132
HIV, 283
HM, 112
hormone(s), 23, 62, 79, 87, 170, 245
hospitalization, 117, 134
host, xi, xiii, 2, 4, 6, 7, 11, 22, 23, 24, 34, 47, 57, 58, 61, 102, 115, 118, 123, 126, 138, 209, 210, 211, 212, 213, 214, 215, 217, 220, 224, 225, 226, 228, 229, 230, 272, 273, 283
human body, 51, 102
human health, ix, 2, 16, 18, 50, 51, 102, 236
human immunodeficiency virus, 271
human leukemia cells, 175
human milk, ix, xi, 10, 13, 20, 28, 39, 41, 45, 87, 88, 89, 90, 98, 100, 101, 105, 106, 110, 111, 112, 113, 115, 116, 117, 119, 122, 124, 125, 128, 130, 134, 135, 136, 137, 138, 139, 140
human subjects, 15, 33, 78, 89, 90, 91, 92, 93
humidity, 14
Hunter, 76
hydrogen, 109, 113, 257
hydrolysis, x, xi, xiii, 4, 24, 30, 31, 67, 68, 70, 71, 77, 81, 85, 86, 87, 88, 89, 90, 91, 92, 93, 99, 102, 107, 136, 156, 173, 175, 177, 178, 182, 196, 204, 235, 239, 240, 259
hydroxyl groups, 71, 95, 96, 211, 212
hyperactivity, 21
hypersensitivity, 104, 107
hypertension, 4, 50
hypertrophy, 18
hypothesis, 18, 20, 23, 103, 128, 131, 132, 146

I

ibuprofen, 214, 216, 218, 226, 227, 228
ideal, 13, 51, 92, 238, 242
identification, 75, 97, 99, 123, 198, 207
IFN, 21, 23
IL-13, 21
ileostomy, 71, 82, 90
ileum, 18, 21, 38, 90, 257
immune defense, 272
immune function, 3, 19, 31, 38, 47
immune modulation, 19
immune response, 20, 21, 22, 133, 236, 268, 273
immune system, ix, 1, 7, 10, 19, 20, 21, 23, 37, 41, 103, 105, 113, 117, 129, 132, 133, 138, 139
immunity, xv, 20, 22, 25, 40, 131, 140
immunization, 273
immunoglobulin, xv, 19, 21, 41, 122, 267, 268, 270, 278
immunomodulation, 19, 23, 47, 133
immunomodulatory, 21, 22, 23, 34, 110, 112, 130, 136
immunomodulatory agents, 34, 112
immunostimulatory, 22, 207
immunotherapy, 270
improvements, 257
impurities, 227
in situ hybridization, 12
in transition, 47
in vitro, xii, 4, 7, 20, 35, 44, 67, 72, 74, 76, 79, 80, 81, 92, 94, 100, 101, 108, 111,

116, 117, 119, 122, 125, 126, 128, 136, 183, 206, 257, 258
in vivo, xii, 8, 19, 24, 40, 71, 72, 77, 91, 92, 93, 99, 108, 118, 128, 139, 144, 146, 258
incidence, 20, 64, 100, 106, 107, 117, 130, 132, 133, 146, 269, 272, 279
income, 52
India, 233
indirect effect, 129, 133
individuals, xii, 3, 8, 14, 15, 31, 51, 57, 68, 73, 92, 104, 144, 147, 148, 161, 162, 164, 165, 166, 167
induction, 21, 24, 37, 38, 46, 61, 91, 212
industrialization, 3
industrialized countries, 116, 135
industry(ies), x, 2, 22, 31. 175, 212, 245, 246, 252
infancy, 21, 116, 134, 135
infants, ix, xi, 10, 12, 13, 19, 20, 31, 32, 36, 38, 40, 43, 47, 100, 102, 105, 106, 107, 111, 113, 115, 116, 117, 119, 120, 122, 126, 127, 128, 131, 132, 134, 135, 138, 139, 140, 239
infection, xi, 9, 10, 115, 118, 121, 126, 128, 133, 137, 140, 175, 270, 272, 277, 278, 279, 283
infectious agents, 271
inflammation, xii, 39, 58, 144, 165, 257
inflammatory bowel disease, 74
inflammatory cells, 21
inflammatory disease, 22, 165
inflammatory mediators, 20, 22
inflammatory responses, 20
influenza, 272, 283
ingestion, 15, 29, 32, 65, 73, 236
ingredients, x, 2, 4, 12, 25, 28, 31, 42, 78, 82, 100, 106, 130, 223, 234, 252, 253
inhibition, xv, 8, 24, 25, 33, 122, 124, 126, 128
inhibitor, 146, 171, 261, 263
injections, 219
injury, 285
innate immunity, 22
insulin, 51, 68, 77
insulin resistance, 80
insulin sensitivity, 68, 77
integration, 175
integrin, 135
integrity, 272
interference, 94, 118
internalization, 109
intervention, xi, 12, 22, 25, 107, 115, 119, 127, 128, 130, 132, 133
intestinal flora, 106, 107
intestinal tract, 107, 241
intestine, 3, 5, 7, 18, 36, 57, 66, 73, 79, 89, 90, 92, 103, 236, 255
ionization, 45, 70, 140, 181, 189
ions, 103, 190, 191, 197, 198, 203
Ireland, 115, 134
iron, 34, 63, 81, 103, 235, 238, 241, 243
irritable bowel syndrome, 104, 107, 112
isoflavone, xiv, 234, 240, 241, 248
isolation, xiii, 173, 204, 207
isoleucine, 260
isomers, 106
issues, xiii, 41, 64, 118, 144, 169, 238
Italy, 281

J

Japan, 149, 243, 245, 248, 269
jejunum, 21
Jordan, 146, 170

K

kidney, 39, 40, 69
Korea, 243, 244, 267
koumiss, xv

L

labeling, 225
lactase, 5, 90, 92
lactate dehydrogenase, 7
lactation, 134, 137, 138
lactic acid, xiv, 9, 36, 39, 233, 236, 240, 241, 242, 243, 248

Lactobacillus, xiv, xvi, 9, 10, 11, 12, 36, 39, 42, 47, 58, 59, 88, 136, 234, 236, 239, 240, 242, 245
lactose, xiii, 10, 31, 38, 66, 90, 102, 105, 117, 118, 120, 121, 130, 139, 233, 234, 238, 242
lactose intolerance, 239
large intestine, 4, 7, 14, 19, 45, 48, 50, 57, 61, 62, 65, 66, 82, 104, 175, 239, 257
LC-MS/MS, 110
LDCs, 61
LDL, 72, 235
leaching, 66, 239
lead, xiii, xiv, 5, 6, 27, 51, 56, 71, 144, 168, 169, 209, 233, 255, 258, 274
lecithin, 239
legionella, 126
legume, x, 49, 50, 68, 77, 82
lesions, 25, 35, 78
ligand, 225
lignin, 62
limb weakness, 270, 276
linoleic acid, 29, 62, 245, 252
lipid metabolism, 75, 103
lipid oxidation, 242
lipids, 5, 60, 62, 68, 105, 113, 117, 170, 237, 255, 276
lipodystrophy, 80
liquid chromatography, 70, 75, 97, 109, 110, 134, 136, 224
liquid phase, 174, 176, 177, 182
liquids, 14
Listeria monocytogenes, 100
liver, 6, 242
liver disease, 169
localization, 285
loci, 285
locus, 274, 275
low-density lipoprotein, 29
lumen, xi, 15, 18, 26, 86, 89, 93, 94, 104, 255, 256, 257
lymph node, 146
lymphocytes, 20
lymphoid tissue, 42, 257
lysine, 245, 260

M

machinery, 65
macromolecules, 39
macrophages, 9, 19
magnesium, 15, 16, 34, 35, 45, 48, 103, 112, 241, 243
majority, 65, 66, 90, 145, 252, 277
malignancy, 145
malnutrition, 55
maltose, 89, 180
mammals, 235
man, 73, 82
management, 6, 174, 281
mannitol, 61, 67, 78
manufacturing, 25, 29, 40, 243
marketing, 58
Maryland, 143, 147
mass, 14, 33, 45, 51, 70, 76, 96, 97, 99, 109, 134, 138, 140, 181, 189, 193, 194, 197, 198, 199, 208, 210, 283
mass spectrometry, 45, 70, 97, 134, 138, 189, 198, 208, 283
materials, 91, 242
matrix, 29, 45, 62, 68, 94, 95, 97, 103, 140
matter, 261
MB, 108, 111
MCP, 98
measurement(s), 17, 51, 74, 91, 94, 96, 170, 178, 179, 182, 207, 214, 215
meat, 262
mechanical stress, 92
media, 67, 116, 117, 129, 224, 225, 226, 230
median, 270, 271
medical, 14, 52, 246
medical care, 249
medicine, 22, 35
Mediterranean, 35, 44
melatonin, 219, 229
membranes, xv, 167, 177, 243, 267, 268, 272, 276, 280
meninges, 137
meningitis, 118, 124
menopause, 239

metabolic, 40, 43, 44, 60, 85
metabolic disorders, 75
metabolic pathways, 87
metabolic syndrome, 61, 77
metabolism, ix, xi, 2, 6, 16, 34, 36, 37, 40, 42, 44, 51, 80, 85, 91, 102, 103, 108, 240, 257
metabolites, 7, 8, 19, 58, 147, 171, 240
metabolized, 6, 254, 257
metal ion(s), 103, 107, 213, 222
methanol, 177, 178, 182, 183, 185, 186
methylation, 6
mice, 8, 20, 21, 25, 33, 35, 38, 41, 42, 46, 47, 76, 78, 80, 104, 112, 273
microbial community, 57
microbiota, ix, 2, 3, 4, 6, 7, 9, 10, 32, 36, 37, 38, 57, 58, 59, 61, 65, 66, 73, 74, 75, 80, 91, 104, 105, 117, 130, 133, 139, 206, 257
microheterogeneity, 169, 170
micronutrients, x, 3, 49, 68
microorganism(s), xiv, 2, 9, 15, 30, 57, 58, 126, 233, 235, 239, 242, 243, 257, 271
microscopy, 96
microstructure, 29, 81
migration, 276
mimicry, 283
mineral aborption, ix, 1
mineralization, 15, 16
miniature, 243
mitogen, 21
modelling, xi, 86, 90
models, xi, 16, 17, 24, 33, 86, 90, 91, 92, 93, 94, 97, 99, 122, 285
modifications, 145, 217, 241
moisture, 29, 30, 62
mole, 157, 166
molecular mass, 71, 211
molecular mimicry, 272, 280, 283
molecular sensors, 226
molecular structure, 74, 210
molecular weight, ix, xiii, 2, 9, 10, 67, 100, 122, 145, 173, 177, 178, 182, 183, 185, 204, 207, 256, 272
molecules, xi, 68, 86, 93, 101, 105, 115, 117, 119, 120, 122, 124, 130, 133, 185, 189, 210, 212, 213, 218, 229, 235, 255, 259, 272, 284
monolayer, 98, 136, 139
monomers, 24, 119
monosaccharide, xii, 2, 11, 71, 144, 154, 156, 157, 161, 165, 168, 181, 182, 185, 189, 260
morbidity, 4, 52, 104, 124, 127, 132
Morocco, 49
mortality, 4, 52, 104, 124, 170
MR, 107, 111, 112
MRI, 286
mRNA, 26, 28
mucin, 25, 140, 257
mucosa, xi, 21, 85, 86, 89, 90, 256
mucus, 23, 118
mumps, 272
muscles, 277
mutant, 35
mutation, 27
mutations, 274
myocardial infarction, 80

N

NaCl, 151
naphthalene, 222
nasopharynx, 118, 124
National Academy of Sciences, 46, 73
National Research Council, 253
nerve, 276, 277, 285
nervous system, 276
Netherlands, 264
neural network, 229
neuropathy, 270, 271, 275, 277, 278, 282
neutral, xiii, 10, 19, 28, 47, 88, 90, 95, 98, 100, 105, 120, 167, 174, 175, 177, 178, 181, 182, 183, 186, 189, 191, 193, 195, 204, 205, 207, 210
New England, 74, 76, 150
Nigeria, 79
nitric oxide, 9
nitrogen, 222

NK cells, 19
NMR, 70, 71, 181, 198, 204, 223
North America, 54, 56
NRC, 253, 264
nuclear magnetic resonance, 70, 273
nucleotide sequence, 275
nucleotides, 131
nucleus, 23
nutraceutical, 3, 175, 246
nutrient, ix, xi, xiv, 8, 9, 33, 57, 78, 85, 86, 93, 104, 251, 253, 255, 256, 258, 259, 260, 263
nutrients, x, 19, 35, 48, 49, 63, 81, 82, 86, 131, 236, 244, 255
nutrition, x, 3, 13, 28, 31, 32, 33, 34, 35, 36, 37, 38, 39, 40, 41, 42, 43, 44, 45, 46, 47, 48, 51, 56, 57, 72, 76, 85, 86, 105, 116, 135, 234
nutritional status, 3

O

obesity, x, 3, 29, 49, 50, 51, 52, 53, 55, 56, 61, 68, 72, 75, 76, 80, 81, 82
obstacles, 71
oculomotor, xv, 267, 268, 276, 277
oculomotor nerve, 268
ODS, 181
oil, 174, 175, 223, 237, 243, 252, 254
oil production, 176
oilseed, 252
oleic acid, 62
oligomers, 24, 109, 175, 182, 195
olive oil, 174, 176, 206, 207, 208
omeprazole, 89, 92
oncogenesis, 35
opportunities, 139, 273
optic nerve, 276
optimization, 175
oral cavity, 87, 91
organ(s), 31, 35, 57, 123, 130
organic matter, 256
osmolality, 256
osmotic pressure, 14, 257
osteoporosis, 4, 17, 18, 236, 238
otitis media, 19, 116, 117, 127, 128, 135, 140
outpatient, 128
ovarian cancer, 170
overweight, 51, 52, 55, 56, 73, 75, 77
oxalate, 241
oxidation, 198
oxidative stress, 236
oxygen, 4

P

p21WAF1/CIP1, 24
p53, 24, 35, 36
Pacific, 1, 45
pain, 14, 22, 104, 235
paresis, 281
participants, 76
pasta, 62, 68
pathogenesis, xv, 120, 130, 267, 268, 269, 272, 273, 278, 279, 280, 283
pathogens, xi, xiv, 7, 9, 10, 18, 23, 27, 33, 101, 102, 105, 106, 116, 117, 120, 122, 125, 126, 128, 138, 140, 234, 236, 271
pathophysiological, 51, 57, 60, 145
pathophysiology, 277, 285
pathways, 7, 9, 39
PCR, 12, 28, 36, 47
PCT, 206
pepsin, 91
peptide, 60, 74, 146, 154, 155
peptide(s), 5, 60, 74224, 225, 234, 240, 241, 242, 246
perfusion, 123
permeability, 18, 98, 109
permission, 55, 56, 147
permit, 93
pertussis, 272, 283
petroleum, 252
pH, xiv, 5, 8, 13, 15, 18, 23, 25, 33, 40, 45, 47, 71, 87, 88, 90, 91, 97, 102, 104, 149, 151, 153, 167, 169, 180, 219, 234, 257
pharmaceutical(s), 3, 22, 31, 38, 46, 112, 175, 212, 213, 214, 216, 218, 219, 220, 222, 225, 226, 227, 228, 229

pharmacokinetics, xi, 86, 93
pharmacology, 34, 39, 40, 93
pharyngitis, 118
phenol, 197, 205
phenol residues, xiii, 174
phenolic compounds, xiii, 174, 175, 177, 182, 183, 186, 189, 204, 205, 207, 208
phosphate, 89
phosphates, 241
phosphorescence, 215, 222, 226, 227
phosphorylation, 28
physical activity, 3, 80
physical properties, 210, 211, 223
physicochemical properties, 28
Physiological, 107
physiology, 9, 42, 44, 247, 255, 265
phytosterols, 245
pigs, 81, 101, 106, 253, 256, 261, 263
placebo, 15, 17, 66, 106, 107, 112, 127, 128, 132, 137, 140
plants, x, 2, 67, 235, 239, 240, 244, 256
plasma levels, 145, 168
plasma membrane, 225
plasma proteins, xii, 144, 146
PM, 112
pneumonia, 101, 118, 122, 123, 126, 137
Poland, 170
polarity, 94, 210, 213
polarization, 45, 110
polymer(s), 9, 41, 67, 99, 102, 175, 229
polymerization, xvi, 2, 9, 66, 71, 175, 179, 189
polymorphism, 275, 285
polyphenols, 44, 93
polysaccharide(s), 2, 26, 33, 46, 51, 61, 67, 70, 109, 111, 175, 182, 206, 207, 208, 235, 236, 240, 254, 257
pools, 195
poor performance, 257, 258
population, 3, 4, 9, 31, 41, 53, 55, 56, 78, 145, 147, 148, 155, 157, 159, 160, 161, 162, 163, 164, 165, 166, 167, 168, 234, 238, 245, 253, 257
positive correlation, 165, 168, 257
potassium, 149, 235, 279, 286
potato, 70, 79
potential benefits, 25, 44
poultry, 252, 253, 254, 255, 257, 259, 261, 263
prebiotic carbohydrates, x, 49, 64, 70, 71, 72
precipitation, 96
pregnancy, 37
preparation, v, 29, 71, 83, 95, 96, 129, 134, 181, 216, 218, 241
preterm infants, 12, 19, 32, 40, 47, 123, 139
prevention, 18, 19, 20, 26, 27, 31, 34, 36, 43, 51, 65, 100, 101, 102, 107, 111, 116, 119, 127, 236, 238, 245
prevention of infection, 19, 102, 116
principles, 226
probe, 137
probiotic, xiv, 19, 20, 29, 30, 32, 34, 43, 131, 132, 137, 234, 242, 245
probiotics, xv, 4, 19, 20, 42, 43, 48, 73, 75, 77, 79, 80, 102, 119, 130, 133, 241
producers, 243
professionals, 40
progesterone, 170
prognosis, 50, 145, 164, 168, 170, 268, 269, 281
pro-inflammatory, 9
project, 32, 46, 81
proliferation, 6, 18, 24, 27, 28, 40, 45
prophylaxis, 128
propranolol, 216, 218, 228
prostate cancer, 238
proteasome, 39
protection, 7, 9, 24, 26, 105, 110, 117, 120, 122, 137, 138, 171, 175
protective role, xi, 115, 117
protein hydrolysates, 246
proteins, 5, 24, 26, 105, 113, 145, 161, 164, 170, 171, 224, 225, 234, 242
proteolysis, 29
proteolytic enzyme, 241
protons, 182
prototype, 176
Pseudomonas aeruginosa, 120, 123, 125, 136, 137, 139, 140

ptosis, 270
public health, 104
pulp, xiii, 173, 175, 176, 182, 191, 197, 198, 206, 208
purification, 133, 177
purity, 96
pyridoxine, 3

Q

quality of life, 14, 41, 52, 104, 107
quantification, 12, 70, 71, 95, 96, 97, 208, 213, 214, 216, 220, 222, 228
quartz, 149
quaternary ammonium, 227
questionnaire, 14

R

reactions, 6, 16, 78, 95, 96, 207, 229
reactivity, 169, 275
reagents, 95, 207
receptors, xi, 22, 23, 105, 116, 118, 120, 121, 122, 125, 131
recognition, 118, 224, 226
recommendations, 20, 48, 166, 252
recovery, 89, 96, 174, 176, 204, 207, 270, 282
rectum, 104
recurrence, 105, 166
reducing sugars, 95
refractive index, 70, 185
regenerate, 151
regenerated cellulose, 177
regions of the world, 72
regression, 180, 186
regression model, 186
rejection, 72
relevance, 40, 88, 102
relief, 22, 105
remission, 166, 270
renewable fuel, 252
requirements, 51, 55, 58, 61
researchers, xiv, 6, 65, 102, 118, 122, 124, 129, 234, 240, 255, 256, 258
reserves, 51
residues, xiii, 62, 64, 66, 95, 109, 130, 154, 155, 160, 161, 167, 174, 190, 191, 273, 274, 278
resistance, xii, 5, 9, 10, 18, 22, 68, 71, 87, 89, 140, 144, 146, 167, 168, 169
resolution, 70
respiratory tract disease, 116
response, 3, 9, 13, 21, 22, 41, 68, 74, 77, 164, 165, 166, 257, 260
rheumatoid arthritis, 170, 171
riboflavin, 3, 235
rights, v
risk(s), xii, 3, 24, 26, 30, 42, 50, 51, 52, 60, 61, 65, 68, 72, 76, 78, 91, 100, 103, 104, 107, 117, 126, 129, 132, 134, 138, 140, 144, 147, 155, 157, 161, 162, 163, 164, 165, 166, 167, 168, 175, 235, 238, 239, 244, 285
risk factors, 50, 51, 68, 72, 78, 103, 285
room temperature, 151, 152, 153, 177, 242
root(s), xv, 15, 22, 54, 56, 64, 276
rotavirus, 43
Royal Society, 77

S

safety, 132, 137
saliva, 88, 91
salmonella, 22, 39, 100, 110
salts, 14, 97, 243, 255
saturated fat, 234, 238, 244
scandium, 225
science, 35, 38, 39, 42, 45, 46, 223
scope, 92
secrete, 45, 87
secretion, 21, 32, 37, 81, 87, 89, 92, 255
sedative, 22
seed, 62, 66, 70, 81
selective estrogen receptor modulator, 146
selectivity, 5, 11, 94, 98, 213, 214
selenium, 81
sensation, 74, 270

sensitivity, 70, 94, 167, 213, 214, 215
sequencing, 80
serology, 283
serum, 27, 29, 45, 60, 61, 62, 79, 150, 171, 214, 215, 218, 219, 220, 227, 271, 273, 286
serum albumin, 170
showing, 26, 106
sialic acid, 105, 120, 157, 160, 171, 275, 276
side effects, 165
silica, 178, 225
skeleton, 16
skin, 176
small intestine, ix, 2, 5, 21, 22, 25, 50, 57, 65, 66, 68, 79, 87, 89, 92, 93, 104, 107, 255, 256
smoking, 103
socioeconomic status, 102
sodium, 40, 43, 71, 97, 149, 197, 198, 203, 216, 219
sodium hydroxide, 71
solid state, 223
solubility, 103, 109, 211, 217, 222, 261
solution, 72, 99, 140, 151, 152, 153, 154, 177, 181, 214, 215, 217, 219, 223
soybean(s), xiv, 27, 235, 236, 237, 241, 243, 244, 245, 251, 254, 255, 258, 260, 263
soymilk, xiv, 234, 235, 236, 237, 239, 240, 241, 246, 247, 249
Spain, 173, 174, 176, 205
species, xi, 6, 10, 11, 12, 36, 47, 57, 58, 59, 65, 86, 87, 88, 91, 92, 93, 101, 103, 107, 119, 198, 236, 239, 240, 253
spectrophotometric method, 215
spectroscopy, ix, xvi, 70, 167, 213, 215, 216, 222, 228
spinal cord, 276
Sprague-Dawley rats, 25
Spring, 265
sputum, 272
stability, 30, 88, 89, 96, 102, 212
stabilization, 29
staphylococci, 118
starch, ix, x, 2, 28, 29, 33, 50, 61, 62, 67, 68, 69, 70, 71, 73, 74, 75, 76, 77, 79, 81, 82, 83, 210, 236, 253, 254
starch polysaccharides, 61, 253, 254
starvation, 52
state(s), 14, 145
statistics, 249, 263
steel, 176
sterile, 118
sterols, 235
stimulation, 25, 36, 92, 103, 133, 236
stock, 152, 153
stoma, 82
stomach, ix, 2, 5, 14, 57, 87, 88, 90, 91, 92
storage, 29, 30, 51, 73, 77, 238
streptococci, 118, 122
stroke, 52, 108
structural changes, 277
structural variation, 273
structure, xii, xv, 4, 22, 64, 66, 68, 80, 81, 82, 95, 97, 106, 109, 119, 144, 160, 166, 197, 198, 199, 203, 211, 255, 268, 273, 274, 275, 278, 279, 284, 285
subacute, 270
subgroups, 59, 155
substitutes, 30, 46, 88, 234
substitution, 96, 218
substrate(s), xi, xiv, 2, 3, 5, 10, 11, 15, 18, 24, 31, 36, 39, 57, 59, 86, 87, 88, 93, 102, 103, 233, 239, 240, 241, 256, 257
sucrose, xiii, 5, 16, 22, 30, 31, 62, 66, 67, 87, 130, 233, 255, 260, 263
Sudan, 209
sugar alcohols, x, 50, 51, 61, 62, 64, 67, 69, 70, 82
sulfate, 243
Sun, 42, 62, 75, 230
supplementation, xiv, 4, 6, 10, 13, 19, 21, 27, 31, 34, 47, 104, 107, 134, 242, 251, 259, 260
suppression, 16, 24
surface area, 18
surfactant, 216
survival, 19, 51, 234
susceptibility, xi, 86

sustainability, x, 49
Sweden, 180
sweeteners, 28, 30, 37, 42, 54, 67, 73, 108
swelling, 14
Switzerland, 263
symptoms, 14, 100, 104, 112, 235, 239, 245, 269, 270, 271, 278
syndrome, xv, 60, 104, 267, 268, 269, 270, 271, 272, 273, 274, 275, 276, 277, 278, 279, 280, 281, 282, 283, 284, 285, 286
synergistic effect, 131, 185
synovial fluid, 171
synthesis, 10, 18, 103, 120, 133, 227
systemic immune response, 21

T

T cell(s), 19, 21
tachycardia, 270
Taiwan, 269, 281
tamoxifen, xii, 144, 145, 146, 147, 148, 153, 155, 157, 160, 161, 162, 163, 164, 165, 166, 167, 168, 170, 171
tannins, 62
target, 4, 7, 22, 23, 28, 38, 68, 70, 96, 97, 101, 120
techniques, xi, 11, 12, 71, 86, 95, 97, 98, 283
technology(ies), 95, 96, 106, 133, 206, 246
temperature, 8, 98, 99, 176, 178, 181, 222, 226, 227
tendon, 271
testing, 102, 223, 281
tetanus, 272, 283
texture, 29, 30, 48, 244
T-helper cell, 23
therapeutics, 34, 40
therapy, xii, 38, 144, 146, 147, 148, 161, 164, 165, 166, 167, 168, 170, 246
thermal treatment, xiii, 173
threats, 50
threonine, 260
thyroxin, 238
TIMP-1, 35
tissue, 21
TNF-α, 22, 23
tofu, 241, 243
toluene, 181
Tonga, 35
tonic, 22
toxic effect, 101
toxic substances, 103
toxicity, xiii, 102, 144, 168, 169, 174
toxin, 101, 111, 131
trachea, 118, 123
traits, 72
transcription, 23, 38
translocation, 23
transmission, 118
transport, 10, 12, 18, 21, 40, 98, 99, 110, 136
treatment, xii, 14, 17, 22, 26, 28, 70, 75, 92, 105, 129, 137, 140, 144, 145, 146, 164, 166, 167, 168, 174, 175, 176, 177, 178, 182, 186, 206, 208, 235, 261, 269
triacylglycerides, 62
trial, 32, 47, 48, 67, 80, 106, 132, 137, 140
trifluoroacetic acid, 149
triggers, 112
triglycerides, 51, 61
trisaccharides of glucose, xiii, 174
trypsin, 261, 263
tryptophan, 161, 167, 170
tumor(s), 9, 24, 25, 42, 33
tumor cells, 38
tumor necrosis factor, 9
turnover, 16
type 2 diabetes, 50, 175

U

United Kingdom (UK), 85, 103, 112, 149, 150
United Nations, 50, 52, 81
United States, 68, 70, 73, 79, 82, 228, 252, 256
untreated breast cancer, xii, 144, 155, 159, 161, 162, 164, 165, 167
upper respiratory tract, 19, 118, 126, 127, 131

urban, 47
urbanization, 3
urinary tract infection, 19
urine, 99, 126, 135, 139, 216, 218, 219, 220, 222, 228, 229
urticaria, 107
USDA, 63, 82
UV, 95, 181, 186, 198, 229

V

vaccine, 272, 283
vacuum, 151, 153, 154, 155, 177
validation, 71, 216
valine, 260
valorization, 174
valuation, 225
valve, 176
variables, 8
variations, 72, 120, 138, 145, 160, 276
varieties, xiv, 79, 82, 88, 102, 245, 251, 258, 260, 263
vegetable oil, 50, 54, 56, 252
vegetables, xv, 64, 68, 245
vein, 39
versatility, 71
viruses, 23, 101
viscosity, 12, 40, 47, 76, 255, 257
vitamin B1, 235
vitamin B12, 235
vitamin C, 63
vitamin D, 45
vitamin E, 239
vitamin K, 3, 63, 219, 226, 229
vitamins, 3, 4, 48, 50, 58, 62, 63, 75, 105, 235, 242, 244, 245
volunteers, 34

W

Washington, 46, 264
waste, 57, 102, 205, 207, 217
waste water, 217
water, xiii, 3, 5, 6, 14, 51, 57, 62, 66, 75, 104, 132, 150, 152, 153, 154, 155, 174, 175, 177, 178, 180, 181, 183, 185, 186, 198, 210, 214, 215, 217, 218, 220, 221, 225, 233, 237, 239, 241, 243, 244, 254, 255, 256, 260
water absorption, 57
weakness, 270, 271
wealth, 57, 102
weight control, x, 2
weight gain, 259, 263
weight loss, 245
weight management, 78
well-being, 4, 10, 14, 51, 57
wellness, 12, 14, 18
Western countries, 279
wheezing, 107
white matter, 276
wool, 153
workers, 213
working population, 14
World Health Organization (WHO), 31, 35, 37, 116, 139, 141
worldwide, xv

Y

yield, 72, 96, 97, 198, 258, 262
young adults, 124

Z

zinc, 34, 63, 81, 103, 221, 230, 235, 241